The Organic Chemistry of Tellurium

The Organic Chemistry of Tellurium

KURT J. IRGOLIC

Department of Chemistry

Texas A&M University

College Station, Texas

GORDON AND BREACH SCIENCE PUBLISHERS

New York London Paris

Gordon and Breach Science Publishers Inc.
1 Park Ave.
New York, N.Y. 10016

Editorial office for Great Britain
Gordon and Breach Science Publishers, Ltd.
42 William IV Street
London W.C. 2

Editorial office for France
Gordon and Breach
7-9 rue Emile Dubois
Paris 75014

Library of Congress Catalog card number 73-84669.
ISBN 0 677 04110 1. Printed in the United Kingdom.

PREFACE

The element tellurium was discovered by Franz Joseph Mueller von Reichenstein in 1783. Its first organic derivatives were prepared by Wöhler in 1840. Although organic tellurium chemistry is almost as old as organic chemistry, many areas - some of which are mentioned in the introduction - remain almost unexplored. This slow development was at least partly caused during the early years by the scarcity of elemental tellurium from which all preparations started. Later, tellurium chemistry was pursued - with few exceptions - as an extension of selenium chemistry. Work presently in progress in some of the unexplored areas of organic tellurium chemistry soon convinces every investigator that there are at least as many dissimilarities as there are similarities in the chemical behavior of selenium and tellurium. Much remains to be accomplished in this field and challenging problems face the resourceful synthetic chemists. This volume should serve as the starting point for further investigations, making available the thus far accumulated knowledge of organic tellurium chemistry in concentrated form.

The author expresses his appreciation for financial support of his research endeavors in the area of organic tellurium chemistry and related fields during the time this book was written to the Selenium-Tellurium Development Association, Inc. of New York and the Robert A. Welch Foundation of Houston, Texas. The Selenium-Tellurium Development Association generously sponsored this book and made its early publication possible. Special thanks go to Mrs. Catherine Mieth, who patiently performed the arduous task of typing the manuscript.

Kurt J. Irgolic

CONTENTS

I. INTRODUCTION 1

II. NOMENCLATURE OF ORGANIC TELLURIUM COMPOUNDS 4

III. ELEMENTAL TELLURIUM, STRUCTURE AND PROPERTIES 13

IV. METHODS FOR THE INTRODUCTION OF TELLURIUM INTO ORGANIC MOLECULES 24

A) Elemental Tellurium 24

1) Reaction with organic halides 24

2) Insertion reactions into carbon-metal bonds . . . 26

3) Reactions with organic radicals 27

4) Replacement of mercury, SO_2 groups and hydrogen . . 28

B) Tellurium Halides 29

1) Condensation reactions of tellurium tetrachloride with elimination of hydrogen chloride . . 29

2) Addition of tellurium tetrachloride to carbon-carbon multiple bonds . 36

3) Reactions of tellurium halides with organometallic compounds . 37

C) Alkali Metal Tellurides and Other Inorganic Tellurium Compounds . . 41

D) Organic Tellurium Compounds 43

V. TELLUROCYANIC ACID, TELLUROCYANATES, TELLURIUM DICYANIDE, TELLURIUM DI(THIOCYANATE), BIS(N,N-DIISOPROPYLTHIOCARBAZOYL) DITELLURIDE AND TELLURIUM DERIVATIVES, XCTe . . 46

A) Tellurocyanates and Tellurocyanic Acid 46

B) Tellurium Dicyanide and Tellurium Di(thiocyanate) . . . 50

C) Bis(N,N-diisopropylthiocarbazoyl) Ditelluride 52

D) Tellurium Derivatives, X=C=Te (X=O,S,Se) 52

VI. COMPOUNDS CONTAINING A SINGLE CARBON-TELLURIUM BOND . . . 56

A) Tellurols 56

B) Tellurenyl Compounds 59

C) Organyl Tellurium Compounds, $RTeX_3$ 67

1) Synthesis of organyl tellurium trihalides . . . 67

2) Reactions of organyl tellurium trihalides . . . 80

D) Tellurinic Acids and Their Derivatives 84

E) Diorganyl Ditellurides 91

1) Synthesis of diorganyl ditellurides 91

2) Reactions of diorganyl ditellurides 100

VII. COMPOUNDS CONTAINING A CARBON-TELLURIUM-CARBON MOIETY . . 105

A) Diorganyl Tellurides 105

1) Symmetric diorganyl tellurides 105

2) Unsymmetric diorganyl tellurides 121

3) Tellurides with two tellurium atoms in the molecule . 138

4) Reactions of diorganyl tellurides 139

B) Diorganyl Tellurium Compounds, R_2TeX_2 143

1) Synthesis of symmetric and unsymmetric diorganyl tellurium compounds, R_2TeX_2 and $RR'TeX_2$ . . 147

2) Diorganyl tellurium compounds, R_2TeXY 187

3) Reactions of diorganyl tellurium dihalides . . . 188

C) Diorganyl Telluroxides 192

D) Diorganyl Tellurones 196

VIII. TRIORGANYL TELLURONIUM COMPOUNDS, $[R_3Te]^+X^-$ 199

1) The synthesis of triorganyl telluronium salts . . 199

2) Racemic and optically active triorganyl telluronium salts $[RR'R''Te]^+X^-$. 224

3) Reactions of triorganyl telluronium compounds . . 229

IX. TETRAORGANYL TELLURIUM COMPOUNDS, R_4Te 234

X. TELLURIUM ANALOGS OF ALDEHYDES AND KETONES 242

XI. ORGANIC TELLURIUM COMPOUNDS CONTAINING A TELLURIUM-METAL OR A TELLURIUM-METALLOID BOND . . 244

A) Organic Compounds of Tellurium with Metals of Group I, II and III . 245

B) Organic Compounds of Tellurium Containing a Tellurium-Group IV Element Bond . 247

C) Organic Compounds Containing a Tellurium-Phosphorus or a Tellurium-Arsenic Bond . 255

D) Organic Tellurium Compounds Containing a Tellurium-Selenium Bond . . . 257

E) Organic Tellurium Compounds as Ligands in Transition Metal Complexes . . 257

XII. HETEROCYCLIC TELLURIUM COMPOUNDS 277

A) Five-membered, Heterocyclic Tellurium Compounds . . . 277

1) Telluracyclopentane and its derivatives 277

2) Tellurophene and its derivatives 279

3) Benzotellurophene and its derivatives 289

4) Dibenzotellurophene and its derivatives 290

B) Six-membered, Heterocyclic Tellurium Compounds with Tellurium as the Only Heteroatom . . 296

1) Telluracyclohexane 296

2) 1-Tellura-3,5-cyclohexanedione and its derivatives . 297

3) Telluroisochroman 305

C) Tellurium Containing, Six-Membered Heterocyclic Ring Systems with Oxygen, Sulfur or Tellurium as Additional Heteroatoms . . 305

1) 1-Oxa-4-telluracyalohexane 305

2) Phenoxtellurine 307

3) 1-Thia-4-telluracyclohexane 320

4) Thiophenoxtellurine 321

5) Telluranthrene 322

6) 1,3,5-Tritelluracyclohexane 322

XIII. TELLURIUM CONTAINING POLYMERS 323

XIV. PHYSICOCHEMICAL INVESTIGATIONS OF ORGANIC TELLURIUM COMPOUNDS 325

A) Infrared, Ultraviolet and Visible Spectroscopy . . . 325

B) Nuclear Magnetic Resonance Spectroscopy 341

C) Structural Investigations 356

D) Mass Spectrometry of Organic Tellurium Compounds . . . 362

XV. ANALYTICAL TECHNIQUES 366

XVI. BIOLOGY OF ORGANIC TELLURIUM COMPOUNDS 369

APPENDIX 1: PATENTS COVERING ORGANIC TELLURIUM COMPOUNDS 380

REFERENCES 386

SUBJECT INDEX 409

FIGURES

Fig. I - 1: The Yearly Publication Rate for Papers Dealing with Organic Tellurium Compounds . . 2

Fig. III - 1: The Unit Cell of Hexagonal Tellurium 16

Fig. III - 2: The Synthesis of Inorganic Tellurium Compounds from Elemental Tellurium . . 18

Fig. IV - 1: The Synthesis of Organic Tellurium Compounds from Elemental Tellurium . 25

Fig. IV - 2: Organic Tellurium Compounds from Tellurium Tetrachloride . . . 29

Fig. IV - 3: Reactions of Organic Tellurium Compounds Proceeding with Formation of Carbon-Tellurium Bonds . 45

Fig. VI - 1: Preparation of Tellurenyl Compounds, RTeX . . . 60

Fig. VI - 2: Partial Molecular Structure of a) Phenyl Bis(thiourea) Tellurium Chloride b) Phenyl Thiourea Tellurium Bromide c) Phenyl Thiourea Tellurium Chloride . . . 63

Fig. VI - 3: Preparation and Reactions of 2-Naphthyl Tellurium Iodide . 66

Fig. VI - 4: Preparation of Organyl Tellurium Trihalides . . 68

Fig. VI - 5: Reactions of Organyl Tellurium Trichlorides . . 81

Fig. VI - 6: Tellurinic Acids and Tellurinic Acid Derivatives . 90

Fig. VI - 7: Syntheses and Reactions of Diorganyl Ditellurides . 104

Fig. VII - 1: Syntheses and Reactions of Diorganyl Tellurides . 120

Fig. VII - 2: The Molecular Structure of Diorganyl Tellurium Dihalides . 144

Fig. VII - 3: Syntheses and Reactions of Diorganyl Tellurium Dihalides . 149

Fig. VII - 4: Halogen Exchange and Replacement in Diorganyl Tellurium Dihalides . . 178

Fig. VIII- 1: Syntheses and Reactions of Triorganyl Telluronium Salts . 207

Fig. VIII- 2: Anion Exchange Reactions of Triorganyl Telluronium Compounds . 210

Fig. XI - 1: Syntheses and Reactions of Compounds Containing Tellurium-Group IV Element Bonds . . . 246

Fig. XII - 1: Synthesis and Reactions of Telluroisochroman . . . 306

Fig. XII - 2: 1-Oxa-4-telluracyclohexane and Its Derivatives . . . 308

TABLES

Table II - 1: Nomenclature of Organic Tellurium Compounds . . . 12

Table IV - 1: Properties of Tellurium Halides 20

Table V - 1: The Calculated and Observed Infrared Absorption Frequencies for the Compounds X=C=Te . . 55

Table VI - 1: Alkanetellurols 56

Table VI - 2: Tellurenyl Compounds, R-Te-X 62

Table VI - 3: Organyl Tellurium Trihalides. 70

Table VI - 4: Tellurinic Acids and Their Derivatives 85

Table VI - 5: Diorganyl Ditellurides 92

Table VII - 1: Symmetric Diorganyl Tellurides, R_2Te 108

Table VII - 2: Unsymmetric Diorganyl Tellurides, RTeR' 122

Table VII - 3: Alkyl Aryl Tellurides Prepared by Rogoz 133

Table VII - 4: Trialkyl Telluronium Alkyltetrahalotellurates(IV) . 147

Table VII - 5: Symmetric Diorganyl Tellurium Compounds, R_2TeX_2 . . . 150

Table VII - 6: Unsymmetric Diorganyl Tellurium Compounds, $RR'TeX_2$ . . . 161

Table VII - 7: Diorganyl Tellurium Dihalides from Diorganyl Tellurides and Organic Bromides . . 176

Table VII - 8: Diorganyl Tellurium Compounds, RR'TeXY 179

Table VII - 9: Diorganyl Telluroxides, RR'TeO 194

Table VIII- 1: Triorganyl Telluronium Compounds, $[R_3Te]^+X^-$ . . . 200

Table VIII- 2: Triorganyl Telluronium Compounds, $[R_2R'Te]^+X^-$ . . 213

Table VIII- 3: Triorganyl Telluronium Compounds, $[RR'R''Te]^+X^-$ . . 225

Table XI - 1: Organic Compounds Containing a Tellurium-Group IV Element Bond, R_3M-Te-R' . . . 248

Table XI - 2: Organic Phosphorus Compounds Containing Tellurium . 256

Table XI - 3: Organic Tellurium Compounds as Ligands in Transition Metal Complexes . . . 259

Table XI - 4: Diorganyl Telluride-Mercuric Halide Adducts $RR'Te \cdot HgX_2$ . . . 275

Table XII - 1: Telluracyclopentane and Its Derivatives . . . 280

Table XII - 2: Tellurophene and Its Derivatives 283

Table XII - 3: Benzotellurophene and Its Derivatives 290

Table XII - 4: Dibenzotellurophene and Its Derivatives . . . 292

Table XII - 5: 2,2'-Biphenylylene Organyl Telluronium Compounds . . . 295

Table XII - 6: Telluracyclohexane and Its Derivatives . . . 298

Table XII - 7: 1-Tellura-3,5-cyclohexanedione and Its Derivatives . . . 301

Table XII - 8: Oximes, 1,1-Dibromides and 1,1-Diiodides of 1-Tellura-3,5-cyclohexanediones . . . 304

Table XII - 9: Phenoxtellurine and Its Derivatives 311

Table XII -10: Phenoxtellurine Complexes 318

Table XII -11: Complexes formed by Phenoxtellurine and Its Derivatives in the Solid State . . . 319

Table XIV - 1: Infrared Absorption of Methyl Tellurium Compounds . . . 327

Table XIV - 2: Infrared Absorption Bands of Alkyl Tellurium Compounds . . 330

Table XIV - 3: Infrared Absorption Bands of Phenyl Tellurium Compounds . . 334

Table XIV - 4: Infrared Absorption Bands of Compounds $R_nTeX_{4-n}\cdot L$ . . . 340

Table XIV - 5: The Ultraviolet and Visible Absorption Spectra of Organic Tellurium Compounds . . . 342

Table XIV - 6: Nuclear Magnetic Resonance Studies of Organic Tellurium Compounds . . 347

Table XIV - 7: Bond Distances and Bond Angles in Organic Tellurium Compounds . . . 358

Table XIV - 8: Mass Spectral Data for Diethyl Telluride and Diethyl Ditelluride . . . 364

Table XIV - 9: Relative Abundances of Ions in the Mass Spectra of Diaryl Ditellurides, Ar_2Te_2 . 365

Table XVI - 1: The Bactericidal Action of Organic Tellurium Compounds . . 370

I. INTRODUCTION

Organic tellurium compounds have a history dating back to 1840. Woehler, who founded synthetic organic chemistry, was also successful in preparing dialkyl tellurides, the first organic compounds of tellurium. Only a few isolated investigations were performed during the next 70 years. Lederer enriched the knowledge in this area of chemistry through his fruitful work in the decade between 1910 and 1920. His synthetic efforts produced a large number of aromatic tellurium derivatives. Morgan, Drew and their coworkers beginning in 1920 explored the chemistry of five- and six-membered ring systems, concentrating on 1-tellura-3,5-cyclohexanedione and its derivatives.

More recently organic tellurium compounds were investigated by Petragnani, de Moura Campos and their coworkers. Russian workers (Egorochkin, Nefedov, Radchenko, Vyazankin) have published many papers in this area with an emphasis on compounds containing tellurium-metal bonds. Schumann and coworkers were also active in this field. Organic tellurium compounds containing radioactive tellurium isotopes have been synthesized recently by decay of organic compounds of 125-iodine[57], 125-antimony[312,315,439a] or from 132-tellurium tetrachloride[3,246,247]. The distribution of publications over a time span of 130 years is presented in Fig. I-1. A minor information explosion during the past decade is clearly evident.

Since 1840 less than 600 publications have appeared dealing with organic tellurium compounds. Compared to organic selenium chemistry, organic tellurium chemistry lies still in its infancy. Selenocarbohydrates are well known substances. The only attempt to prepare a corresponding tellurium compound had failed[381]. The chemistry of tellurophene is almost unexplored. Tellurium containing amino acids or peptides are unknown.

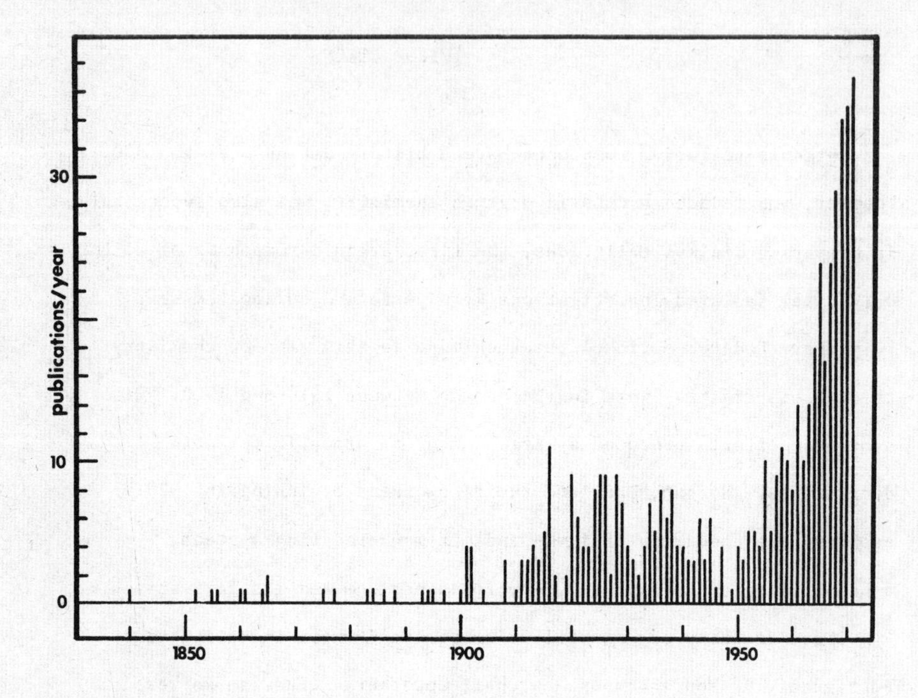

Fig. I-1: The Yearly Publication Rate for Papers Dealing with Organic Tellurium Compounds.

Heterocyclic tellurium compounds with nitrogen atoms as ring members have not yet been prepared. Even many "facts" reported in the literature cannot be considered as the final truth. Many contradictory statements will be found in the following sections. These doubtful reports have been included in the discussion, because there was in many cases no way of selecting the reliable results. It is our hope that new investigations will soon resolve the open questions.

The chapters in this book have been organized to provide an easy access to the information available on the topics treated. The first part of each section describes the preparation of the compounds. The second part dealing with the properties and chemical reactivity is usually short containing cross-references to the synthetic sections, where a particular reaction is treated in more detail. The synthetic methods for a certain class of organic tellurium compounds are outlined in general terms. Literature references and data for specific compounds are collected in tables. This arrangement should make it possible, to locate easily a specific compound, its melting or boiling point, the methods available for its synthesis and the yields as reported in the literature.

An attempt has been made to completely present the field of organic tellurium compounds. The literature has been searched from 1840 through 1971. The papers abstracted in Chemical Abstracts, Volume 76, No. 1 to 11, have been included.

The symbol R is employed throughout this book to represent aliphatic as well as aromatic groups. The meaning of R is either stated in the text or explained in a note following a chemical equation or a structure.

The following review articles have been very helpful in preparing this chapter: J. N. E. Day[95] (1928, organic compounds of tellurium), G. T. Morgan[299] (1935, heterocyclic tellurium compounds), A. N. Nesmeyanov[320] (1945, organic tellurium compounds), H. Rheinboldt[362] (1955, organic tellurium compounds), S. C. Abrahams[2] (1956, stereochemistry), M. de Moura Campos[303] (1960, vicinal group participation in organic compounds), N. Petragnani and M. de Moura Campos[347] (1967, new topics in organotellurium chemistry), E. W. Abel and D. A. Armitage[1] (1967, compounds with a tellurium-group IV element bond) and K. J. Irgolic and R. A. Zingaro[185b] (1970, reactions of organotellurium compounds).

II. NOMENCLATURE OF ORGANIC TELLURIUM COMPOUNDS

It certainly would be desirable to have a set of nomenclature rules which would make it possible to name all organometallic and organometalloidal compounds in a simple and consistent manner. Such a set of official rules does not yet exist. Substitutive nomenclature which has been used extensivel in the literature seems to be the most adaptable method of forming names. Substitutive nomenclature has, however, the disadvantage that the metallic and metalloidal part is in most cases an appendix to the organic parent compound. Such an arrangement might not be acceptable to some organometallic chemists, who would like to see their favorite element in a more prominent position.

The nomenclature of organic tellurium compounds should follow the rules set forth for the sulfur and selenium derivatives in order to have a consistent way of naming the organic compounds of the group VI elements. The suggestions for naming organic tellurium compounds presented here are patterned according to the rules established for sulfur derivatives by IUPAC, which are published in the book "Nomenclature of Organic Chemistry" (Butterworth, London, 1965). The acceptable names for organotellurium compounds have been compiled in the following pages. Not all of these names have actually been used in the literature. Table II-1 contains the suggested prefixes, suffixes and functional class names for quick reference.

Suffixes are available only for those organotellurium compounds, which contain not more than one carbon-tellurium bond. For compounds with two or more carbon-tellurium bonds one is presently forced to use substitutive nomenclature with prefixes or radicofunctional nomenclature in conjunction with functional class names. Functional class names are available for only a few types of compounds in which tellurium is bonded to only one carbon atom. The names for the organic parts of the organometallic compounds are formed according to the nomenclature rules of organic chemistry.

It should be pointed out that in substitutive nomenclature the entire name of a compound is written as one word, while in the radicofunctional naming method the functional class name and the different radicals are separate words. The latter method seems to give rise to less confusion than substitutive names unless liberal use of parentheses is made.

R-Te-H: Using substitutive nomenclature, the suffix for the principal group -TeH is *tellurol*. Combined with the parent compound one obtains names like *alkanetellurol*. Should the alkane be substituted, there exists in cases the choice between using prefixes for the substituent or conjunctive nomenclature.

$Cl-CH_2CH_2-TeH$	*c*-$C_6H_{11}CH_2-TeH$
chloroalkanetellurol	*cyclohexylmethanetellurol* *cyclohexanemethanetellurol*

Should the -TeH group not be the important part of the compound, the prefix *hydrotelluro* is employed, e.g., *hydrotelluroalkane*. *Tellurol* is not a generic name like *alcohol*. Therefore, names like *methyl tellurol* are incorrect. Tellurium analogs of aromatic phenols have also been named by placing the prefix *telluro*, which indicates the replacement of oxygen by tellurium, in front of the name for the phenol, e.g., *tellurophenol*. This practice should be discontinued.

Salts of tellurols can be named in two ways as shown for the sodium derivative:

RTeNa	*sodium alkanetellurolate* *sodium alkyl telluride*

<u>*R-Te-X:*</u> In these derivatives of divalent tellurium X represents the following groups: Cl, Br, I, ClO_4, $-S-\overset{O}{\underset{O}{S}}-CH_3$, $-S-\overset{O}{\underset{O}{S}}-C_6H_5$, $-S-\overset{S}{C}-OCH_3$ and $-S_2O_3Na$. These compounds can be looked upon as anhydrides of the unknown *alkanetellurenic acid,* R-Te-OH, with the acids whose anions are given above. According to this view names like *alkanetellurenyl halide* or *alkanetellurenyl methanethiosulfonate* are formed. However, since the compounds $RTeCl_3$ are called *alkyl tellurium trichlorides,* it is reasonable to propose *alkyl tellurium chloride* for RTeCl. The term *tellurenyl* should then be used only as a general name for these compounds. As a prefix in substitutive nomenclature *nitratotelluro, chlorotelluro, (methyl xanthatotelluro)* may be used.

<u>*R-Te-R:*</u> Radicofunctional nomenclature is generally employed for this class of compounds. *Telluride* is the functional class name.

	R-Te-R	$C_6H_5-Te-CH_3$
radicofunctional:	*dialkyl telluride*	*phenyl methyl telluride*
substitutive:	*alkyltelluroalkane*	*methyltellurobenzene*

Although IUPAC favors substitutive nomenclature, the radicofunctional names are preferred by most authors unless a simple RTe- group is attached to a complicated organic molecule, e.g.:

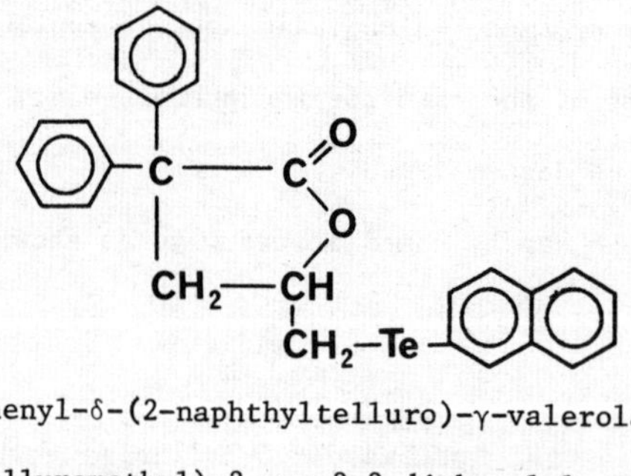

α,α-diphenyl-δ-(2-naphthyltelluro)-γ-valerolactone

5-(2-naphthyltelluromethyl)-2-oxo-3,3-diphenyl-1-oxacyclopentane

The term *telluroether* should not be used. The tellurides containing several tellurium atoms in an aliphatic chain can be named using replacement nomenclature according to the following example:

$$CH_3\text{-Te-}(CH_2)_5\text{-Te-}C_2H_5$$
2,8-ditelluradecane

The prefix *telluro* designates divalent tellurium and has been employed in naming symmetric tellurides. Names like *tellurodiacetic acid* for $Te(CH_2COOH)_2$ are confusing and should not be used. The radicofunctional name *bis(carboxymethyl) telluride* is preferred.

R-Te-Te-R: The functional class name of *ditelluride* together with the appropriate names for organic radicals give the almost exclusively used name of *dialkyl ditelluride*. The group R-Te-Te- is called *alkylditelluro* when employed as a prefix. *Ditelluro* designates the bivalent -Te-Te- group.

$Te\text{-}CH_2COOH$	*bis(carboxymethyl) ditelluride*
$Te\text{-}CH_2COOH$	*(carboxymethylditelluro)acetic acid*

Names like *ditellurodiacetic acid* which are sometimes found in the literature should not be used.

R_2TeX_2: These compounds are named according to radicofunctional nomenclature rules. *Tellurium dichloride* is considered to be the functional class name with the groups bonded to tellurium added as radicals.

$$(C_6H_5)_2TeCl_2$$

functional class name:	*tellurium dichloride*
radical:	*phenyl*
complete name:	*diphenyl tellurium dichloride*

In the case of a compound with two difference organic groups, both are written as separate words. The radicals should be arranged alphabetically according to the rules of organic nomenclature.

$C_6H_5\text{-}TeCl_2\text{-}CH_2Cl$	*chloromethyl phenyl tellurium dichloride*

So far the compounds with X representing F, Cl, Br, I, CN, OH, NO_3, C_2O_4, SO_4, SO_4H, picrate, carboxylate and OR have been prepared. They are named *dialkyl tellurium dinitrate, dialkyl tellurium oxalate,* etc. Compounds are known, which are represented by the general formula R_2TeXY. $R_2Te(OH)Cl$ would be called *dialkyl tellurium chloride hydroxide.* There is, however, some experimental evidence, that the hydroxides are telluronium compounds, $[R_2Te(OH)]^+Cl^-$ and should therefore be named *dialkyl hydroxy telluronium chloride.* The respective anhydrides, $[R_2Te\text{-}O\text{-}TeR_2]^{++}2Cl^-$, could be called *tetraalkyl μ-oxo ditelluronium dichloride.*

R_2TeO: The correct functional class name is *telluroxide.* Using the preferred radicofunctional nomenclature, names like *dialkyl telluroxide* and *alkyl aryl telluroxide* are obtained. Different organic radicals are again written as separate words. The prefix for *RTe(O)-* is *alkyltellurinyl.*

$R\text{-}\overset{O}{\overset{\uparrow}{Te}}\text{-}C_6H_5$

(alkyltellurinyl)benzene

$O{\leftarrow}Te(CH_2\text{-}COOH)_2$

(carboxymethyltellurinyl)acetic acid
bis(carboxymethyl) telluroxide

Ring compounds having the group -Te(O)- as a ring member are named as tellurium oxide.

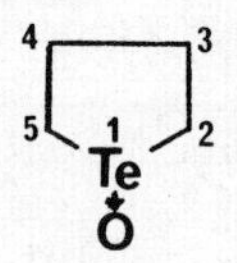

1-telluracyclopentane Te-oxide
1-telluracyclopentane 1-oxide

R_2TeO_2: *Tellurone* is the functional class name for these compounds. The names are formed in the same way as discussed for *telluroxides.*

$R\overset{O}{\underset{O}{Te}}$ - as prefix is called *alkyltelluronyl.* The TeO_2 group as a ring member is designated as *Te, Te-dioxide.*

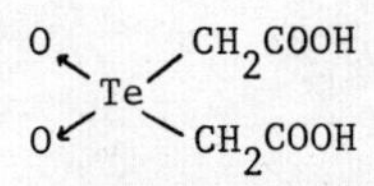

bis(carboxymethyl) tellurone
(carboxymethyltelluronyl)acetic acid

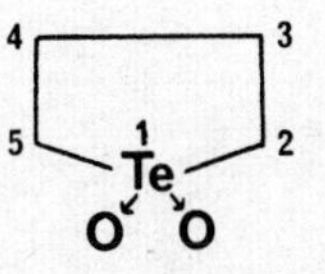

1-telluracyclopentane Te,Te-dioxide
1-telluracyclopentane 1,1-dioxide

$RTeX_3$: Using $RTeCl_3$ as an example, the composite functional class name is *tellurium trichloride*. Adding the radical the name becomes *alkyl tellurium trichloride*. So far compounds with X representing F, Cl, Br, I, OH and $-SC(S)NR_2$ have been prepared. *Trichlorotelluro* is used as a prefix for $-TeCl_3$.

RTe(O)OH: Substitutive nomenclature is employed for this class of compounds with *tellurinic acid* as a suffix. The organic component is considered to be the parent compound.

$C_4H_9Te(O)OH$ — *butanetellurinic acid*, *1-tellurinobutane*

The prefixes *tellurino and alkyltellurinyl* designate the groups -Te(O)OH and R-Te(→O)-, respectively.

$RTeO_3H$: The *alkanetelluronic acids*, not yet prepared, are named using *telluronic acid* as suffix, and *tellurono* and *alkyltelluronyl* as prefixes for the groups $-TeO_3H$ and R-Te(→O)(→O)-, respectively.

RTe(O)X: Compounds with X representing Cl, Br, I and NO_3 are known. *Tellurinyl halide* is the correct suffix generating names of the type *alkanetellurinyl iodide*. The anhydrides [X= RTe(O)-] are named *alkanetellurinic anhydrides*.

R_3TeX: Telluronium salts with X representing the anions F^-, Cl^-, Br^-, I^-, OH^-, picrate$^-$, CrO_4^{--}, $Cr_2O_7^{--}$, $PtCl_6^{--}$ and other complex anions derived from metal halides are known. An unsymmetric telluronium salt would be named *ethyl methyl phenyl telluronium halide*. *Telluronium halide* is considered to be the functional class name to which are added as separate words the radical groups bonded to tellurium. The following names should serve as examples.

$[(CH_3)_3Te]^+I^-$	*trimethyl telluronium iodide*
$[(CH_3)_2C_6H_5Te]^+OH^-$	*dimethyl phenyl telluronium hydroxide*
$[(C_6H_5)_2CH_3Te]_2^+[PtCl_6]^{--}$	*bis(methyl diphenyl telluronium) hexachloroplatinate(IV)*

If the cation is cited as a prefix *telluronio* is used, e. g., *(dimethyltelloronio)benzene hydroxide* for $[(CH_3)_2TeC_6H_5]^+OH^-$.

R_4Te: Tellurium compounds with four carbon-tellurium bonds are named using *tellurium* as a functional class name with the organic groups added as radicals.

$(C_4H_9)_4Te$	*tetrabutyl tellurium*
$(C_6H_5)_2(CH_3)_2Te$	*dimethyl diphenyl tellurium*

R_2CTe: The tellurium analogs of ketones are either named by indicating the substitution of tellurium for oxygen by adding *telluro* to the functional class name ketone, or in substitutive nomenclature by using the prefix *telluroxo* (not *telluro*) to indicate a tellurium atom formally double bonded to a carbon atom.

$CH_3-\overset{Te}{\overset{\|\|}{C}}-C_2H_5$	*ethyl methyl telluroketone* *2-telluroxobutane*

$RC(H)=Te$: Aldehydes in which oxygen is replaced by tellurium, are named by using *telluroformyl* as prefix, *carbotelluraldehyde* as suffix or the term *telluro* together with the name for the *O*-aldehyde.

$C_3H_7-C(=Te)-H$	*telluroformylpropane* *propanecarbotelluraldehyde* *tellurobutanal*

<u>*Ring compounds:*</u> Replacement nomenclature with *tellura* indicating the substitution of -Te- for $-CH_2-$ is generally employed for naming heterocyclic tellurium compounds.

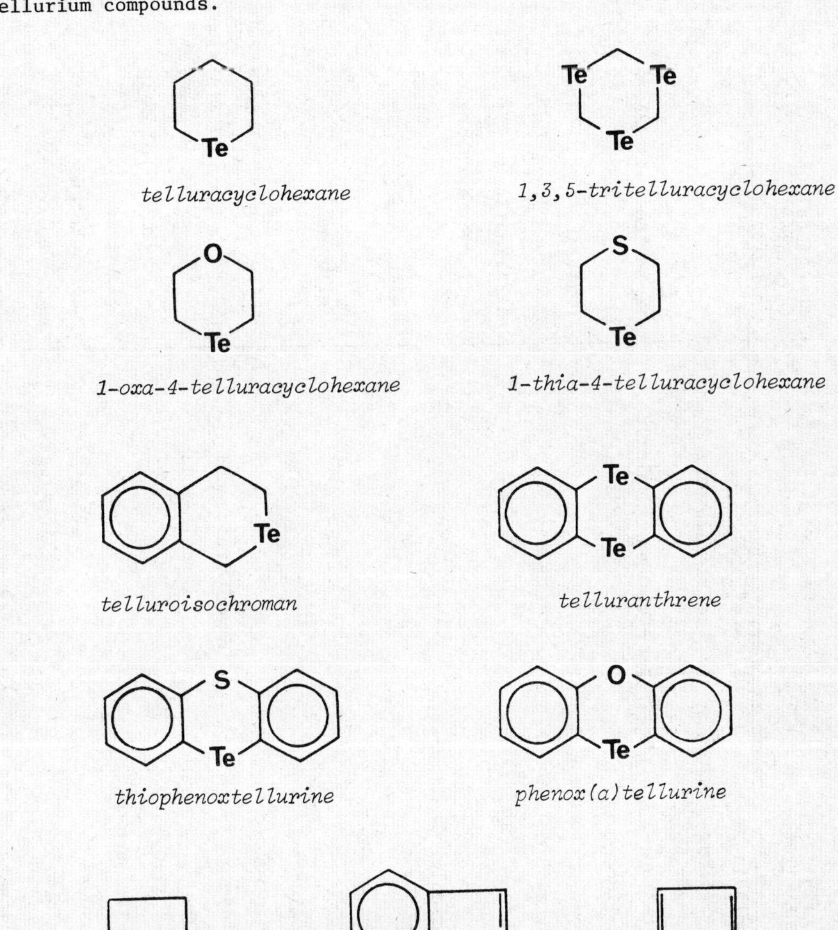

telluracyclohexane — *1,3,5-tritelluracyclohexane*

1-oxa-4-telluracyclohexane — *1-thia-4-telluracyclohexane*

telluroisochroman — *telluranthrene*

thiophenoxtellurine — *phenox(a)tellurine*

Te Te Te

telluracyclopentane — *benzotellurophene* — *tellurophene*

TABLE II-1 Nomenclature of Organic Tellurium Compounds

Group * (or compound)	Prefix (substitutive)	Suffix (substitutive)	Functional class name (radicofunctional)
-TeH	hydrotelluro-	-tellurol	
-TeOH	hydroxytelluro- tellurenyl-	-tellurenic acid	
-TeCl	chlorotelluro-	-tellurenyl chloride	tellurium chloride
-TeR	alkyltelluro-		
-Te-	telluro-, tellura-†		telluride
-TeTeR	alkylditelluro-		
-TeTe-	ditelluro-		ditelluride
$\gt TeCl_2$			tellurium dichloride
$R\dot{T}eCl_2$	alkyl(dichloro)telluro-		
$\gt TeO$			telluroxide
RTeO	alkyltellurinyl-		
$\gt TeO_2$			tellurone
$R\dot{T}eO_2$	alkyltelluronyl-		
$-TeX_3$	trichlorotelluro-		tellurium trichloride
-Te(O)OH	tellurino-	-tellurinic acid	
$-TeO_3H$	tellurono-	-telluronic acid	
R_2Te-(R)	trialkyltelluronio		telluronium (halide)
R_3Te-(R)	trialkyltelluro-		tellurium
$\dot{C}$=Te	telluroxo-		telluroketone
-C(Te)H	telluroformyl-	-carbotelluraldehyde	telluroaldehyde

* The free valences indicated on the tellurium atom are connected to organic groups *via* carbon atoms.

†

III. ELEMENTAL TELLURIUM, STRUCTURE AND PROPERTIES

The abundance of tellurium in the earth's crust has been determined to be 2×10^{-7} per cent by weight. Tellurium is thus as rare as gold. Tellurium can replace sulfur in sulfidic ores and may become enriched to 5×10^{-3} per cent by weight. Although a number of tellurium minerals are known, they do not form large deposits. Tellurium has always been obtained as a by-product of mining and processing operations of other ores. Today tellurium accumulates as a by-product of the electrolytic copper refining process. The anode sludge can contain up to 8 per cent tellurium in addition to selenium, copper, silver, gold and the platinum-group metals.

The sludge is roasted in an oxidizing atmosphere. The roasted product is melted with sodium carbonate or with a sodium nitrate-sodium hydroxide mixture converting tellurium into water soluble sodium tellurite. The aqueous extract of the cooled melt precipitates tellurium dioxide upon neutralization with sulfuric acid. Sodium selenite remains in solution. An alternate method treats the roasted sludge with hot 15 per cent sulfuric acid. The undissolved material is extracted with 10 per cent aqueous sodium hydroxide. The extraction residue may then be treated with sodium carbonate to recover the last amounts of tellurium. The tellurium dioxide obtained in this manner reacts with sulfur dioxide in concentrated hydrochloric acid solution or with elemental carbon at 400-450° to produce tellurium.

This crude tellurium contains as impurities selenium, copper, arsenic, antimony, bismuth and traces of other elements. The purification is achieved by employing one, but generally several of the following methods[151,201]. Distillation under a hydrogen atmosphere separates tellurium from all non-volatile substances. Only traces of hydrogen telluride are formed. The separation from selenium can be accomplished by dissolving the impure element in a hot, concentrated hydrochloric acid-nitric acid mixture. After removing all the nitric acid selenium is preferentially reduced by sulfur dioxide at high acid concentration. Upon dilution tellurium precipitates [16,151,179]. Melting the impure tellurium with potassium cyanide in a reducing atmosphere converts selenium and sulfur into potassium seleno- and thiocyanates, respectively, while tellurium forms potassium telluride. The melt is taken up with water. When air is passed through this solution elemental tellurium is precipitated. In order to remove traces of metals like iron and aluminum, the tellurium is treated with dilute hydrochloric acid. A boiling aqueous potassium cyanide solution is used to dissolve gold and residual selenium. Tellurium can also be purified by recrystallization of the basic tellurium nitrate $2TeO_2 \cdot HNO_3$, which is formed upon dissolution of finely divided tellurium at 70° in nitric acid of density 1.25g/ml. Careful heating of the nitrate produces pure tellurium dioxide[151,179]. The methods available for the extraction and purification of tellurium have been reviewed[186a,320a]. It has been stated[420a], that distillation is the best purification method for elemental tellurium.

While sulfur and selenium can exist in several allotropic modifications, only one crystalline and an amorphous form of tellurium have thus far been detected. Amorphous tellurium is reported to precipitate upon reduction of tellurous acid, H_2TeO_3, with sulfur dioxide or hydrazine hydrate as brown to grey powder[368]. A tellurium melt, which had been rapidly cooled, was claimed to be a mixture of amorphous and crystalline tellurium[151,383]. The X-ray powder patterns of tellurium samples prepared according to methods, which should have given amorphous products, were identical to those of crystalline tellurium[383]. Tellurium evaporated in vacuum onto a collodion membrane gave electron diffraction maxima similar to those produced by crystalline material[368]. However, tellurium precipitated by reduction of tellurous acid by hydrazine hydrate, gave a diffraction pattern characteristic of an amorphous phase[368]. Recently, rapidly solidified samples of tellurium were found to give a ^{125}Te nuclear magnetic resonance spectrum, which indicated the presence of amorphous parts in the specimen[194e]. The values reported for the enthalpy of transformation of amorphous to crystalline tellurium vary between -24 kcal/mole and +2.63 kcal/mole[383]. Powder patterns of vacuum deposited tellurium gave no indication of a structure change in the temperature range from -192° to +360°[100,326].

Crystalline tellurium possesses a hexagonal unit cell containing three atoms. The tellurium atoms are arranged in helices of three-fold symmetry, the prism edge serving as the screw axis. The fourth atom in the helix is positioned exactly above the first atom. The radius of the helix, the perpendicular distance from the center of each atom to the axis, is 1.20Å. The shortest distance between two atoms in the same helix is 2.68Å, while 3.46Å was found to be the shortest distance between atoms in neighboring helices[40,402,403,447]. Figure III-1 shows the unit cell for hexagonal tellurium.

A three-fold screw axis is either right-handed or left-handed. Cleavage planes of tellurium crystals were exposed to hot sulfuric acid. The etch pits showed, that nearly equal numbers of right- and left-handed crystals were formed in these experiments[194d]. The observed anisotropy of the ^{125}Te chemical shifts was in good agreement with the calculated one for right- and left-handed crystals[194d].

Crystalline tellurium is a silver-white substance with metallic luster. It melts at 449.8° to a dark liquid and boils at 1390°[151]. Gaseous tellurium is diatomic from 1400-1800° with an internuclear distance of 2.59Å[265,266]. This value obtained from electron diffraction experiments agrees well with that of 2.61Å derived from spectroscopic data[184]. The density of crystalline tellurium is 6.25g/ml, while a density range of 5.8-6.1 g/ml is reported for the amorphous modification[151].

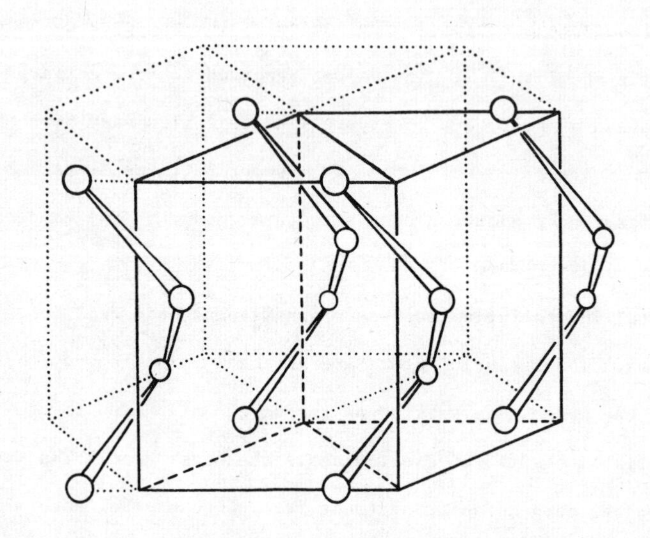

Fig. III-1: The Hexagonal Unit Cell of Tellurium

The electrical conductivity of tellurium at 500° is only 1 per cent that of cold mercury[202]. Light causes a small increase in the conductivity[151,17] Tellurium is insoluble in all solvents which do not react with it. Its extreme brittleness allows it to be easily powdered. A detailed discussion of the physical properties of tellurium is available in the literature[21a].

The industrial demand for tellurium is small. The 69 patents claiming applications for organic tellurium compounds are listed in Appendix I. Small quantities of tellurium are employed as secondary vulcanizing agents for natural rubber[16]. The addition of 0.02-0.085 per cent tellurium to lead increases the corrosion resistance of lead. Steel can be worked more easily if it is alloyed with tellurium. Tellurium added to glass and ceramics imparts a blue to brown color, but in larger concentrations a purple color caused by the presence of polytellurides is obtained. The industrial applications of tellurium and its compounds have been described by Champness[65a], Nachtman[308a], Aborn[1a] and Cooper[83c]. Tellurium and inorganic tellurium compounds show great promise as catalysts in the oxidation of organic materials. High selectivity, high yields and resistance towards poisoning have been claimed in recent years for tellurium containing catalysts. Kollonitsch and Kline[194c] discuss these applications in detail. The annual tellurium production amounts to approximately 100 tons[179,194c].

In the following paragraphs only those chemical reactions of tellurium are discussed in any detail, which produce inorganic starting materials for the synthesis of organic tellurium compounds. These reactions are schematically presented in Figure III-2. A survey of the reactions of elemental tellurium with organic substances is included in section IV.

Elemental tellurium combines with the halogens to produce tellurium halides. Fluorine easily forms tellurium hexafluoride over a wide temperature range with ditellurium decafluoride being present as a by-product. Tellurium tetrafluoride can be obtained by heating the hexafluoride with elemental tellurium in an alumina tube to 200°[428]. The tellurium fluorides have not yet been employed in the synthesis of organic tellurium compounds. Tellurium difluoride is unknown[13]. Chlorine and bromine combine with tellurium to produce in almost quantitative yield tellurium tetrachloride[413] and tetrabromide[44], respectively. These crystalline solids can be distilled in presence of the respective halogens[413] and can be sublimed in vacuum[44,113]. Chlorinating agents like SO_2Cl_2, $SOCl_2$[183,242] and S_2Cl_2[241] produce at room temperature tellurium tetrachloride in good yields (eq. 1-3).

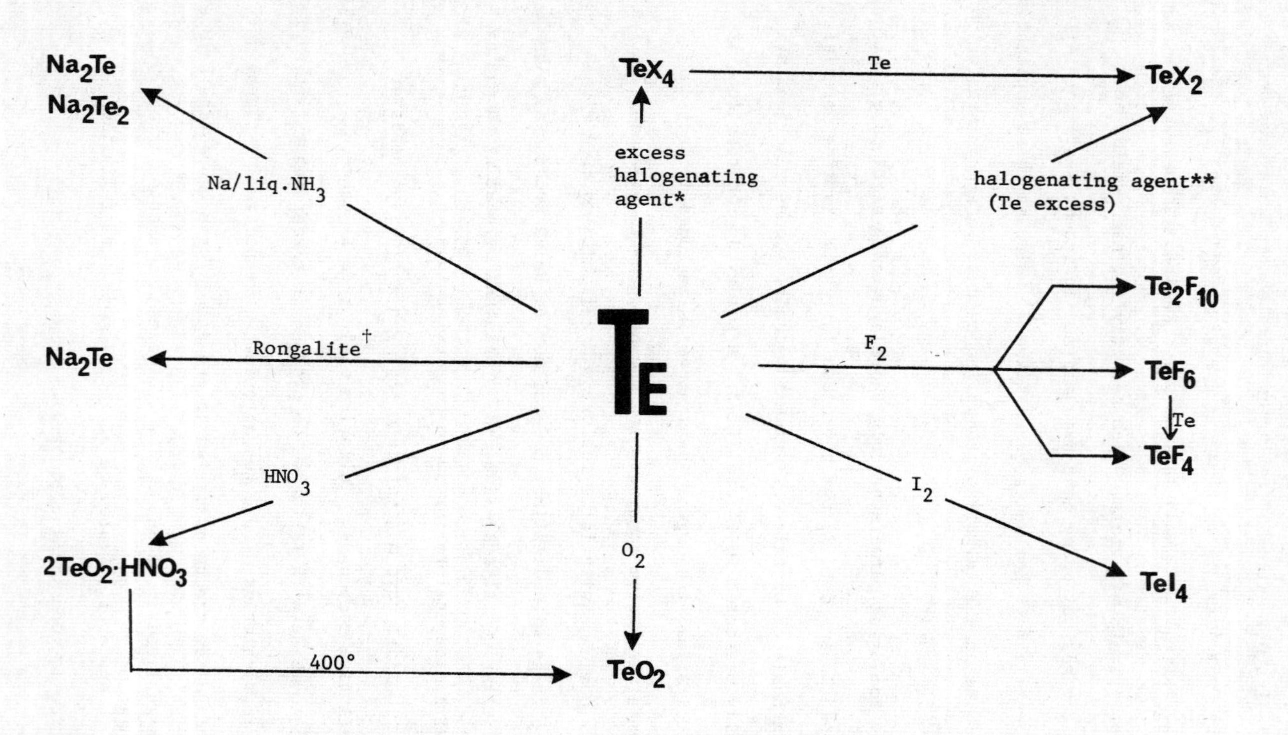

Fig. III-2: The Synthesis of Inorganic Tellurium Compounds from Elemental Tellurium

* Cl_2, Br_2, $SOCl_2$, SO_2Cl_2, S_2Cl_2, IBr ** Cl_2, Br_2, $SOCl_2$, SO_2Cl_2, S_2Cl_2, $BrCF_3$, CCl_2F_2 † $HOCH_2S(O)ONa$

(1) $Te + 2SO_2Cl_2 \rightarrow TeCl_4 + 2SO_2$

(2) $Te + 2SOCl_2 \rightarrow TeCl_4 + SO_2 + S$

(3) $Te + 2S_2Cl_2 \rightarrow TeCl_4 + 4S$

Molten iodine bromide and tellurium powder give a quantitative yield of tellurium tetrabromide, which is isolated by extraction with carbon tetrachloride[164].

In these halogenation reactions the product first formed is a tellurium dihalide[183,200,241,242,413], which then combines with additional reagent to the tetrahalide as shown by equation (4) for the reaction with chlorine.

(4) $Te \xrightarrow{Cl_2} TeCl_2 \xrightarrow{Cl_2} TeCl_4$

When tellurium and iodine were heated together and the crushed melt was extracted with carbon tetrachloride, only tellurium tetraiodide was isolated[91]. Attempts to prepare tellurium diiodide have failed[90,163].

Tellurium dihalides are best prepared by reducing the tetrachloride[241] or tetrabromide[44,93] with tellurium. Aynsley[13,14] obtained the dichloride by passing gaseous difluoiodichloromethane over molten tellurium. Similarly, the dibromide was formed from tellurium and bromotrifluoromethane at 500°[15]. The tellurium dihalides disproportionate in the solid state at room temperature to tellurium and the tetrahalide[13,15,92,93]. The decomposition proceeds also in solution and is instantaneous when traces of water are present[13,15,93]. Damiens[92] reviewed the tellurium halide work published prior to 1923. Table III-1 summarizes some of the properties of the tellurium halides.

Sodium telluride and sodium ditelluride are convenient starting materials for the preparation of organic tellurium compounds. Elemental tellurium when added to a solution of sodium metal in liquid ammonia yields sodium tellurides. The potentiometric titration of a tellurium

Table III - 1

Properties of Tellurium Halides

Compound	m.p. (°C)	b.p. (°C)	density g/ml (°C)	hydrolysis	soluble in	insoluble in
TeF_4	129.6	---		fast		
TeF_6	-38.9° (subl.)	---		slow		
Te_2F_{10}	-33.7	59.0	2.836 (25)	very slow	hydrocarbon, chlorinated hydrocarbons	
$TeCl_2$	208	328		very unstable	disproportionation in ether, dioxane, pyridine	CCl_4
$TeCl_4$	225	390	2.559 (232) 3.21 (18)	in moist air	acetic acid, toluene, benzene aliphatic alcohols chloroform, concd. aq.tartaric acid	CS_2, cyclohexane, CCl_4
$TeBr_2$	---	---	---	very unstable	ether, chloroform	
$TeBr_4$	363, 380	414-427	4.310 (15)		ether, glac. acetic acid, aqueous HBr $pH < 4.5$	benzene
TeI_4	280° (sealed tube)	---	5.403 (15)	slow (cold H_2O) fast (hot H_2O)	slightly in ether, ethanol, acetone	acetic acid, $CHCl_3$, CCl_4, CS_2

suspension with a solution of sodium in liquid ammonia gave evidence of the formation of Na_2Te, Na_2Te_2 and Na_2Te_4[452]. Kraus and coworkers investigated this reaction in detail. They found that Na_2Te is the initial product[201]. Since Na_2Te is only slightly soluble in liquid ammonia, it precipitates. With additional tellurium the soluble Na_2Te_2 is formed, which can incorporate more tellurium to give Na_2Te_4. The enthalpies for the reactions (5)-(9) were determined[203].

(5) $2Na\cdot(NH_3)_x + Te \longrightarrow Na_2Te$ $\Delta H = -86.9$ kcal

(6) $Na_2Te + Te \longrightarrow Na_2Te_2\cdot(NH_3)_x$ $\Delta H = -21.9$ kcal

(7) $Na_2Te_2 + xNH_3 \longrightarrow Na_2Te_2\cdot(NH_3)_x$ $\Delta H = -4.40$ kcal

(8) $Na_2Te_2\cdot(NH_3)_x + 2Te \longrightarrow Na_2Te_4\cdot(NH_3)_x$ $\Delta H = -1.90$ kcal

(9) $Na + xNH_3 \longrightarrow Na\cdot(NH_3)_x$ $\Delta H = +1.45$ kcal

From these data the enthalpies of formation of Na_2Te and Na_2Te_2 from the elements were calculated to be -84.0 and -101.5 kcal/mole, respectively[203]. Voronin[429a] obtained -84.3 kcal/mole for ΔH_f of Na_2Te at 850°K from data for liquid tellurium-sodium alloys.

It is possible to prepare under the proper conditions almost exclusively the monotelluride. When one mole of tellurium is added in portions to two moles of sodium in liquid ammonia, a colorless suspension of Na_2Te is formed[42]. When equimolar amounts of the two elements are combined in liquid ammonia at -78° a dark green solution is produced containing Na_2Te_2 among other products[76]. Sodium telluride, Na_2Te, has also been synthesized[122a] in 80 per cent yield by heating the elements in an evacuated quartz ampoule to 300°.

Tschugaeff and Chlopin[420] obtained tellurides by adding tellurium to an aqueous sodium hydroxide solution containing sodium formaldehyde sulfoxylate (Rongalite-C). The polytellurides initially produced imparted a permanganate-like color to the solution which became pale pink after the reduction to the monotelluride was completed. The molar ratios of

Rongalite to tellurium vary between 6.5:1[420] and 2.4:1[17,180,269]. Bergson[26] found, that tellurium with Rongalite in a 2:1 molar ratio produced predominantly the ditelluride, with monotelluride and polytellurides being present as impurities. All these reactions are performed under a nitrogen atmosphere to protect the tellurides from oxidation.

Tschugaeff[420] described the reactions by the following equations (10)-(13):

$$3Te + 6NaOH \longrightarrow 2Na_2Te + Na_2TeO_3 + 3H_2O \tag{10}$$

$$Na_2Te + Te \longrightarrow Na_2Te_2 \tag{11}$$

$$Na_2TeO_3 + 3HOCH_2SO_2Na + 3NaOH \longrightarrow Na_2Te + 3CH_2O + 3Na_2SO_3 + 3H_2O \tag{12}$$

$$Na_2Te_2 + HOCH_2SO_2Na + 3NaOH \longrightarrow 2Na_2Te + CH_2O + Na_2SO_3 + 2H_2O \tag{13}$$

After completion of the reaction the solution contains only Na_2Te, NaOH and Na_2SO_3. It is, however, possible, that the tellurium does not disproportionate according to equation (10), but is converted to the telluride as described in equation (14).

$$HOCH_2\overset{O}{\underset{O}{S}}\text{-Na} \xrightarrow{Te} HOCH_2\overset{O}{\underset{O}{S}}\text{-TeNa} \xrightarrow{NaOH} Na_2Te + CH_2O + Na_2SO_3 \tag{14}$$

Sodium dithionite, $Na_2S_2O_4$, in basic solution also reduces elemental tellurium to sodium telluride[405,420]. It is possible that the sodium sulfoxylate obtained according to equation (15) is the active reagent.

$$Na_2S_2O_4 + 2NaOH \longrightarrow Na_2SO_3 + Na_2SO_2 + H_2O \tag{15}$$

The reduction might then proceed as described above for Rongalite-C. The preparation of sodium telluride using $Na_2S_2O_4$ is inferior to the synthesis employing Rongalite. Woehler prepared impure potassium[444] and sodium[443] telluride by heating tellurium with the carbonaceous materials obtained either by thermal decomposition of potassium tartrate or by reduction of sodium carbonate with carbon. This method is only of historic interest.

Tellurium dioxide is prepared by thermal decomposition of $2TeO_2 \cdot HNO_3$ at 400-430°. The latter compound is formed, when tellurium reacts with concentrated nitric acid of density 1.42 g/ml[262].

IV. METHODS FOR THE INTRODUCTION OF TELLURIUM INTO ORGANIC MOLECULES

Only a general outline of the reactions leading to the formation of tellurium-carbon bonds will be given here. Specific cases will be dealt with in the appropriate sections describing the different classes of organic tellurium compounds.

A number of organic tellurium compounds can be synthesized directly from elemental tellurium. The inorganic tellurium compounds most often employed as starting materials are tellurium tetrahalides and alkali metal tellurides. Diphosphorus pentatelluride, aluminum telluride, tellur dioxide, hydrogen telluride, tetramethyoxy tellurium and hexamethyoxy tellurium have found only very limited application. Certain organic tellurium compounds can react further with organic reagents to yield compounds with more organic groups bonded to the tellurium atom than in the starting material.

A) Elemental Tellurium

Elemental tellurium undergoes the following reactions with formation of at least one carbon-tellurium bond. These reactions are summarized in Fig. IV-1.

1) Reactions with organic halides

Only one aryl iodide and a few alkyl iodides have been reacted with elemental tellurium to produce in yields of 50 per cent or less the dialkyl tellurium diiodides[83,96,391,423,425] (eq. 16).

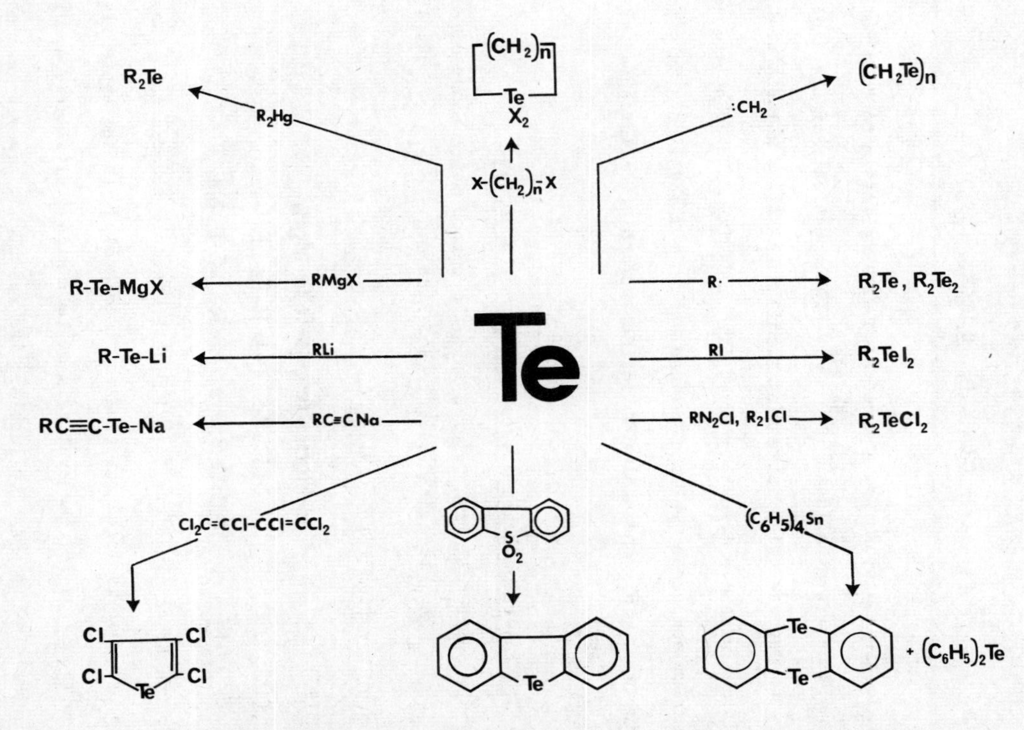

Fig. IV-1: The Synthesis of Organic Tellurium Compounds from Elemental Tellurium

(16) $$2RI + Te \longrightarrow R_2TeI_2$$

$$(R = CH_3, C_2H_5, C_6H_5CH_2, C_6F_5)$$

Organic dihalides, $X-(CH_2)_n-X$, combine with tellurium to form telluracycloalkane 1,1-dihalides[121,294,298]. The reactivity of the dihalides seems to decrease with decreasing atomic mass of the halogen atom. Dihalides with n = 4 and 5 produce five and six-membered heterocyclic tellurium compounds, while dihalides with n = 2 and 3 gave only indefinite products[121]. With hexachlorobutadiene tetrachlorotellurophene was produced[256].

2) <u>Insertion reactions into carbon-metal bonds</u>

Elemental tellurium inserts into the carbon-metal bonds of compounds of the type RMgBr[164a,323d,343], RLi[343,348b,c,d] and R-C≡C-Na[38a,41,43] as shown by equations (17)-(19).

(17) $$R\text{-}MgBr + Te \longrightarrow R\text{-}Te\text{-}MgBr$$

(18) $$R\text{-}Li + Te \longrightarrow R\text{-}Te\text{-}Li$$

(19) $$R\text{-}C{\equiv}C\text{-}Na + Te \longrightarrow R\text{-}C{\equiv}C\text{-}Te\text{-}Na$$

The products of equations (17)-(19) are unstable substances. They are, therefore, not isolated but treated immediately with the appropriate reagent required for the synthesis of tellurides, ditellurides, diorganyl tellurium dihalides and tellurols. The reactions, however, are not as simple as, for instance, represented by equations (17) and (20) for the synthesis of a diorganyl telluride.

(20) $$C_6H_5\text{-}Te\text{-}MgBr + RI \longrightarrow C_6H_5\text{-}Te\text{-}R + MgBrI$$

Petragnani and de Moura Campos[343] reported the isolation of phenyl tellurium trichloride, diphenyl tellurium dichloride and elemental tellurium after treating the reaction mixture, which originally contained tellurium and phenyl magnesium bromide or phenyl lithium, with SO_2Cl_2. They did not obtain an unsymmetric telluride upon addition of ethyl iodide to the reaction mixture, while Bowden and Braude[39] successfully synthesized ethyl phenyl telluride under similar conditions. Petragnani and de Moura Campos[343] proposed a radical mechanism and the reaction sequence given in equation (21), which both explain the results of their investigations.

(21)

$$R\text{-}MgBr + Te \longrightarrow R\text{-}Te\text{-}MgBr$$

$$R\text{-}Te\text{-}MgBr + Te \longrightarrow R\text{-}TeTe\text{-}R + Te(MgBr)_2 \xrightarrow{H_2O} H_2Te + Mg(OH)Br$$

$$R\text{-}TeTe\text{-}R \xrightarrow{SO_2Cl_2} \underline{\underline{RTeCl_3}} \qquad H_2Te \xrightarrow{O_2} \underline{\underline{Te}} + H_2O$$

$$R\text{-}TeTe\text{-}R + RMgBr \longrightarrow R\,Te\text{-}R + R\,Te\text{-}MgBr$$

$$R\,Te\text{-}R \xrightarrow{SO_2Cl_2} \underline{\underline{R_2TeCl_2}}$$

------- secondary reactions   isolated products (double underlined) $R = C_6H_5$

The detailed mechanism of this heterogeneous reaction must, however, still be considered unsolved. It has been shown, that elemental tellurium and a number of aromatic Grignard reagents in tetrahydrofuran produce in the presence of oxygen diaryl ditellurides in yields as high as 80 per cent[164a]. Aliphatic Grignard reagents did not react.

Sodium acetylides in liquid ammonia react with tellurium according to equation (19). The alkynyl sodium telluride without being isolated produced with alkyl halides the unsymmetric, unsaturated tellurides in yields of approximately 50 per cent[38a,41,43].

Equimolar amounts of bis(pentafluorophenyl) thallium bromide and tellurium were heated in a sealed tube for 3 days at 190°. Bis(pentafluorophenyl) telluride was obtained in 18 per cent yield[95a].

3) Reactions with organic radicals

Mirrors of elemental tellurium were removed by radicals like $CH_3\cdot$[366,367], $CF_3\cdot$[24], $C_3H_7\cdot$ and $CH_2:$[365] generated by thermal decomposition of methane[23,367], butane[366], acetone[148,149,366] diethyl ether[366], diisopropyl, methyl ethyl, methyl propyl, and methyl butyl ketone[148,149], dipropyl ketone[334], benzophenone[150], diazomethane[365,441] and hexafluoroacetone[24]. Diorganyl tellurides, ditellurides and polytelluroformaldehyde[365] were obtained as products. These reactions are unimportant as synthetic methods, since the organic tellurium compounds formed can be prepared much easier and in higher yields by other methods. Waters[437] proposed, that diphenyl tellurium dichloride, obtained from elemental tellurium and benzenediazonium chloride in acetone, is formed through the attack of phenyl radicals, generated by decomposition of the diazonium salt, on tellurium. This method was employed by Taniyama[415] for the synthesis of a number of diaryl tellurium dichlorides in yields of approximately 10 per cent. Diphenyliodonium chloride is claimed to react like the diazonium salts[375,376]. Free radicals might also be the intermediates in the formation of diphenyl telluride[35a,379] and telluranthrene[379] from tellurium and tetraphenyl tin.

4) Replacement of mercury, SO_2 groups and hydrogen

Diaryl mercury[3,83,171,173,199,254,407,429,449] reacts with tellurium upon heating in absence of a solvent according to equation (22).

(22) $$R_2Hg + 2Te \longrightarrow R_2Te + HgTe$$

The yields in these reactions range from 53-100 per cent.

Tellurium replaces the SO_2 group in biphenylylene sulfone[87] and thianthrene 5,5,10,10-tetroxide[330] upon heating, to give in both cases dibenzotellurophene (eq. 23). No reaction took place between diphenyl sulfone and tellurium at 380°[198]. Gurshovich[161] reported that 1,3-dimethylbenzimidazoline kept with tellurium at 100-200° produced 1,3-dimethyl-2-telluroxobenzimidazoline.

(23)

B) Tellurium Halides

Among the tellurium tetrahalides the tetrachloride is used almost exclusively. This section will therefore be devoted mainly to the reactions between tellurium tetrachloride and organic reagents, which proceed under formation of at least one carbon-tellurium bond. Only very few reports on the reactivity of the tetrabromide and tetraiodide are available. Fig. IV-2 summarizes the reactions of tellurium tetrachloride, which lead to organic tellurium compounds.

1) Condensation reactions of tellurium tetrachloride with elimination of hydrogen chloride

Tellurium tetrachloride combines with substances containing activated hydrogen atoms to form cyclic and linear condensation products. 1,3-Diketones, monoketones, carboxylic acid anhydrides, dimethyl sulfite[260a] and substituted aromatic compounds have thus far been investigated as organic components. Tellurium tetrabromide and tetraiodide do not undergo condensation reactions to the same extent as the tetrachloride,

a) 1,3-Diketones: 1,3-Diketones of the general formula *(1)*, with

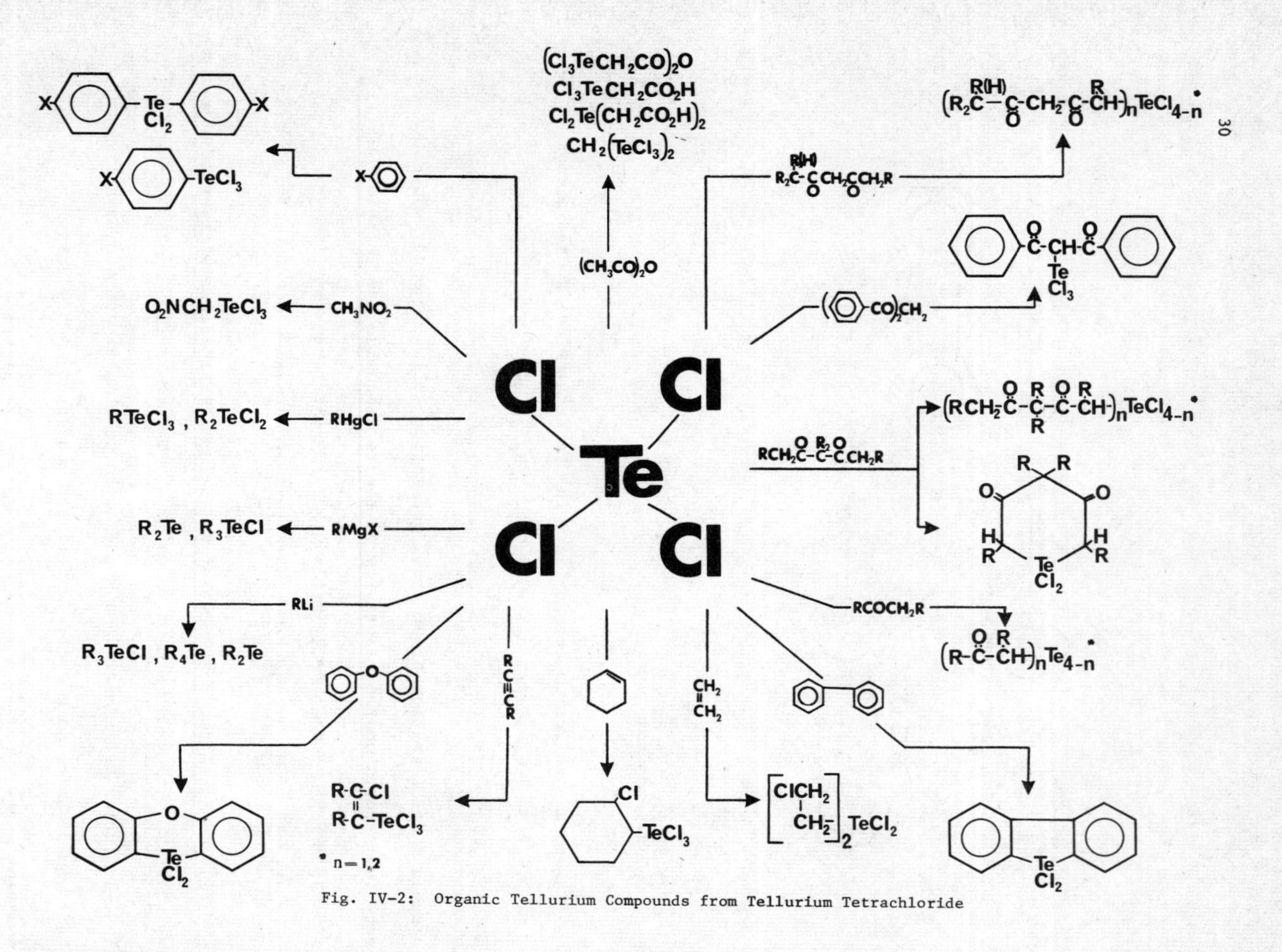

Fig. IV-2: Organic Tellurium Compounds from Tellurium Tetrachloride

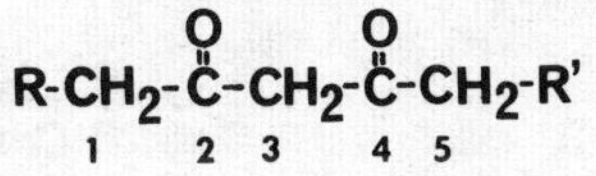

$$R\text{-}CH_2\text{-}\overset{O}{\overset{\|}{C}}\text{-}CH_2\text{-}\overset{O}{\overset{\|}{C}}\text{-}CH_2\text{-}R'$$

1 2 3 4 5

(1)

R and R' representing hydrogen atoms or the same or different aliphatic groups, condense with tellurium tetrachloride producing 1-tellura-3,5-cyclohexanedione 1,1-dichlorides *(2)* (eq.24).

(24) $RHC(H)\text{-}C(=O)\text{-}CH_2\text{-}C(=O)\text{-}CHR'(H) + Cl\text{-}TeCl_2\text{-}Cl \xrightarrow{-2HCl}$ (2)

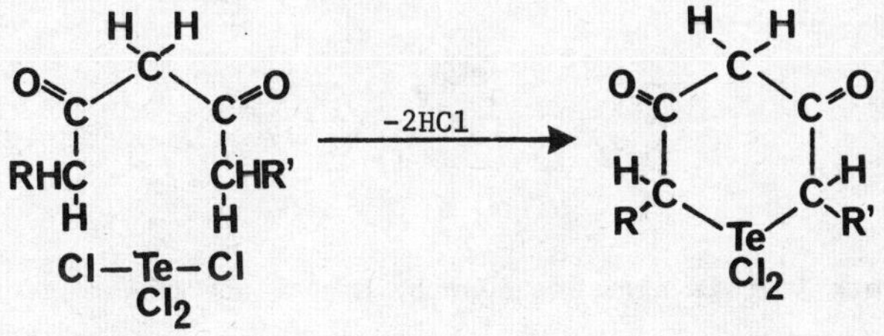

(2)

Condensation takes place at the 1 and 5 position of the ketone and not at the more activated 3 position. Substitution of alkyl groups for these methylene hydrogen atoms does not prevent cyclization. A phenyl group or two benzyl groups in 3 position, however, do not allow a cyclic product to be formed.

1,3-Diketones of the formula *(3)* or *(4)* produce compounds of structure *(5)*. Similar linear condensation products were obtained as by-products of the reactions of tellurium tetrachloride with the diketones *(1)*. Only when the carbon atoms in the 1 and 5 position do not bear any hydrogen atoms, will tellurium tetrachloride attack the methylene group in the 3 position. Thus, dibenzoylmethane yielded the very unstable (dibenzoyl)methyl tellurium trichloride *(6)*.

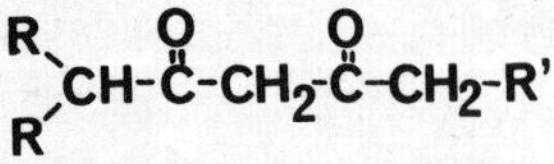

$$R_2CH-\overset{O}{\overset{\|}{C}}-CH_2-\overset{O}{\overset{\|}{C}}-CH_2-R'$$

(3)

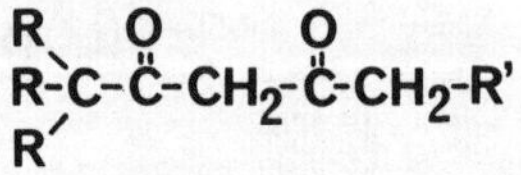

$$R_3C-\overset{O}{\overset{\|}{C}}-CH_2-\overset{O}{\overset{\|}{C}}-CH_2-R'$$

(4)

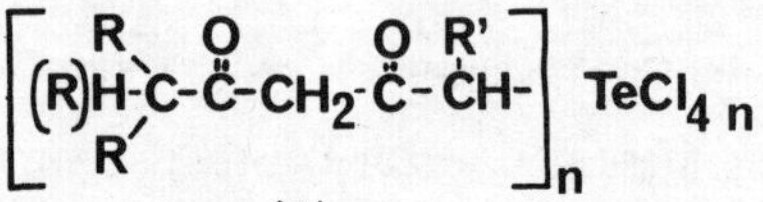

$$\left[(R)H\text{-}C(R)(R)-\overset{O}{\overset{\|}{C}}-CH_2-\overset{O}{\overset{\|}{C}}-\overset{R'}{CH}-\right]_n \cdot TeCl_{4n}$$

(5)

R,R' = alkyl; n = 1 and/or 2

$$C_6H_5-\underset{O}{\underset{\|}{C}}-\overset{TeCl_3}{CH}-\underset{O}{\underset{\|}{C}}-C_6H_5$$

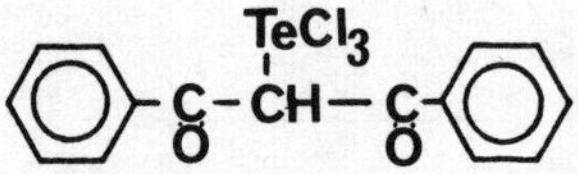

(6)

Most of the work in this area was done by Morgan and coworkers in the years 1920-1930. Pertinent literature references are cited in the sections dealing with specific compounds. Some of these condensation reactions between tellurium tetrachloride and 1,3-diketones[260] were reinvestigated in 1967; the earlier results were confirmed.

Morgan and coworkers[289] proposed, that enolization using hydrogen atoms in positions 1 and 5 is followed by addition of tellurium tetrachloride to the thus generated carbon-carbon double bonds. Hydrogen chloride is then eliminated from the intermediate chlorohydroxy compound restoring the keto function (eq. 25). Addition reactions of tellurium tetrachloride to carbon-carbon double bonds are known. They are discussed in section IV-B-2.

b) Monoketones: Monoketones react with tellurium tetrachloride in chloroform in a similar manner as 1,3-diketones yielding, however, only linear condensation products (eq. 26). The addition of the tetrachloride to the enol form of the ketone was discussed as a possible mechanism[290].

(25) $R-CH_2-\overset{O}{\overset{\|}{C}}-CH_2-\overset{O}{\overset{\|}{C}}-CH_2-R \rightleftarrows R-CH=\overset{OH}{\overset{|}{C}}-CH_2-\overset{OH}{\overset{|}{C}}=CH-R$

$\xrightarrow{TeCl_4}$

(26) $nR-CH_2-\overset{O}{\overset{\|}{C}}-CH_2-R' + TeCl_4 \longrightarrow \left[R-CH_2-\overset{O}{\overset{\|}{C}}-\overset{R'}{CH}-\right]_n TeCl_{4-n} + nHCl$

R,R' = aliphatic or aromatic group, n = 1 or 2

Organyl tellurium trichlorides and diorganyl tellurium dichlorides were present together in the reaction mixtures. However, only the less soluble compound was usually isolated[290]. The presence of one methylene group in α-position to the carbonyl group seems to be a requirement for a successful condensation, since 1,1,3,3-tetramethylacetone did not react with tellurium tetrachloride[290].

c) Carboxylic acid anhydrides: Acetic anhydride and tellurium tetrachloride in a 6:1 molar ratio in chloroform solution yielded bis(carboxymethyl) tellurium dichloride and methylene bis(tellurium trichloride) *(7)*. The reactants mixed in a molar ratio of 2:1 gave carboxymethyl tellurium trichloride and *(7)*. The primary condensation products were derivatives of acetic anhydride. Hydrolysis to the acid derivatives occurred during isolation of the products[286] (eq. 26a).

$$(26a)\quad (CH_3CO)_2O \xrightarrow{TeCl_4} \begin{cases} \xrightarrow{6:1,\ \text{then } H_2O} (HOOCCH_2)_2TeCl_2 + CH_2(TeCl_3)_2 \\ \xrightarrow{2:1} HOOCCH_2\text{-}TeCl_3 + \underset{(7)}{CH_2(TeCl_3)_2} \end{cases}$$

The reactivity of the higher anhydrides towards tellurium tetrachloride is markedly less than that of acetic anhydride[293]. Nitromethane is reported to yield nitromethyl tellurium trichloride[331].

d) Aromatic compounds: Benzene derivatives activated by an alkoxy[293,393], phenoxy[103,422a], thiophenoxy[339], hydroxy[358], dialkylamino[294], acetamido[358], 2-quinolyl[360] or 9-acridinyl[360] group condense with tellurium tetrachloride in refluxing chloroform or carbon tetrachloride or upon heating on the water bath without solvent to give aryl tellurium trichlorides. These trichlorides are insoluble in the solvents employed and precipitate during the reaction. Diaryl tellurium dichlorides are not obtained under these conditions, but will form upon prolonged heating of tellurium tetrachloride with a large excess of the organic reagent[293,297] (eq.27).

$$(27)\quad X\text{-}C_6H_5 \xrightarrow{TeCl_4} \begin{cases} X\text{-}C_6H_4\text{-}TeCl_3 \\ (X\text{-}C_6H_4)_2TeCl_2 \end{cases}$$

Ogawa[323a] reported in 1970 that the not activated benzene derivatives toluene and *o*-xylene, condense with tellurium tetrachloride at the boiling temperatures of these aromatic compounds to produce aryl tellurium trichlorides in good yields. Naphthyl alkyl ethers[293,360], naphthols[374], and monoethers of hydroquinone and resorcinol[293] did not give crystalline substances.

A systematic investigation with respect to the position taken by the trichlorotelluro group as a function of the substituents already in the aromatic ring has not been carried out. It seems, however, that the rules of aromatic substitution hold. A number of amines and heterocyclic nitrogen compounds form only adducts with tellurium tetrachloride[240,252,273,293,332,358,360].

Tellurium tetrabromide and phenetole heated at 180-90° produced only bromophenetoles and metallic tellurium[297]. Montignie[273,274] obtained elemental analyses for products isolated from reactions between tellurium tetrabromide and benzoquinone, naphthoquinone, anthraquinone and pyrrole, which suggest, that organic tellurium compounds had been formed.

Biphenyl heated with tellurium tetrachloride at 140-65° produced dibenzotellurophene dichloride[84] (eq. 28). Tellurium tetrabromide gave a low yield of the corresponding dibromide[84].

(28) $C_6H_5-C_6H_4-TeCl_3$ $\xrightarrow{-2HCl}$ dibenzotellurophene $TeCl_2$

Diphenyl ether[103,105], bis(4-methoxyphenyl) ether[146], bis(4-chlorophenyl) ether[422], 4-fluorodiphenyl ether[422a], and 4-methyl-4'-chlorodiphenyl ether[55] heated with tellurium tetrachloride without solvent condensed to derivatives of phenoxtellurine 10,10-dichloride *(8)*. Diphenyl sulfide did not react in this way[339].

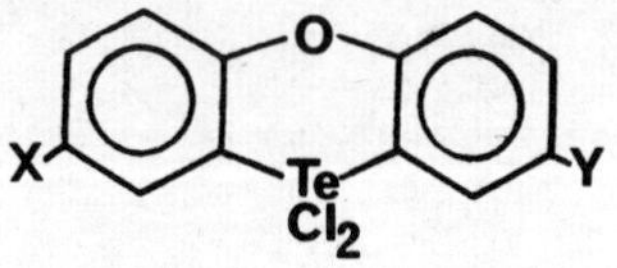

X, Y, O, Te, Cl_2

(8)

Concerning the mechanism of these reactions consult the section on heterocyclic tellurium compounds (XII-C-2).

2) Addition of tellurium tetrachloride to carbon-carbon multiple bonds

Addition of one mole tellurium tetrachloride to at least two moles of cyclohexene[140,323], ethylene[140], propene[12a,323b], 1-butene[323,323b] and 2,2-diphenyl-4-pentenoic acid[304] in carbon tetrachloride or acetonitrile gave diorganyl tellurium dichlorides. With equimolar quantities of cyclohexene[300,323,323b], with phenylacetylene and diphenylacetylene[305] only tellurium trichlorides were obtained. Diisobutylene, styrene, 1,4-diphenyl-1,3-butadiene and stilbene produced tellurium and unidentified viscous substances[304]. Elmaleh and coworkers, however, obtained from styrene and tellurium tetrachloride in diethyl ether a compound, whose analysis agreed with the formula $C_{16}H_{14}TeCl_3$. Diarylethylenes, however, gave with $TeCl_4$ adducts of the type $(R_2C{=}CH_2)_2 \cdot TeCl_2$[119a].

Tellurium tetrabromide did not react with ethylene[121], but gave with carbethoxymethylenetriphenylphosphorane the tellurium dibromide *(9)* as shown in equation (29)[345]. The addition reactions of tellurium tetrahalides have not been exhaustively investigated. There are reports, that ethylene does not combine with tellurium tetrachloride[121] and that 2-chlorocyclohexyl tellurium trichloride is stable in excess cyclohexene[304]. Further work in this area should clarify the disputed points and show whether tellurium tetrachloride will add to other types of unsaturated compounds.

(29) $$2(C_6H_5)_3P{=}CH{-}\overset{O}{\overset{\|}{C}}{-}OC_2H_5 + TeBr_4$$

$$\longrightarrow \left[\begin{array}{c} (C_6H_5)_3\overset{\oplus}{P}{-}CH{-}\overset{O}{\overset{\|}{C}}{-}OC_2H_5 \\ | \\ TeBr_2 \\ | \\ (C_6H_5)_3\underset{\oplus}{P}{-}CH{-}\underset{O}{\underset{\|}{C}}{-}OC_2H_5 \end{array}\right]^{++} 2Br^-$$

(9)

3) <u>Reactions of tellurium halides with organometallic compounds</u>

Tellurium tetrahalides have been reacted with Grignard reagents, organyl mercury chlorides, organic lithium compounds, diethyl zinc and bis(pentafluorophenyl)thallium bromide. Tellurium tetrabromide and tetraiodide gave the same yields in reactions with Grignard reagents as tellurium tetrachloride[364].

a) Grignard reagents: The reaction of tellurium tetrachloride with an excess of a Grignard reagent in ether or an ether-benzene mixture is complex. It is not possible to prepare, for example, a diorganyl tellurium dichloride in good yield by adding two moles of Grignard reagent to one mole of tellurium tetrachloride according to equation (30).

(30) $$2RMgX + TeCl_4 \longrightarrow R_2TeCl_2 + 2MgXCl$$

Elemental tellurium, diorganyl tellurium dichlorides, diorganyl tellurides, and triorganyl telluronium chlorides have been observed as products of such reactions. Rheinboldt and Petragnani[364] observed, that the formation of elemental tellurium by reduction of tellurium tetrachloride is largely prevented, when the etheral Grignard solution diluted with twice the volume of benzene is added slowly to an ice-cold suspension of the tetrachloride in ether. Refluxing such a mixture gave diorganyl tellurides in yields of more than 80 per cent[364]. When the reaction mixture was kept cold and then hydrolyzed[316] triorganyl telluronium halides were isolated in moderate yields together with tellurides. The molar ratios of Grignard reagents to tellurium tetrachlorides employed in these reactions varied between 5:1 and 10:1.

Since it is known that diorganyl tellurium dichlorides are reduced by Grignard reagents to diorganyl tellurides[229,232-236,238], according to equation (31), it is likely that the diorganyl tellurium dichlorides are intermediates in the direct preparation of the tellurides from tellurium tetrachloride.

$$(31) \qquad R_2TeCl_2 + 2CH_3MgI \longrightarrow R_2Te + C_2H_6 + MgI_2 + MgCl_2$$

The conversion of the diorganyl tellurium dichlorides to tellurides probably proceeds *via* a tetraorganyl tellurium compound. Wittig and Fritz[442] and Hellwinkel and Fahrbach[171-173] have synthesized such tetraorganyl tellurium compounds and found, that their thermal decomposition leads to diorganyl tellurides. The tetraalkyl tellurium compounds decompose already in solution at room temperature producing dialkyl tellurides[172]. In the reactions between tellurium tetrachloride and a Grignard reagent enough of the latter was always used, to convert $TeCl_4$ into tetraorganyl tellurium (eq. 32). The thermal decomposition and hydrolysis of these compounds would give the observed product (eq. 33). The elucidation of the exact mechanism of the reaction of Grignard reagents with tellurium tetrahalides will come only after a more detailed study in this area. No definite conclusions can be drawn from the data now available.

Lederer[219,222-226,232-236] used "tellurium dihalides", together with a four-fold molar excess of Grignard reagents in preparing organic

(32) $$TeCl_4 + 4RMgX \longrightarrow R_4Te + 4MgXCl$$

(33) $$R_4Te \longrightarrow R_2Te + R_2$$
$$R_4Te \xrightarrow{H_2O/HX} R_3TeX + RH$$

tellurium compounds. He wrote the following equation (34) for these reactions.

(34) $$TeX_2 + 2RMgX \longrightarrow R_2Te + 2MgX_2$$

After hydrolysis of the reaction mixture, elemental tellurium, hydrocarbons, diorganyl tellurides and ditellurides were isolated. It is almost certain, that these reactions were run with a mixture of tellurium and tellurium tetrahalide. Aynsley and coworkers[13,14,15] showed that tellurium dihalides disproportionate into tellurium and tellurium tetrahalide in ethereal solution, especially in presence of traces of moisture. The Grignard reagent will then react with elemental tellurium and tellurium tetrahalide. The reactions with tellurium have been discussed in section IV-A-2

b) Aryl mercury chlorides: Aryl mercury chlorides, which can be prepared from mercuric chloride and diazonium salts, are employed to form carbon-tellurium bonds, when a direct condensation between tellurium tetrachloride and an aromatic compound is not possible. Equimolar quantities of the aryl mercury chloride and tellurium tetrachloride in chloroform, acetonitrile or dioxane give exclusively the aryl tellurium trichloride[55,122,197,304,339,357,363] in yields ranging from 60-90 per cent.

(35) $$RHgCl + TeCl_4 \longrightarrow RTeCl_3 + HgCl_2$$

Aryl mercury chlorides can also be used to synthesize diorganyl tellurium dichlorides as has been shown in the case of the reaction of 2,2-diphenyl-5-chloromercuri-4-pentanolactone and tellurium tetrachloride in a 2:1 molar ratio[304]. This reaction should be capable of more extensive application.

c) Lithioalkanes: Organic lithium compounds combine with tellurium tetrachloride with elimination of lithium chloride. Again one would expect the reactions to proceed according to equation (36) depending on the molar ratios employed.

(36)
$$TeCl_4 \begin{cases} + 1RLi \longrightarrow RTeCl_3 + LiCl \\ + 2RLi \longrightarrow R_2TeCl_2 + 2LiCl \\ + 3RLi \longrightarrow R_3TeCl + 3LiCl \\ + 4RLi \longrightarrow R_4Te + 4LiCl \end{cases}$$

All reported reactions with one exception used at least 4 moles of the lithium compound for each mole of tellurium tetrachloride. The products recovered in these cases were diorganyl tellurides[45,83], tetraorganyl tellurium[83,171,172,442] and triorganyl telluronium compounds[172]. The equations (37)-(39) outline the postulated reaction steps.

(37)
$$4RLi + TeCl_4 \longrightarrow R_4Te + 4LiCl$$

(38)
$$R_4Te + HX \longrightarrow R_3TeX + RH$$

(39)
$$R_4Te \longrightarrow R_2Te + R_2$$

Reaction (39) proceeds very easily at room temperature with tetraalkyl tellurium compounds. The triorganyl telluronium compounds are probably formed by hydrolysis of R_4Te according to equation (38).

Only Kostiner and coworkers[197] used 3 moles of an organolithium compound per mole of tellurium tetrachloride. Tris(perfluorophenyl) telluronium chloride was thus obtained from perfluorophenyl lithium.

C) Alkali Metal Tellurides and Other Inorganic Tellurium Compounds

Alkali metal tellurides are easily alkylated by organic chlorides, bromides and iodides (eq. 40). The alkylation reactions are carried out

(40)
$$Na_2Te + 2RX \longrightarrow R_2Te + 2NaX$$
$$Na_2Te_2 + 2RX \longrightarrow R_2Te_2 + 2NaX$$

either in an aqueous medium, when the telluride was prepared according to Tschugaeff and Chlopin[420], or in liquid ammonia[42], when the telluride was synthesized from the alkali metal and elemental tellurium. Although the aqueous systems are easier to handle, liquid ammonia is the appropriate solvent for higher alkyl halides, which are not appreciably soluble in an aqueous medium. Diorganyl tellurides[17,18,42,121] and a few diorganyl ditellurides[26,76] were prepared in this way. Mack[257] discovered an interesting method to introduce tellurium into an organic molecule. Upon mixing stoichiometric amounts of diacetylenes, R-C≡C-C≡C-R, with sodium telluride in methanol at room temperature and hydrolyzing the reaction mixture, 2,5-disubstituted tellurophenes were isolated[257]. A patent was recently taken out on this process[P-13]. Although the authors do not speculate about the reaction mechanism, one can postulate that sodium telluride adds to the carbon-carbon triple bond forming a 3,4-disodiotelluro-phene derivative, which then undergoes hydrolysis as shown in scheme (41). Hydrolysis experiments with heavy water and attempts to carry out this reaction in an inert solvent should clarify the mechanism.

(41)

C——C   Na–C——C–Na

R-C   C-R   + Na_2Te →   R-C–Te–C-R

↓

HC——CH

R-C–Te–C-R

R = H[133b,257], CH_2OH, $C(CH_3)_2OH$,

C_6H_5, pyrrolidinylmethyl[257]

Aluminum telluride, Al_2Te_3, besides being employed instead of sodium telluride in the synthesis of telluracycloalkanes[294,298], reacted with aliphatic alcohols at 300-350° to produce tellurols[19] and tellurides[310] (eq. 42).

(42)

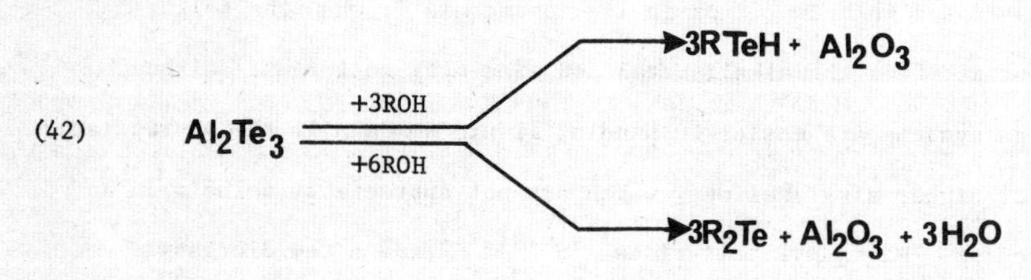

A British patent[P-8] claims, that saturated aliphatic carboxylic acids with at least seven carbon atoms gave with diphosphorus pentatelluride, P_2Te_5, alkanecarboditelluroic acid anhydrides, RC(Te)-Te-C(Te)R. Hydrogen telluride bubbled through a mixture of a ketone with concentrated hydrochloric acid produced telluroketones, $R_2C{=}Te$[255]. The preparation of bis(2-cyanoethyl) telluride by electrolysis of a suspension of tellurium in 1<u>N</u> aqueous Na_2SO_4 in the presence of acrylonitrile at a pH between 7.5 and 9, involves the addition of electrolytically generated hydrogen telluride to the ethylenic bond[191].

Tetra- or hexamethoxy tellurium treated with 2,2'-dilithiobiphenyl yielded bis(biphenylylene) tellurium[171]

D) Organic Tellurium Compounds

Only a very general outline of the reactions of organic tellurium compounds leading to the formation of additional carbon-tellurium bonds will be given here. For more detailed information the sections dealing with the reacting compound and the product should be consulted.

Alkylation of naphthyl tellurium iodide with Grignard reagents was used to synthesize unsymmetric tellurides[305,427]. Tetraorganyl tellurium compounds, which were hydrolyzed to triorganyl telluronium compounds, and diorganyl tellurides were the products of the reaction of Grignard reagents with diorganyl tellurium dichlorides[237] and diorganyl ditellurides[342], respectively. Diphenyl tellurium dichloride[442], triphenyl telluronium chloride[442] and trimethyl telluronium iodide[171] with the appropriate organic lithium compounds formed tetraorganyl tellurium compounds. The aliphatic derivatives could not be isolated. Triethyl telluronium chloride when heated with diethyl zinc to 100° yielded diethyl telluride, tetraethyl tellurium again being a likely intermediate[261]. Organyl mercury chlorides reacted with organyl tellurium trichlorides and produced dialkyl tellurium dichlorides[363], which can also be obtained by addition of trichlorides to carbon-carbon double bonds[300,302,304]. Organyl tellurium trichlorides and tribromides condensed with acetone, acetophenone, 4-phenoxy-, 4-methoxy-, 4-ethoxy-, 4-dimethylamino- and 2,4-dihydroxybenzene to form diorganyl tellurium dihalides[340]. 2-Phenoxyphenyl tellurium trichloride[103] and the corresponding thio derivative[339] condensed intramolecularly to phenoxtellurine and thiophenoxtellurine, respectively. Trichlorides with ditellurides gave diorganyl tellurium dichlorides and tellurium[341] (eq. 43).

(43) $$2R_2Te_2 + 2RTeCl_3 \longrightarrow 3R_2TeCl_2 + 3Te$$

Diorganyl tellurides combine with organic halides to give triorganyl telluronium halides. Diorganyl ditellurides are cleaved by methyl iodide to diorganyl methyl telluronium iodide and organyl methyl tellurium diiodide[288], while *vic*-dibromides transform ditellurides into diorganyl tellurium dibromides with elimination of tellurium[341] (eq. 44).

(44) $$R{-}TeTe{-}R + R{-}CHBrCHBr{-}R \longrightarrow R_2TeBr_2 + Te + RCH{=}CHR$$

Ditellurides decompose thermally to tellurides and tellurium[341]. Fig. IV-3 summarizes the reactions of organic tellurium compounds, which proceed with formation of at least one additional carbon-tellurium bond.

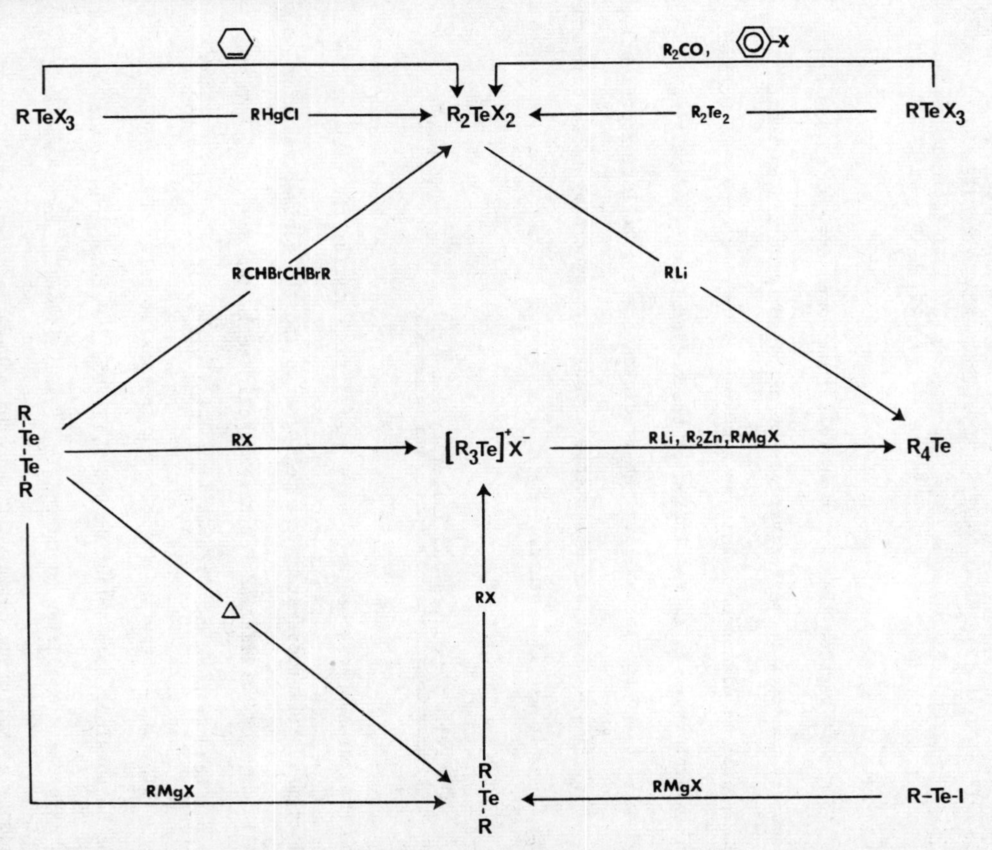

Fig. IV-3: Reactions of Organic Tellurium Compounds Proceeding with Formation of Additional Carbon-Tellurium Bonds.

V. TELLUROCYANIC ACID, TELLUROCYANATES, TELLURIUM DICYANIDE, TELLURIUM DI(THIOCYANATE), BIS(N,N-DIISOPROPYLTHIOCARBAZOYL) DITELLURIDE AND TELLURIUM DERIVATIVES, XCTe

In this section the few known tellurium compounds, which one can formally consider to be derivatives of carbon dioxide and carbonic acid, will be discussed. This area of tellurium chemistry has not been fully explored. By extrapolation of thermodynamic data of pertinent group VI compounds instability was predicted for the tellurium derivatives. A number of contradictory statements especially concerning tellurocyanates are found in the literature.

A) Tellurocyanates and Tellurocyanic Acid

Tellurocyanates would be very useful reagents for the preparation of organic tellurocyanates, for tellurourea and its derivatives, for tellurosemicarbazides and for heterocyclic tellurium compounds. Telluro-cyanic acid is unknown. Its melting point was estimated to be -45°[34].

Berzelius[28] reported in 1821 that potassium tellurocyanate could not be obtained from tellurium and $K_4[Fe(CN)_6]$, although selenium reacted as expected. According to Oppenheim[324] and Shimose[393] only very small amounts of tellurium dissolve in aqueous potassium cyanide solutions. Sulfur and selenium formed the respective cyanates. Telluropentathionate, contrary to the selenium compound (eq. 45), hydrolyzed according to equation (46) and did not produce any tellurocyanate[124]. A very small amount of tellurium dissolved in various basic solvents like pyridine and ethylene-diamine containing hydrogen cyanide. Upon dilution with ether the reaction mixtures deposited the dissolved tellurium[268]. Greenwood and coworkers[157] were unable to isolate tellurocyanates after refluxing a suspension of

(45) $$Se(S_2O_3)_2^{--} + 2CN^- + 2OH^- \longrightarrow S_2O_3^{--} + SeCN^- + SCN^- + SO_4^{--} + H_2O$$

(46) $$2Te(S_2O_3)_2^{--} + 6OH^- \longrightarrow 4S_2O_3^{--} + Te + TeO_3^{--} + 2H_2O$$

tellurium in an aqueous or ethanolic solution of potassium cyanide. The reactants failed to combine when suspended in acetone. The desired compound could not be obtained by fusing tellurium with potassium cyanide or a KCN-NaCN eutectic[157]. Tellurium dicyanide and potassium telluride[157] did not yield potassium tellurocyanate (eq. 47).

(47) $$Te(CN)_2 + K_2Te \not\longrightarrow 2KTeCN$$

Tellurium dissolved very slowly in liquid ammonia containing aluminum or potassium cyanide. Evaporation of these solutions to dryness caused deposition of tellurium. Potassium tetratelluride was unreactive towards potassium cyanide[27]. Wassermann, however, claimed to have obtained the respective tellurocyanates from a tellurium-sodium cyanide or tellurium-potassium cyanide melt[P-2].

Predictions of thermodynamic properties and spectral characteristics of the tellurocyanate ion based on extrapolations in the series XCN^- (X = O,S,Se) are found rather frequently in the literature. The standard free energy of formation of $TeCN^-$ was estimated to be +54 kcal/mole[165]. Birckenbach and Kellermann[32] measured the reducing power of pseudohalide anions. From experiments with aqueous and ethanolic solutions prepared from tellurium and potassium cyanide the chalcogenocyanate anions were arranged in the following order of increasing reducing power (decreasing stability).

$$OCN^- < SCN^- \ll SeCN^- \ll TeCN^-$$

These solutions were not 0.1M with respect to KTeCN as erroneously stated in reference 32*. Gusarsky and Treinin[162] measuring the ultraviolet spectra of the anions XCN^- (X = S, Se, Te) detected a shift of the absorption bands to longer wavelength with decreasing electronegativity of the group VI atoms. The long wavelength maximum at ~260mμ observed with a solution of unknown concentration obtained by shaking tellurium in a 0.1M sodium cyanide solution, was attributed to a charge transfer-to-solvent process. A shoulder was located at ~235mμ.

Vibrational force constants and the location of ir-absorption bands for $TeCN^-$ were calculated. It was assumed, that in the series XCN^- (X = O, S, Se, Te) the force constants, k_{C-X}, are linearly dependent upon the ionization potential of the atom X. Thus, the band corresponding to the Te-C stretching vibration should be found at 455 cm^{-1}, while the bending mode should give rise to a band at ~400 cm^{-1} [157]. Wagner[436] assuming a carbon-tellurium bond length of 2.03Å in $TeCN^-$ calculated π-bond orders for this anion using an internally consistent procedure based on "unsymmetrical group orbitals". The mean amplitudes of vibrations for the tellurocyanate ion were calculated by Nagarajan and Hariharan[309].

Downs[101] reported in 1968 the isolation of a salt of the tellurocyanate anion. He found, that acetone, a powerful solvent for potassium selenocyanate but not for potassium cyanide, when warmed with tellurium and potassium cyanide, produced a solution having an absorption band at 2079 cm^{-1}. Greenwood and coworkers had predicted 2083-2086 cm^{-1} as the location of the C-N stretching mode in the $TeCN^-$ anion[157]. A shift to lower wavenumbers was observed, when tellurium was replaced by selenium and sulfur in the solution. Addition of water to the acetonic, tellurium containing sylution or evaporation of the solution caused tellurium to precipitate. An attempt to prepare Cs(TeCN) was equally unsuccessful[101].

* See correction in reference 33, p. 10, footnote 2.

From these observations it can be concluded that tellurocyanates exist in solution, but that cations even as large as Cs^+ cannot stabilize the tellurocyanate anion in the solid state.

Tetraethylammonium cyanide reacted in dimethylformamide with tellurium. The solution deposited upon evaporation pale yellow crystals having the formula *(10)*. Austad[12b] prepared tetramethylammonium and

$$[(C_2H_5)_4N]^+ TeCN^- \cdot H\overset{O}{\overset{\|}{C}}\text{-}N(CH_3)_2$$

(10)

tetraphenylarsonium tellurocyanate in good yields from the corresponding cyanides and excess tellurium in acetonitrile. The tellurocyanates could be recrystallized from acetone. The tetraphenylarsonium salt decomposed only slowly in humid air. It was stable for months, even in direct sunlight, when stored in a closed bottle. The tetramethylammonium derivative is less stable. It slowly darkens on storage and decomposes rapidly in humid air. The Raman spectrum[101] in dimethylformamide showed a band at 2080 cm^{-1}, while the corresponding ir-absorption[12b] in acetonitrile was located at 2081 cm^{-1}. Maxima in the uv-region were found at 247 and 273 mμ[12b].

Solutions of tellurocyanates rapidly deposit tellurium in presence of oxygen, on treatment with water and on addition of non-aqueous solutions of the first row transition metal salts and of silver, lead and mercury[101]. The dry salts were unaffected by oxygen[12b].

Organic tellurocyanates, R-TeCN, according to claims by Borglin[P-31 - P-35, P-36 - P-39, P-40, P-41, P-43, P-44, P-51] and Wasserman[P-2] were prepared from metal tellurocyanates and appropriate organic reagents. The first concrete directions for the preparation of an organic tellurocyanate were given by Nefedov[319]. He heated a carefully ground mixture of equimolar amounts of potassium cyanide and tellurium for 3 hours by increasing the temperature from 100-250°. The cooled melt was extracted with acetonitrile. The resulting solution after addition of copper(II) acetate gave with azulene the tellurocyanate *(11)* as violet needles with a melting point of 80-1° in 7 per cent yield.

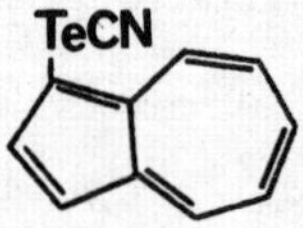

TeCN

(11)

Attempts to prepare isotellurocyanates, RNCTe, from isonitriles and tellurium failed[131,245]. Thayer[418] was unable to synthesize $R_3SiNCTe$ from the corresponding isonitrile and tellurium, although selenium had given the expected compound.

B) Tellurium Dicyanide and Tellurium Di(thiocyanate)

Tellurium dicyanide was first prepared by Cocksedge[81] according to equation (48).

$$TeBr_4 + 3AgCN \longrightarrow Te(CN)_2 + 3AgBr + BrCN \quad (48)$$

The reactants were heated in purified, dry benzene for three days. Tellurium dicyanide precipitated. It was dissolved in ether after decantation of the benzene. Upon evaporation of the solvent colorless crystals of the adduct $2Te(CN)_2 \cdot (C_2H_5)_2O$ were obtained. The coordinated ether was released upon heating on a water bath. The solvent-free compound can be purified by vacuum distillation, but approximately half of the material is lost by decomposition[81]. A rapid high vacuum sublimation at 120° gave a slightly pink colored product in 80 per cent yield[136]. The thermal decomposition of $Te(CN)_2$ was slow at 100° but at 190° a sudden gas evolution and partial sublimation of tellurium dicyanide occurred.[81]

Tellurium dicyanide decomposes in air in a few minutes with formation of elemental tellurium[81]. Water hydrolyzed the compound[33,81] according to equation (49).

$$2\,Te(CN)_2 + 3\,H_2O \longrightarrow Te + H_2TeO_3 + 4HCN \quad (49)$$

Tellurium dicyanide dissolves in cold methanol, but on warming hydrolysis takes place. It is slightly soluble in chloroform, carbon tetrachloride and benzene. A saturated solution in diethyl ether contains 1g in 60 ml[81].

Challenger and coworkers[60] obtained diphenylcyanobismuthine and diphenyl ditelluride by combining tellurium dicyanide and triphenylbismuthine in ether at room temperature. The scheme (50)-(51) was offered as a possible reaction sequence.

$$R_3Bi + Te(CN)_2 \longrightarrow R_2BiCN + RTeCN \quad (50)$$

$$RTeCN + 2H_2O \longrightarrow RTeH + HCNO$$

$$2\,RTeH + 0.5O_2 \longrightarrow R_2Te_2 + H_2O \quad (51)$$

$R = C_6H_5$

Rheinboldt[362], however, pointed out, that there are alternate and more likely mechanisms which would account for the observed products.

Fritz and Keller[136] reported the infrared absorption bands of tellurium dicyanide: 2181(m,sh), 2179(m), 1316(w), 1086(w), 460(vw) and 403(s) cm^{-1}. They assigned the 2181 and 2179 cm^{-1} bands to C≡N modes, while the 403 cm^{-1} band was assigned to the antisymmetric Te–C stretching vibration. Greenwood and coworkers[157] on the basis of their calculations believe the 403 cm^{-1} band to be caused by the bending mode, while the stretching vibration is responsible for the 459 cm^{-1} absorption.

Montignie[274] prepared tellurium di(thiocyanate), $Te(SCN)_2$, from silver thiocyanate and tellurium tetrabromide in benzene. Ditellurium dicyanide,

$Te_2(CN)_2$, could not be obtained by electrolysis of a solution containing potassium tellurocyanate[32].

C) Bis(N,N-diisopropylthiocarbazoyl) Ditelluride

Anthoni, Larsen and Nielsen[12] bubbled a fast stream of hydrogen telluride through a 10 per cent ethanolic solution of N-isothiocyanato-diisopropylamine *(12)* and isolated in 70 per cent yield brick-red crystals of the ditelluride *(14)* (eq. 52).

$$\text{(52)}\quad \underset{(12)}{R_2N\text{-}N{=}C{=}S} + H_2Te \longrightarrow \underset{(13)}{R_2N\text{-}NH\text{-}C(=S)TeH} \xrightarrow{O_2} \tfrac{1}{2}\,\underset{(14)}{R_2N\text{-}NH\text{-}\overset{S}{\overset{\|}{C}}\text{-}TeTe\text{-}\overset{S}{\overset{\|}{C}}\text{-}NH\text{-}NR_2}$$

$R = i\text{-}C_3H_7$

The expected tellurothiocarbazic acid *(13)* could not be isolated. The extreme sensitivity of the hydrotelluro group toward oxidation facilitated the conversion of the acid *(13)* to the ditelluride *(14)*. The authors did not state whether an inert, oxygen free atmosphere besides excess hydrogen telluride was employed in this synthesis. The ditelluride structure *(14)*, although probably correct, has not yet been confirmed by experimental data. The compound *(14)* decomposed at 110-3° without melting. In air it deposited elemental tellurium. Its infrared absorption bands are reported.

D) Tellurium Derivatives, X=C=Te (X = O, S, Se)

The tellurium compounds X=C=Te, with X representing a group VI element, have all been prepared with the exception of carbon ditelluride, CTe_2. The physical and chemical properties of these rather unstable tellurides are largely unknown. Carbon oxide telluride, O=C=Te, a gas at room temperature, was prepared from carbon monoxide and tellurium in very small yields in 1944[369]. No further reports concerning this substance were found in the literature.

Carbon sulfide telluride, S=C=Te, is the best known compound in this series. Stock and Praetorius[411] first prepared it by passing an electric arc under carbon disulfide between a graphite cathode and a tellurium-10 per cent graphite anode. The difficult separation from the by-product C_3S_2 was accomplished by distillation and addition of naphthylamine, which converted C_3S_2 into thiomalonnaphthylamide. Other investigators[398,438] using the same method, kept the reaction mixture at 0°[438], maintained a CO_2 atmosphere above the reaction mixture to minimize explosion hazards[438], and performed the distillation of the product at -78°[398].

Pure carbon sulfide telluride decomposes even at -50° under very weak illumination. Carbon disulfide solutions of a concentration smaller than 5 per cent, however, are stable enough at room temperature under normal incandescent light to permit infrared studies[438]. The telluride, a red solid at -80°, with a density of 2.9g/ml at -50°[411], melts at -54° and decomposes at higher temperatures[438]. Recently Steudel[408] prepared carbon sulfide telluride from carbon monosulfide and tellurium. Carbon monosulfide was generated from carbon disulfide at 0.1 torr in a high frequency discharge. The walls of the discharge tube were covered with tellurium. Similarly, by employing carbon diselenide, an eight per cent yield of carbon selenide telluride was obtained[408]. The thermal instability of this compound prevented its purification by sublimation in high vacuum.

The claim of Stock and Blumenthal[410], that they had obtained carbon ditelluride under conditions similar to the ones used in the carbon sulfide telluride synthesis, was withdrawn by Stock and Praetorius[411]. The enthalpy of formation of carbon ditelluride is expected to be highly positive making its existence doubtful.

	$CO_{2\,(g)}$	$CS_{2\,(l)}$	$CSe_{2\,(l)}$	CTe_2
ΔH_f^o(kcal/mole)	-94.1	+21.4	+34	+?

The attempts to prepare this compound from carbon tetrachloride and hydrogen telluride at 500°, from silver telluride and carbon tetrabromide or tetraiodide, from carbon and tellurium[158] and from methylene chloride and tellurium at 450-80°[186] have been unsuccessful.

Besides the thermal decomposition of X=C=Te compounds only two reactions of SCTe, given in equation (53) and (54), have been reported[411].

(53) $$SCTe + Br_2 \xrightarrow{CS_2} TeBr_4 + Te + \text{other coumpounds}$$

(54) $$2SCTe + Hg \xrightarrow{\text{benzene}} CS_2 + C + 2Te$$

The infrared investigations carried out with X=C=Te compounds are summarized in Table V-1. The instability of these tellurides complicated these experiments. Data are available only for S=C=Te and Se=C=Te. Calculations for the other molecules gave approximate frequencies for their fundamental modes[190,438]. Bond stretching force constants[190,439] and interaction force constants[439] were also derived. Cyvin[88,89] calculated the mean square amplitudes of vibration for S=C=Te. The bond lengths in carbon sulfide telluride were obtained from microwave data as 1.90Å and 1.557Å for C-Te and C-S, respectively[167,398]. The molecular dipole moment was found to be 0.172 ± 0.002 Debye units[167].

The tellurium analog of carbon monoxide, carbon monotelluride, CTe, has not yet been detected. In the theoretical treatment of diatomic heteronuclear molecules this compound has been frequently included. The internuclear equilibrium distance was deduced from Sutherland's potential functions to be 1.949Å[414,421]. For the dissociation energy values between 4.8 and 6.06 eV were obtained[110,196,406]. The force constant for an infinitesimal amplitude is 4.945×10^{-5} dyne/cm[414,421]. Its natural frequency of vibration should be 875 cm^{-1} [196].

Table V-1

The Calculated and Experimental Infrared Absorption Frequencies of the Compounds X=C=Te

Compound		C=Te stretching mode, cm^{-1}	Bending mode, cm^{-1}	C=X stretching mode, cm^{-1}
O=C=Te	calcd	567[1)]	---	2010[3)], 2042[1)]
	exp.	---	---	---
S=C=Te	calcd.	435[1)]	---	1385[1)]
	exp.	423[3),4),6)]	337[3),4),5)]	1347[3),4),6)]
Se=C=Te	calcd.	312[1)]	---	1225, 1222[1)]
	exp.	294[2),8)]	---	1179[2),7)]
Te=C=Te	calcd.	257[1)]	---	1121[1)], 1130[3)]
	exp.	---	---	---

1) reference 190, 2) reference 408, 3) reference 438, 4) reference 439,

5) deduced from first overtone

6) observed directly and calculated from bands corresponding to the sum and difference of these two bands

7) observed directly

8) deduced from combination bands

VI. COMPOUNDS CONTAINING A SINGLE CARBON-TELLURIUM BOND

The compounds treated in this section contain only one carbon-tellurium bond. The second bond to the tellurium atom in divalent tellurium derivatives is formed with the following atoms or groups: H, Cl, Br, I, CNS, ClO_4, NO_3, $-S_2O_3Na$, $-S-\overset{O}{\underset{O}{S}}-CH_3$, $-S-\overset{O}{\underset{O}{S}}-C_6H_5$ and $-S-\overset{S}{\overset{\|}{C}}-OCH_3$. Tellurols, R-Te-H, and tellurenyl compounds, R-Te-X, are the only members of this class of compounds discussed in this section. Tellurium derivatives, in which the second tellurium valence is satisfied by a lithium, sodium, magnesium or a group IV or V element atom, are discussed in section XI. Ditellurides, R-Te-Te-R, are included in this section. Tetravalent tellurium derivatives belonging into this section have the general formula RTe(O)OH and $R-TeX_3$, with X representing F, Cl, Br, I, OH and $-SC(S)NR_2$. The derivatives of tellurinic acids will also be discussed.

Alimarin[5-7] mentioned that he tested benzenetelluronic acid, $C_6H_5-TeO_3H$, as an analytical reagent for the detection of metal ions. However, the literature report describing the synthesis of this compound could not be located. The existence of telluronic acids therefore remains doubtful.

A) Tellurols

Tellurols, R-Te-H, are compounds which are very sensitive towards oxidation[19]. They possess an obnoxious and persistent odor. Purified samples are usually yellow. The color is caused by the presence of ditellurides, which are easily formed by oxidation of the tellurols[401] (eq. 55).

$$(55) \qquad 2RTeH \xrightarrow{O_2} R\text{-}TeTe\text{-}R$$

Alkanetellurols, the only tellurols prepared in pure form, can be synthesized by the following methods:

a) Alcohols added to aluminum telluride, Al_2Te_3, at temperatures between 240-350° under an atmosphere of hydrogen gave alkanetellurols[19,310]. According to Baroni[19] this method is of general applicability.

b) A solution of sodium ethoxide in absolute ethanol after saturation with hydrogen telluride was heated with an alkyl bromide. Methane- and ethanetellurol could not be prepared in this manner[19].

c) Ditellurides, R_2Te_2, and sodium in a 1:2 molar ratio are added alternately to boiling liquid ammonia. Evaporation of the ammonia left a solid residue. Treatment of the residue with dilute sulfuric acid set free the tellurol[401].

Table VI-1 lists the tellurols prepared according to these methods. Yields were not reported in the abstracts of Baroni's[19] and Natta's[310] papers.

Table VI-1

Alkanetellurols

R-Te-H R	Method	b.p. °C exp (calcd)*	Reference
CH_3	a	57.0 (62.9)	19
	c	---	401
C_2H_5	a	90.0 (92.5)	19, 310
C_3H_7	a,b	121.0 (119.5)	19
C_4H_9	a,b	151.0 (144.4)	19

* according to equation (55a)

Anderson[11] calculated the boiling points of the known alkanetellurols using equation (55a).

(55a) $$\text{b.p. } (^\circ\text{C}) = 230.14 \text{ (total b.p. number)}^{1/3} - 543$$

Miller[P-64] employed tellurols in the preparation of photoconducting cadmium films from high purity cadmium. Mercury, silver and lead salts of alkanetellurols could not be prepared by combining the reagents in ethanol as solvent[19].

Aromatic tellurols have not yet been isolated in pure form. They are postulated, however, as reaction intermediates. Giua and Cherchi[147] reacted phenyl magnesium bromide with elemental tellurium. Upon hydrolysis hydrogen telluride, diphenyl telluride and diphenyl ditelluride were observed. The ditelluride was probably formed by oxidation of benzenetellurol, generated from phenyltelluro magnesium bromide by hydrolysis. Lederer[219] reduced diphenyl ditelluride with sodium in ethanol. Besides tellurium and diphenyl telluride a small amount of benzenetellurol was formed. The ether extract shaken with mercuric chloride produced a yellow precipitate, which analyzed for C_6H_5-Te-HgCl. Petragnani[342] prepared unsymmetric tellurides from ditellurides and aromatic Grignard reagents (eq. 56).

(56) $$\text{R-TeTe-R} + \text{R'MgX} \longrightarrow \text{R-Te-MgX} + \text{R-Te-R'}$$

(57) $$\text{R-Te-MgX} + H_2O \longrightarrow \text{R-TeH} + \text{MgX(OH)}$$

The organyltelluro magnesium bromides formed in reaction (56) were hydrolyzed to the tellurols (eq. 57). The tellurols, however, were not isolated.

B) Tellurenyl Compounds

Compounds of the type R-Te-X with X representing Cl, Br, I, CNS, ClO_4, NO_3, $-S_2O_3Na$, $-S-\overset{O}{\underset{O}{S}}-CH_3$, $-S-\overset{O}{\underset{O}{S}}-C_6H_5$ and $-S-\overset{S}{C}-OCH_3$ are known only in the aromatic series (R = phenyl, 2-naphthyl and 4-methoxyphenyl). 2-Naphthyl tellurium iodide is the only simple halide derivative in this class of compounds[427]. Phenyl tellurium chloride, bromide, perchlorate, nitrate and thiocyanate were obtained only as complex compounds with thiourea[127-130].

The compounds R-Te-X were prepared according to one of the following methods, which are summarized in Fig. VI-1.

a) Controlled halogenolysis of ditellurides: Only 2-naphthyl tellurium iodide[427] was prepared in this manner (eq. 58). No further reports concerning this reaction could be located.

$$\text{(58)} \quad R{-}TeTe{-}R + I_2 \longrightarrow 2\,R{-}Te{-}I$$

b) Reduction of aryl tellurium trichlorides: Aryl tellurium trichlorides were reduced by methanethiosulfonate and thiourea to aryl tellurium methanethiosulfonates[125] (eq. 59) and thiourea adducts of aryl tellurium chlorides[127-130] (eq. 60), respectively.

$$\text{(59)} \quad RTeCl_3 + 3CH_3SO_2SNa \longrightarrow RTe{-}S{-}\overset{O}{\underset{O}{S}}CH + NaCl + CH_3\overset{O}{\underset{O}{S}}{-}SS{-}\overset{O}{\underset{O}{S}}CH$$

$$\text{(60)} \quad RTeCl_3 + 3S{=}C(NH_2)_2 \longrightarrow RTeCl\cdot SC(NH_2)_2 + \left[H_2N{=}\underset{NH_2}{C}{-}SS{-}\underset{NH_2}{C}{=}NH_2\right]^{++} 2Cl^-$$

Mixing the trichloride in methanol with thiourea in water in a 1:3 molar ratio resulted in the formation of the 1:1 adduct, $RTeCl\cdot SC(NH_2)_2$, while addition of a drop of concentrated hydrochloric acid to a mixture containing the reagents in a 1:5 molar ratio produced the 1:2 adducts $RTeCl\cdot 2SC(NH_2)_2$[127].

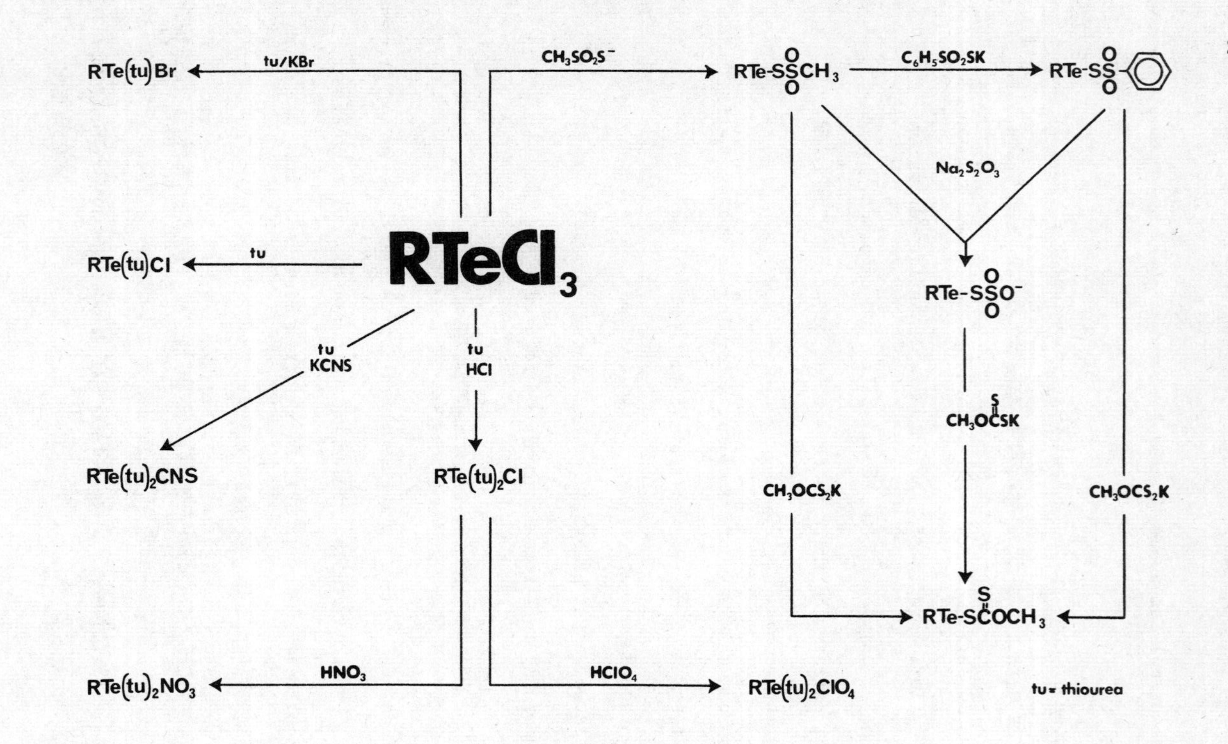

Fig. VI-1: Preparation of Tellurenyl Compounds, RTeX

c) The exchange of X in RTe-X: The exchange of the methanethiosulfonate group in aryl tellurium methanethiosulfonates for a benzenethiosulfonate, thiosulfate or methyl xanthate group proceeds in methanolic or aqueous solution[125].

The thiourea adducts $RTeCl \cdot 2SC(NH_2)_2$ when treated with nitric acid or perchloric acid produce the compounds $RTeX \cdot 2SC(NH_2)_2$ ($X = NO_3$, ClO_4)[127]. When the reduction of $RTeCl_3$ with thiourea (eq. 60) is carried out in the presence of potassium bromide or thiocyanate the adducts $RTeX \cdot SC(NH_2)_2$ (X = Br, CNS) are obtained[127]. It is not known whether these exchange reactions taken place before or after the reduction of the trichloride.

The tellurenyl compounds, which have been described in the literature, are collected in Table VI-2.

The crystal structures of the phenyl tellurium halide thiourea adducts were determined by Foss and coworkers[128-130]. The tellurium atom in the compounds $C_6H_5TeX \cdot SC(NH_2)_2$ (X = Cl, Br) is bonded to a sulfur, a carbon and a halogen atom with all four atoms in one plane. The chlorine or bromine atom from the next molecule approaches within 3.71Å and 3.77Å, respectively. These long distances indicate weak bonding. The angle between the directions of the tellurium-carbon and tellurium-halogen bond is 164° in both cases[128,130]. The tellurium-chlorine bond length of 3.61Å in the adduct $C_6H_5TeCl \cdot 2SC(NH_2)_2$ justifies the ionic formulation $C_6H_5Te \cdot [SC(NH_2)_2]_2^+ Cl^-$. The carbon-tellurium-chlorine bond angle is 163°[128,129].

The arrangement of the ligands around the divalent tellurium atom can be regarded as square-planar with a vacant position *trans* to the phenyl group. Figure VI-2 gives more details about the molecular structures of these three compounds.

Table VI-2

Tellurenyl Compounds, R-Te-X

R-Te-X·L$_n$							
R	L	n	X	Method	Yield, %	mp. °C	Reference
phenyl	tu*)	1	Cl	b	79	166	127
	tu	2	Cl	b	85	168	127
	tu	1	Br	b	99	178	127
	tu	2	ClO_4	b	--	136	127
	tu	2	NO_3	b	--	135	127
	tu	2	CNS	b	58	109	127
4-methoxyphenyl	--	-	$S_2O_3^-$ †)	c	--	---	125
	--	-	$-SS(O)(O)-CH_3$	b	68	107	125
	--	-	$-SS(O)(O)-C_6H_5$	c	--	110	125
	--	-	$-S-C(=S)-OCH_3$	c	--	78	125
2-naphthyl	--	-	I	a	86	130 and 220 (dec)	427
	$(C_6H_5)_3P$	1	I	-	--	--	346

*) tu = thiourea   †) in solution only

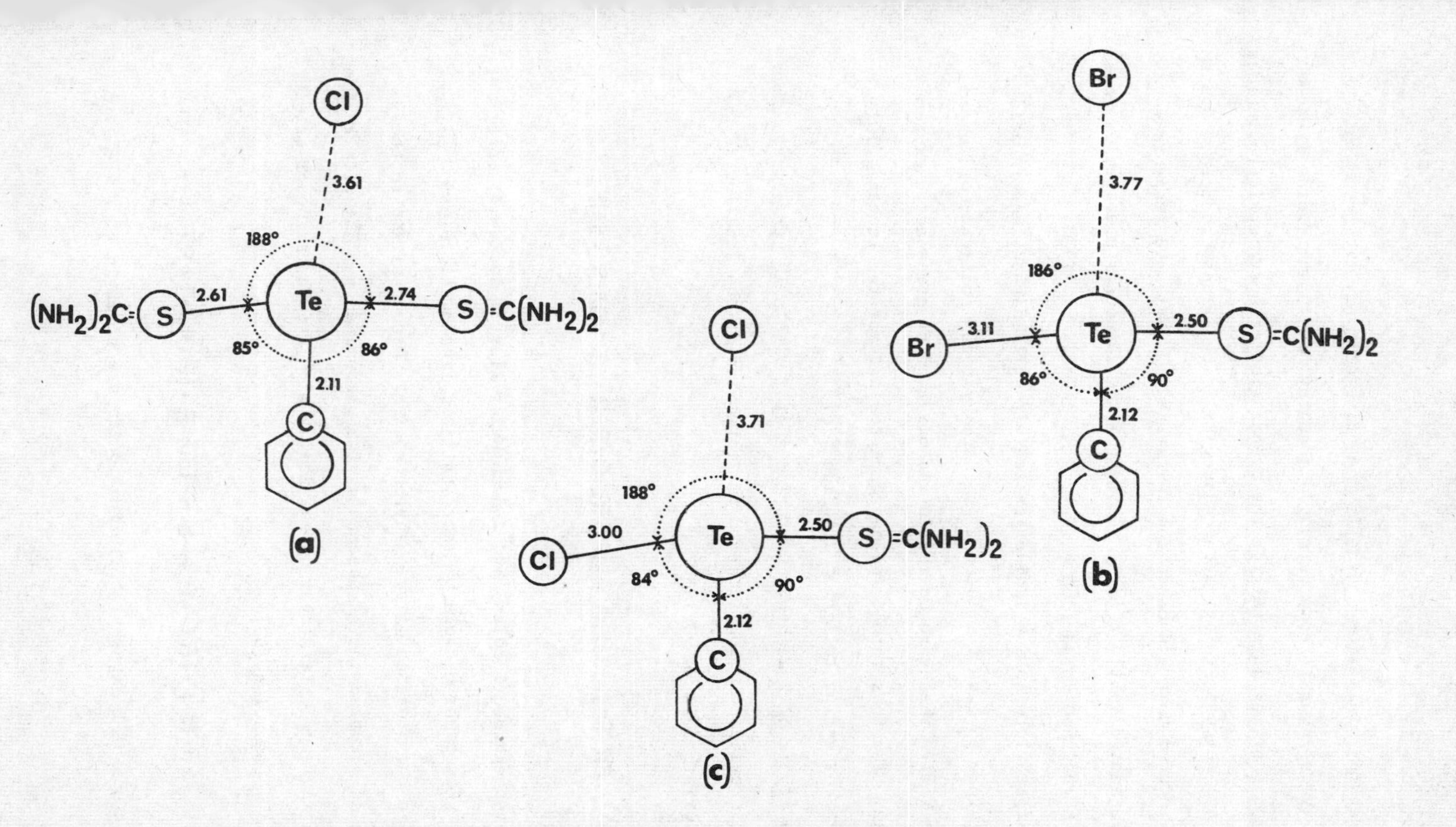

Fig. VI-2: The Structure of (a) Phenyl Bis(thiourea) Tellurium Chloride, (b) Phenyl Thiourea Tellurium Bromide and (c) Phenyl Thiourea Tellurium Chloride

2-Naphthyl tellurium iodide is the only tellurenyl compound whose reactivity is known in some detail. 2-Naphthyl tellurium iodide is a convenient starting material for the preparation of unsymmetric tellurides. With phenyl mercuric chloride the adduct RR'Te·HgClI was produced. Sodium sulfide reduced this adduct to the telluride[427]. Grignard reagents gave the tellurides directly[305,427]. A list of the compounds prepared by this method is included in section VII-A-2.

2-Naphthyl tellurium iodide reacted with allyldiphenylacetic acid in chloroform to give the lactone *(15)* and hydrogen iodide according to equation (61). The hydrogen iodide was oxidized by atmospheric oxygen to iodine. The thus formed iodine converted part of the lactone *(15)* to the corresponding tellurium diiodide *(16)*[302].

Hydrolysis of 2-naphthyl tellurium iodide with a 10 per cent aqueous sodium hydroxide solution lead to the formation of the ditelluride and the tellurinic acid anhydride[427].

$$(62)\quad 6R\text{-}Te\text{-}I + 3H_2O \longrightarrow R\text{-}\overset{O}{\overset{\uparrow}{Te}}\text{-}O\text{-}\overset{O}{\overset{\uparrow}{Te}}\text{-}R + R\text{-}TeTe\text{-}R + 6HI$$

Triphenylphosphine (2 moles) and 2-naphthyl tellurium iodide (1 mole) gave 1:1 adduct which could be recrystallized from a chloroform - benzene mixture in the presence of triphenylphosphine. Decomposition to the ditelluride occurred in the absence of the phosphine[346]. The ditelluride was also isolated after treating the iodide with triethylamine or triethyl phosphite[346] [equation (63)].

$$(63)\quad 2R\text{-}Te\text{-}I + (R'O)_3P \longrightarrow R\text{-}TeTe\text{-}R + (R'O)_2\overset{O}{\overset{\uparrow}{P}}\text{-}I + R'I$$

The reactions of 2-naphthyl tellurium iodide are given in Fig. VI-3.

(61)

$-HI$ O_2 I_2

(15) (16)

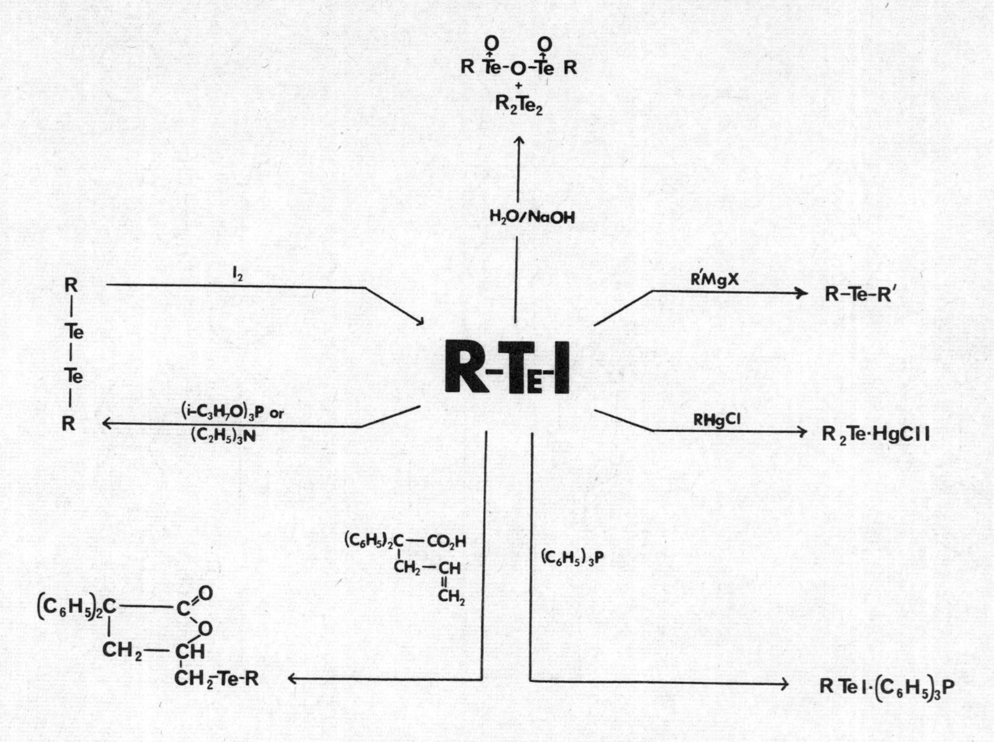

Fig. VI-3: Preparation and Reactions of 2-Naphthyl Tellurium Iodide

C. Organyl Tellurium Compounds, $RTeX_3$

Organic tellurium compounds of the type $RTeX_3$, in which X stands for F, Cl, Br, I, OH and $-SC(S)NR_2$, have been prepared. While a large number of aromatic derivatives are known, only a few unsubstituted aliphatic compounds have been synthesized. The aromatic derivatives possess much greater stability than the aliphatic compounds. Methyl tellurium trichlorides, for instance, is reported to be moisture sensitive and to decompose in acetonitrile and nitrobenzene solution. Solutions in dichloromethane or benzene are stable for a few days. Methyl tellurium trichloride and tribromide are associated in the latter two solvents[447a]. The aromatic tellurium trihalides are stable under these conditions.

According to a single crystal X-ray investigation 2-chloroethyl tellurium trichloride is polymeric in the solid state with the tellurium atom surrounded by five ligands in a square pyramidal arrangement[194b].

1) Synthesis of organyl tellurium trihalides

The reactions which have been developed for the synthesis of organyl tellurium trihalides are displayed in Fig. VI-4. Condensation reactions with aliphatic and aromatic compounds containing activating groups proceed satisfactorily only with tellurium tetrachloride. 1,3-Diketones, monoketones and carboxylic acid anhydrides have been employed as aliphatic components.

a) Tellurium tetrachloride and 1,3-diketones:

With tellurium tetrachloride 1,3-diketones *(17)* give organyl tellurium

$R^1R^2R^3C-C(=O)-CR^7R^8-C(=O)-CR^4R^5R^6$

(17)

$R^1R^2R^3C-C(=O)-CR^7R^8-C(=O)-CR^4R^5-TeCl_3$

(18)

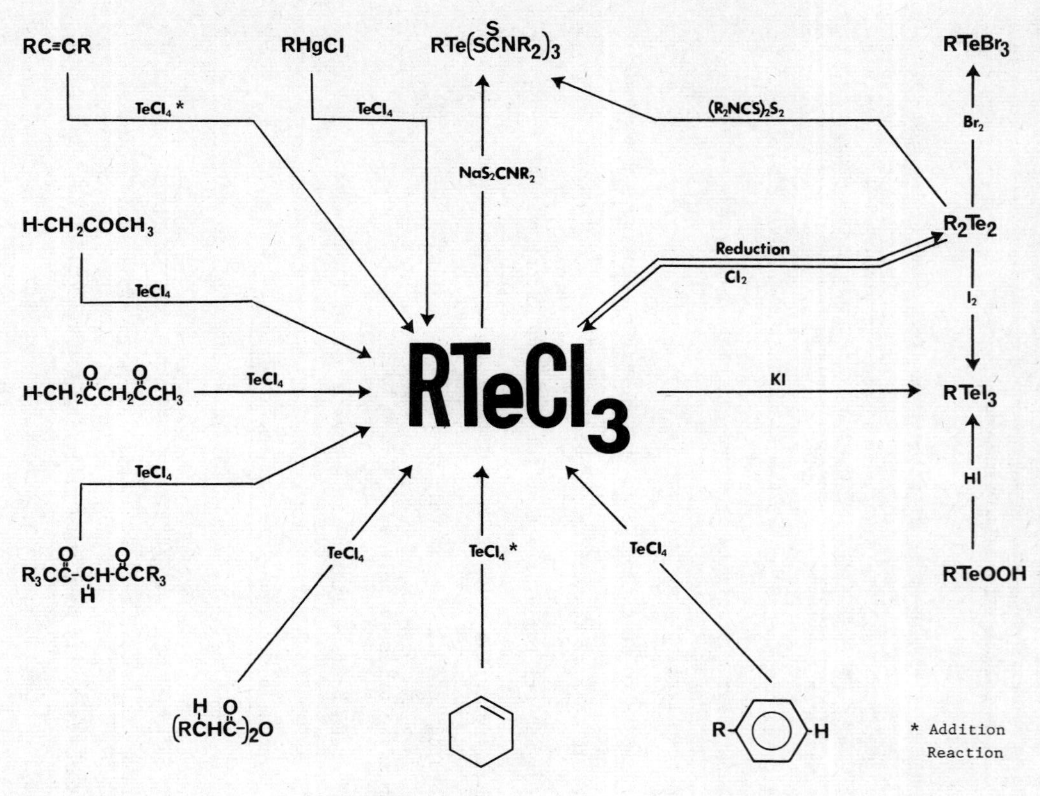

Fig. VI-4: Preparation of Organyl Tellurium Trihalides

trichlorides *(18)* if at least one R-group among R^1–R^6 is a hydrogen atom. When R^1, R^2, R^3 and R^4 are hydrogen atoms telluracyclohexane derivatives are preferentially formed as discussed in section IV-B-1a. Trichlorides *(18)* are expected as by-products in these reactions and have been isolated in a few cases. If two or more organic groups are bonded to the carbon atom 1 while one or more hydrogen atoms are attached to the carbon atom 5, the trichloride *(18)* is the major product when equimolar amounts of the reagents are employed. The hydrogen atoms in position 3 ($R^7 = R^8 = H$) react only when no hydrogen atoms are available on the carbon atoms 1 and 5. The experimental data available indicate that the formation of organyl tellurium trichlorides is not influenced by the nature of the substituents R^7 and R^8 in position 3. Literature references for individual compounds are given in Table VI-3.

Organyl tellurium trichlorides *(18)* with at least one hydrogen atom in position 3 (R^7 and/or R^8 = H) can enolize to *(19)* or *(20)*.

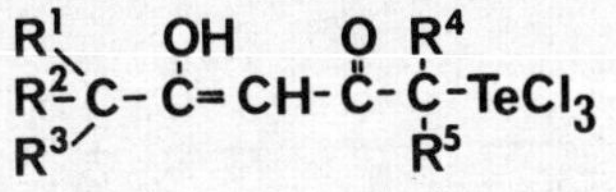

$$R^1R^2R^3C-C(OH)=CH-C(=O)-CR^4R^5-TeCl_3 \quad (19)$$

$$R^1R^2R^3C-C(=O)-CH=C(OH)-CR^4R^5-TeCl_3 \quad (20)$$

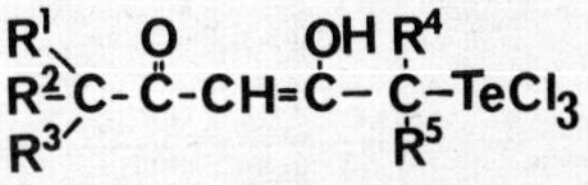

The chloroform employed as a solvent in these reactions contained ethanol, which reacted with the enols producing the enol ethers listed in Table VI-3. Purified chloroform gave the free enols. The experiments described in the literature do not allow a conclusion as to the position of the alkoxy group in the enol ethers. The formulas for these compounds as written in Table VI-3 are therefore tentative.

b) Tellurium tetrachloride and monoketones: Aliphatic and aliphatic-aromatic ketones were generally reacted with tellurium tetrachloride in a 2:1 molar ratio. Organyl tellurium trichlorides and diorganyl tellurium dichlorides were observed as products (eq. 64). The

Table VI-3

Organyl Tellurium Trihalides

RTeX3 R	X	Method	Yield %	mp. °C	Reference
methyl	F	h	-	-	120
	Cl	g	63	137-9 (dec)	447a
	Br	g	100	140 (dec)	76
		i	-	140-56	107
	I	g	76	175-8	447c
		i		100-180	107
nitromethyl	Cl	c	-	>225°	331
carboxymethyl	Cl	c	32	-	286
	Cl	g	-	-	286
	Br	g	-	-	286
HOOC-CH(CH_3)-	Cl	c	-	145-50	293
	Br	g	-	139-41	293
Cl_3Te-CH_2-	Cl	c	14	173	286
	Cl	g	-	173	286

Table VI-3(cont'd)

Br_3Te-CH_2-	Br	g	-	214(dec)	286
ethyl	Cl	g	75	114-5	447b
	Br	g	78	138-40	447c
	I	g	58	178-80	447c
		i	-	135(dec)	249
$ClCH_2CH_2-$	Cl	f	-	116	12a
$CH_3CHClCH_2-$	Cl	f	-	oil	323b
butyl	I	i	-	dec.	17
$CH_3CH_2C(O)CH_2-$	Cl	b	75	101	290
$(CH_3)_3CC(O)CH_2-$	Cl	b	21	114-5	290
$CH_3CH_2C(O)CH(CH_3)-$	Cl	b	84.6	77-8	290
$C_2H_5CH_2C(O)CH(C_2H_5)-$	Cl	b	77	70	290
$C_6H_5C(O)CH(CH_3)-$	Cl	b	61	114.5	290
$C_6H_5C(O)CH(C_2H_5)-$	Cl	b	71	128-9	290
$C_6H_5C(O)CH(C_6H_5)-$	Cl	b	60	142-3	290
$CH_3C(O)CH=C(OC_2H_5)CH_2-$	Cl	a	40	106-7	276
$C_3H_7C(O)CH=C(OC_2H_5)CH_2-$	Cl	a	28	106	282

Table VI-3(cont'd)

$C_6H_{13}C(O)CH{=}C(OH)CH_2-$	Cl	a	30	117 (dec)	282
$C_8H_{17}C(O)CH{=}C(OH)CH_2-$	Cl	a	16	114-5 (dec)	287
$(CH_3)_2CHC(O)CH{=}C(OC_2H_5)CH_2-$	Cl	a	30	103 (dec)	287
$(CH_3)_3CC(O)CH{=}C(OC_2H_5)CH_2-$	Cl	a	80	116	277
$(CH_3)_2CHCH_2C(O)CH{=}C(OC_2H_5)-CH_2-$	Cl	a	42	100-1 (dec)	282
$CH_3CH_2C(O)CH{=}C(OC_2H_5)CH(CH_3)-$	Cl	a	7	110-1 (dec)	282, 298
$(CH_3)_2CHC(O)CH{=}C(OC_2H_5)C(CH_3)_2-$	Cl	a	0.4	87-9 (dec)	287
$C_2H_5CH_2C(O)CH{=}C(OH)CH(C_2H_5)-$	Cl	a	-	87-8	283
$C_2H_5OC(O)CH{=}C(OC_2H_5)CH_2-$	Cl	a	32	90-2	282
$C_6H_5COCH{=}C(OC_2H_5)CH_2-$	Cl	a	80	138-40 (dec)	276
$(C_6H_5CO)_2CH-$	Cl	a	-	-	277
(3,4-diacetyl-5-methylfuryl)methyl[1)]	Cl	a	-	-	282
2-chlorocyclohexyl	Cl	f	-	111-14 (dec)	300, 323, 323b
	I	h	88	175-90 (dec)	304
2-chloro-2-phenylethenyl	Cl	f	94	205-15 (dec)	305
1-phenyl-2-chloro-2-phenylethenyl	Cl	f	77	125-8	305
phenyl	Cl	e	66	215-8	122
	DTC[2)]	g	-	-	126

Table VI-3(cont'd)

2-methylphenyl	Br	g	100	131 (dec)[3)]	164a
	Br	g	75	165-7[4)]	348c
4-methylphenyl	Cl	d	74	196-7 (dec)	323a
	Cl	e	-	181-2	122
3,4-dimethylphenyl	Cl	d	-	194-5	323a
2-chlorophenyl	Br	g	100	178 (dec)	164a
4-fluorophenyl	Br	g	100	291 (dec)	164a
4-methoxyphenyl	Cl	d	100	190	293, 358
	Cl	g	-	192	293
	Br	g	100	188-90 (dec)	293, 339
	I	g	93	131-3 (dec)	339
	DMDTC[5)]	g,h	90	184 (dec)	126
	DEDTC[6)]	g,h	90	156 (dec)	126
	PDTC[7)]	g,h	90	178 (dec)	126
	OH	h	-	-	293
4-ethoxyphenyl	Cl	d	92	183	288, 358
	Br	g	100	195-205 (dec)	339

Table VI-3(cont'd)

4-ethoxyphenyl	I	g	100	133-4 (dec)	339
4-phenoxyphenyl	Cl	d	90	156-7	103
	Br	g	100	207	339
	I	g	100	161 (dec)	339
2-thiophenoxyphenyl	Cl	e	80.6	213-5	339
4-thiophenoxyphenyl	Cl	d	42	165	339
	Br	g	100	207	339
	I	g	100	180 (dec)	339
2-(4'-methylphenoxy)phenyl	Cl	e	80	180-5 (dec)	55
2-(4'-carboxyphenoxy)phenyl	Cl	e	-	205-6 (dec)	55
3-methyl-4-methoxyphenyl	Cl	d	-	232-3	293
	Cl	g	-	-	293
	Br	g	-	-	293
2(or 4)-methyl-4(or 2)-methoxyphenyl	Cl	d	-	154	293
2-methoxy-5-methylphenyl	Cl	d	-	135	293
3-hydroxy-4-methoxyphenyl	Cl	d	-	157-9	288
2,4-dimethoxyphenyl	Cl	d	67	155-6	288

Table VI-3 (cont'd)

?-bromo-?-methoxyphenyl[8)]	Cl	d	-	184	293
	Br	g	-	153	293
4-acetamidophenyl	Cl	d	-	-	358
4-hydroxyphenyl	Cl	d	~100	213	358
	I	h	-	125-50(dec)	339
2-formylphenyl	Cl	g	95	195-200	348c
4-(4'-carboxyquinolyl-2')phenyl	Cl	d	47.6	237	360
4-(4'-aminoquinolyl-2')phenyl	Cl	d	45.1	243 (dec)	360
4-(5',6'-benzo-4'-carboxy-quinolyl-2')phenyl	Cl	d	38	276-80 (dec)	360
4-(acridinyl-9')phenyl	Cl	d	63	254	360
pentafluorophenyl	Cl	e	low	128-30	197, 357
1-naphthyl	Cl	e	96	175-80 (dec)	339
	Br	g	100	160 (dec)	339
	I	g	100	133 (dec)	339
2-naphthyl	Cl	e	85.2	200-2	363
	Br	g	100	212-5 (dec)	427

Table VI-3(cont'd)

1) from 1,1,2,2-tetraacetylethane

2) DTC = dimethyl-, diethyl- and cyclpentamethylenedithiocarbamate

3) recrystallized from methanol

4) recrystallized from hexane $CHCl_3$

5) dimethyldithiocarbamate

6) diethyldithiocarbamate

7) cyclopentamethylenedithiocarbamate

8) from 4-bromomethoxybenzene

less soluble compounds were isolated[290].

(64)

$$R\text{-}\overset{O}{\overset{\|}{C}}\text{-}CH_2\text{-}R \xrightarrow{TeCl_4} \begin{cases} R\text{-}\overset{O}{\overset{\|}{C}}\text{-}\overset{R}{\overset{|}{C}}H\text{-}TeCl_3 \\ (R\text{-}\overset{O}{\overset{\|}{C}}\text{-}\overset{R}{\overset{|}{C}}H)_2TeCl_2 \end{cases}$$

The condensation reactions were carried out in chloroform. 1,1,3,3-Tetramethylacetone did not condense with tellurium tetrachloride, indicating that at least one methylene group in the α-position to the carbonyl function is required for a successful reaction[290]. Table VI-3 lists the tellurium trichlorides prepared in this way.

c) Tellurium tetrachloride and carboxylic acid anhydrides: The condensation reactions of tellurium tetrachloride and acetic acid anhydride are discussed in section IV-B-1c. The isolated products and pertinent data are listed in Table VI-3. Nitromethane refluxed in presence of tellurium tetrachloride gave nitromethyl tellurium trichloride[331]. Trihalotelluromethyl methyl sulfite was obtained from dimethyl sulfite and tellurium tetrachloride[260a] (eq. 65).

(65)

$$CH_3Y + TeX_4 \longrightarrow Y\text{-}CH_2\text{-}TeX_3 + HX$$

Y,X: NO_2, Cl; CH_3OSO_2, Cl; CH_3OSO_2, Br

d) Tellurium tetrachloride and aromatic compounds: Benzene derivatives containing a ring-position activated by an alkoxy, phenoxy, thiophenoxy, hydroxy, dialkylamino, acetamido, 2-quinolyl or 9-acridinyl group condense with tellurium tetrachloride in refluxing chloroform or carbon tetrachloride or upon heating without solvent on a water-bath to yield organyl tellurium trichlorides. More details about these reactions have been presented in section IV-B-1d. Table VI-3 lists the compounds synthesized by this method together with the appropriate literature references. Ogawa[323a] succeeded in condensing tellurium tetrachloride with toluene and *o*-xylene. The trichlorotelluro group entered *para* to the methyl groups. Only unstable products were obtained from *m*- and *p*-xylene. Morgan and Burgess[296] obtained 3-methyl-4-hydroxyphenyl tellurium trichloride by condensing $TeOCl_2$ with 2-methylphenol.

e) Tellurium tetrachloride and aryl mercury chlorides: If the direct condensation of tellurium tetrachloride with an aromatic compound is not feasible the reaction with aryl mercury chlorides will give the aryl tellurium trichlorides[55] (eq. 66).

(66) $$R\text{-}HgCl + TeCl_4 \longrightarrow RTeCl_3 + HgCl_2$$

Equimolar amounts of the reactants refluxed in chloroform, acetonitrile or dioxane for several hours give good yields of the trichlorides. The use of dioxane has the advantage of precipitating mercuric chloride as the dioxane adduct[55]. The organic mercury compounds are generally prepared from the appropriate diazonium salts. For individual compounds consult Table VI-3.

f) Tellurium tetrachloride and unsaturated organic compounds: Tellurium tetrachloride adds across the carbon-carbon multiple bond in propene[323b] cyclohexene[123,300], phenylacetylene and diphenylacetylene[305] to give substituted organyl tellurium trichlorides. These compounds are listed in Table VI-3. For further information consult section IV-B-2.

g) Halogenolysis of diorganyl ditellurides: The conversion of ditellurides, R_2Te_2, to organyl tellurium trichlorides by action of elemental chlorine (eq. 67) is of importance only when the ditelluride can be synthesized from sodium ditelluride and an organyl halide. Ditellurides are generally prepared by reduction of organyl tellurium trichlorides (see section VI-E-1).

(67) $$R{-}TeTe{-}R \underset{\text{Reduction}}{\overset{3Cl_2}{\rightleftarrows}} 2RTeCl_3$$

Since the condensation reactions described in section VI-C-1a through VI-C-1e proceed - as far as is known- only with tellurium tetrachloride and not with the tetrabromide and tetraiodide, the organyl tellurium tribromides and triiodides can not be prepared in this manner. However, the reduction of the trichlorides to the ditellurides, and the subsequent reaction of the ditellurides with bromine or iodine in chloroform, carbon tetrachloride or benzene solution is a convenient way to the organyl tellurium tribromides and triiodides[339,427]. Reduction and halogenolysis proceed almost quantitatively. Further literature references are given in Table VI-3.

Bis(thiocarbamoyl) disulfides show halogen character in their reactions with diorganyl ditellurides in chloroform yielding phenyl and 4-methoxyphenyl tellurium tris(dithiocarbamates) in 90 per cent yield[126] (eq. 68).

(68) $$R_2Te_2 + 3[R_2N{-}CS]_2S_2 \longrightarrow 2RTe[S\overset{S}{C}NR_2]_3$$

Bis(2-benzothiazolyl) disulfide is reported to react similarly[126]. The aryl tellurium tris(dithiocarbamates) possess thermochromic properties[126].

h) Halogen exchange in organyl tellurium trihalides: Halogen exchange reactions for organyl tellurium trihalides have been little investigated. 4-Hydroxyphenyl[339] and 2-chlorocyclohexyl tellurium triiodides[304] were obtained from the respective trichlorides and potassium iodide in aqueous ethanolic solution. The conversion of 4-hydroxyphenyl tellurium trichloride into the tribromide with potassium bromide failed[339]. 4-Methoxyphenyl tellurium trichloride and sodium dithiocarbamates gave the corresponding tris(dithiocarbamates)[126]. Emeleus and Heal[120] prepared methyl tellurium trifluoride from the triiodide and silver fluoride. Equation (69) describes the reactions for halogen exchange.

(69) $$RTeX_3 + 3MY \longrightarrow RTeY_3 + 3MX$$

M = metal, X and Y = halogen

4-Methoxyphenyl tellurium trichloride was hydrolyzed by water. According to the results of elemental analyses the product could have been the trihydroxide.

i) Conversion of tellurinic acids, RTe(O)OH, into organyl tellurium trihalides: Balfe and coworkers[17] treated the complex substances $(C_4H_9)(C_2H_5O(O)CCH_2)_2TeO \cdot 2C_4H_9Te(O)OH$, $(C_4H_9)_2TeO \cdot 3C_4H_9Te(O)OH$ and $2(C_4H_9)_2TeO \cdot C_4H_9TeOOH$, with hydrogen iodide. The tellurinic acids contained in these compounds were converted to organyl tellurium triiodides (eq. 70).

(70) $$C_4H_9TeOOH + 3HI \longrightarrow C_4H_9TeI_3 + 2H_2O$$

Drew[107] isolated methyl tellurium trihalides by reacting "Vernon's β-base", $(CH_3TeO_2)^-(CH_3)_3Te^+$ with hydrogen bromide and iodide. Similarly, ethyl tellurium triiodide was prepared[249]. For details concerning this reaction and the structure of the "β-base" consult section VII-B.

2) Reactions of organyl tellurium trihalides

Only a general discussion of the reactions of organyl tellurium trihalides is given in this section. A more extensive presentation can be found in those sections, in which the particular reaction is employed in the synthesis of other organic tellurium compounds. The reactions of organyl tellurium trichlorides are compiled in schematic form in Fig. VI-5

Organyl tellurium trichlorides are easily and almost always quantitatively reduced to diorganyl ditellurides by sodium sulfide

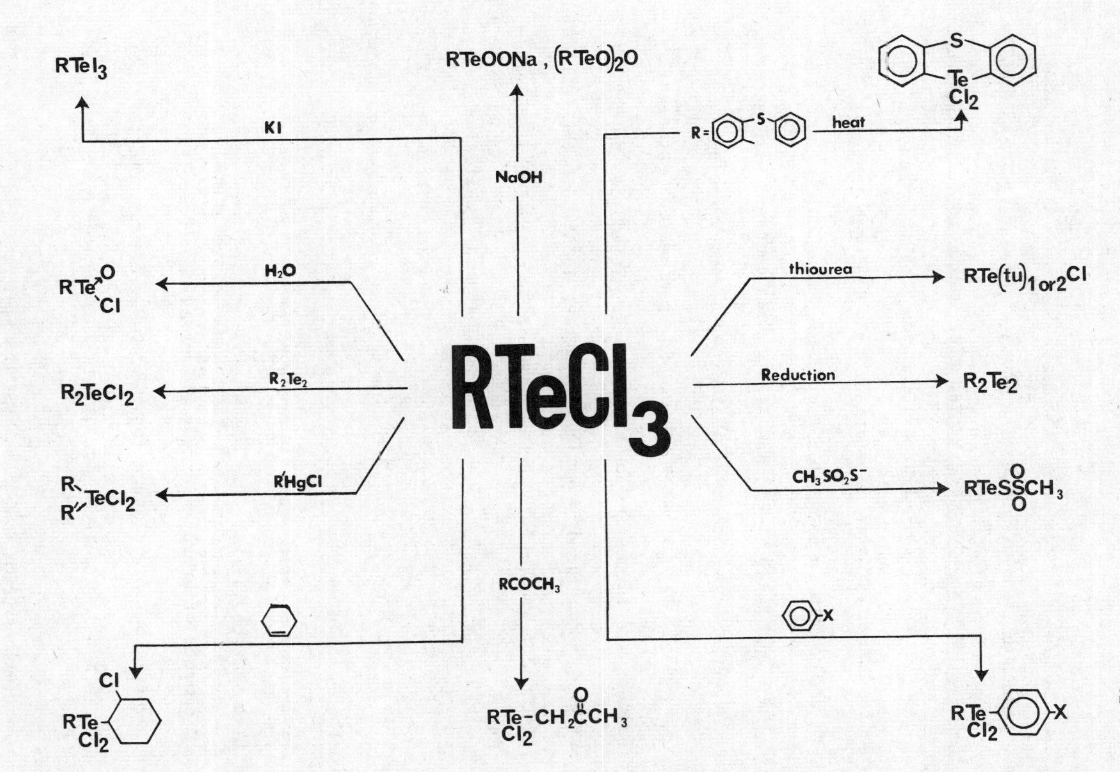

Fig. VI-5: The Reactions of Organyl Tellurium Trichlorides

nonahydrate[339,359], by potassium disulfite, $K_2S_2O_5$[286,293], and by zinc in chloroform or ethanol[197,357] (see section VI-E). However, 2-chloropropyl tellurium trichloride was decomposed by reducing agents to tellurium and propene[323b]. The reduction to tellurenyl compounds by thiourea and methanethiosulfonate was treated in section VI-B. Halogen exchange reactions were dealt with in the preceding section VI-C-1h.

Organyl tellurium trichlorides and tribromides condense with acetone, acetophenone, 4-phenoxy-, 4-methoxy-, 4-ethoxy-, 4-dimethylamino- and 2,4-dihydroxybenzene[340] with elimination of hydrogen halide. Diorganyl tellurium dihalides are formed in these processes. Triiodides do not react with any of the mentioned compounds. Further details can be found in section VII-B. Intramolecular condensation of 2-chloro-1,2-diphenylvinyl[37] 2-biphenylyl[84], and 2-phenoxy-[103] and 2-thiophenoxyphenyl tellurium trichloride[339] produces heterocyclic tellurium compounds (see section XII). Aromatic tellurium trichlorides and 2-chlorocyclohexyl tellurium trichloride add to the carbon-carbon double bond in cyclohexene, 2,2-diphenyl-4-pentenoic acid and 9-allyl-9-carboxy-fluorene[304] forming diorganyl tellurium dichlorides (see section VII-B).

Equimolar amounts of organyl mercury chlorides and organyl tellurium trichlorides react in boiling dioxane (eq. 71) to give symmetric (R = R') and unsymmetric (R ≠ R') tellurium dichlorides[363] (see section VII-B).

$$\text{(71)} \quad RTeCl_3 + R'HgCl \longrightarrow RR'TeCl_2 + HgCl_2$$

Aryl tellurium trichlorides and ditellurides refluxed in toluene produced tellurium and diorganyl tellurium dichlorides[341]. Drew[107] reported that potassium iodide combined with methyl tellurium triiodide to give almost black crystals of $K[CH_3TeI_4]$.

Hydrolysis of organyl tellurium trichlorides and tribromides by water gave tellurinic acid halides, while sodium hydroxide produced the acid or the anhydride (see section VI-D).

(72)

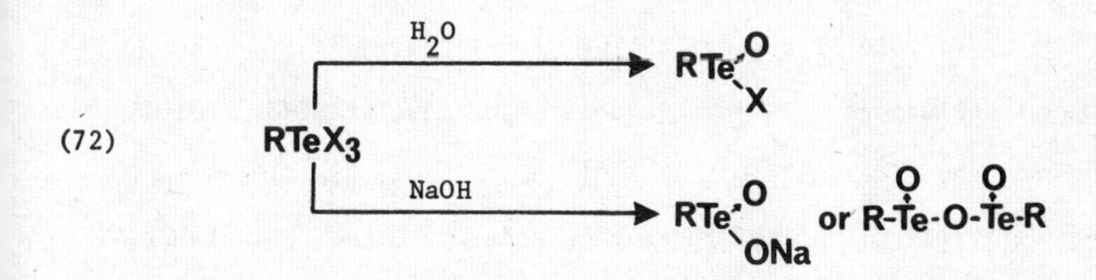

The chloroalkyl tellurium trichlorides, obtained by addition of tellurium tetrachloride to an olefin, are thermally decomposed to an inorganic tellurium compound, to olefins, hydrocarbons, chloroalkanes and chloroalkenes[12a,323b].

Methyl, ethyl and 4-methoxyphenyl tellurium trichlorides reacted with antimony pentachloride in dichloromethane to give moisture sensitive adducts $RTeCl_3 \cdot SbCl_5$ in 61-68 per cent yield, which melted at 125-7° (R = CH_3), 86-8° (R = C_2H_5) and 131-3° (R = 4-$CH_3OC_6H_4$). Methyl tellurium trichloride did not form an adduct with boron trichloride[447b].

Organyl tellurium trihalides, dissolved in benzene, reacted with a dichloromethane solution of thiourea to give the adducts $RTeX_3 \cdot SC(NH_2)_2$[447c] (R, X, % yield, m.p. given):

CH_3, Cl, 92, 123-5°; CH_3, Br, 94, 120-2°; CH_3, I, 86, 97-9°; C_2H_5, Cl, 83, 108-20; C_2H_5, Br, 79, 100-2°; C_2H_5, I, 75, 175-6°; 4-$CH_3OC_6H_4$, Cl, 66, 133-5°.

The tetramethylthiourea adducts of methyl tellurium trichloride and tribromide, which melted at 123° and 120°, respectively, were obtained similarly[447d].

The thin layer chromatographic behavior of phenyl ^{127}Te-tellurium trichloride has been investigated[319b]. The dipole moment of phenyl tellurium trichloride in benzene was determined to be 2.90D[192a].

D) Tellurinic Acids and Their Derivatives

Only eleven tellurinic acids, RTe(O)OH, are known. They have been prepared by hydrolysis of organyl tellurium trichlorides or oxidation of diorganyl tellurides. Tellurinic acid chlorides, bromides, iodides, nitrates and anhydrides, RTe(O)X [X = Cl, Br, I, NO_3, RTe(O)O] comprise the known derivatives. All of these compounds are cited in Table VI-4. The following methods were employed in their synthesis:

a) Oxidation of diorganyl tellurides to tellurinic acids:

Dipentyl telluride, carbethoxymethyl pentyl telluride, carbomenthoxymethyl pentyl telluride[18] and carbomenthoxymethyl butyl telluride[17] are reported to be oxidized by air or the hydrogen peroxide-urea adduct to pentane- and butanetellurinic acid, respectively. Adducts of tellurinic acids with telluroxides, $nRTeOOH \cdot mR_2TeO$, which are discussed in section VII-C were also formed in these oxidation reactions[17].

b) Hydrolysis of organyl tellurium trihalides by water:

Cold water hydrolyzes organyl tellurium trichlorides and tribromides to the respective tellurinic acid halides (eq. 72). The only systematic investigation of these hydrolytic reactions was carried out by Petragnani and Vicentini[338]. These authors found that cold water does not affect organyl tellurium triiodides while boiling water converted them into products of indefinite composition. Only 4-methoxybenzenetellurinic acid iodide was isolated in pure form. There are large discrepancies in the melting points reported for 4-methoxy- and 4-ethoxybenzenetellurinic acid chlorides[338,358] (see Table VI-4).

c) Hydrolysis of organyl tellurium trihalides in alkaline medium:

Organyl tellurium trichlorides were dissolved in 1N, 4N[348c] or 2N[358] sodium hydroxide solutions. Tellurinic acids were isolated upon acidification of these solutions with sulfuric acid,[358] hydrochloric acid or acetic acid.[348c] In some cases a 70-80 per cent sodium hydroxide

Table VI-4

Tellurinic Acids and Their Derivatives

RTe(O)X		Method	Yield %	mp. °C	Reference
R	X				
methyl	OH	*	–	–	107
	Cl	†	–	–	249
	Br	†	–	–	249
	I	†	–	–	249
	RTe(O)O	*	–	~230(dec)	107
ethyl	I	†	–	–	249
butyl	OH	a	16	250(dec)	17
pentyl	OH	a	–	200–20(dec)	18
phenyl	OH	g	–	210–1	219
	Cl	b	88	250	338
	Br	b	86	247–9	338
	NO_3	e	56	232–3	147,219
	RTe(O)O	c	90–5	220–5	338
2-methylphenyl	OH	c	–	180–8	348c

Table VI-4 (cont'd)

4-hydroxyphenyl	OH	c	–	dec.	358
	Cl	b	100	dec	358
	RTe(O)O	c	75-80	200(dec)	338
4-methoxyphenyl	OH	c,g	–	indefinite	358
	Cl	b	67	400-500(dec)	358
	Cl	b	90	225-35(dec)	338
	Cl	h	–	–	340
	Br	b	96	231-4	338
	I	b	67.2	190-5(dec)	338
	RTe(O)O	c	100	200-5(dec)	338
4-ethoxyphenyl	OH	c,g	100	indefinite	358
	Cl	b	88	224-6	338
	Cl	b	96	400-500(dec)	358
	Cl	h	–	–	340
	Br	b	91.5	233-6	338
	RTe(O)O	c	90-5	206-10	338

Table VI-4 (cont'd)

4-phenoxyphenyl	Cl	b	98.5	185-6	338
	Cl	h	-	-	340
	Br	b	100	188-9	338
	RTe(O)O	c	100	276(dec)	338
2-formylphenyl	OH	c	70	~350(dec)	348c
4-(4'-carboxyquinolyl-2')-phenyl	Cl	b	100	179	360
4-(5',6'-benzo-4'-carboxy-quinolyl-2')phenyl	Cl	b	100	302	360
4-(acridinyl-9')phenyl	Cl	b	80	276(dec)	360
4-hydroxy-3-nitrophenyl	OH	f	70	221(dec)	358
4-hydroxy-3,5-dinitrophenyl	OH	g	56	-	358
	NO_3	f	56	-	358
2-naphthyl	Cl	b	71.5	197 and 257	338
	Br	b	85	216	338
	RTe(O)O	c	100	230	338
	RTe(O)O	c,d	100	230	427

* obtained from "Vernon's β-base" (see section VII-B)

† in solution only, not isolated

solution was employed[358]. Petragnani and Vicentini[338] obtained, however, the anhydrides of tellurinic acids by dissolving the respective trihalides in 10 per cent aqueous sodium carbonate or sodium hydroxide solution and acidifying the resulting solution with 10 per cent acetic acid.

d) Disproportionation of tellurenyl compounds in sodium hydroxide solution: 2-Naphthyl tellurium iodide disproportionated into the ditelluride and tellurinic acid anhydride[427] according to equation (62) (section VI-B). Whether other tellurenyl compounds react in the same way is unknown.

e) Oxidation of diorganyl ditellurides by nitric acid to tellurinic acid nitrates: Diphenyl ditelluride warmed with halogen-free nitric acid (65 per cent) was oxidized to the tellurinic acid nitrate[147,219]. No further reports concerning this potentially useful reaction could be located in the literature.

f) Nitration of aromatic tellurinic acids:

Reichel and Kirschbaum[358] prepared hydroxynitrobenzenetellurinic acids by nitration of 4-hydroxybenzenetellurinic acid with a nitric acid sulfuric acid mixture. With excess nitric acid the tellurinic acid nitrate *(21)* was isolated as shown in equation (73).

(73) $HO-C_6H_4-TeOOH \xrightarrow{HNO_3} HO-C_6H_3(NO_2)-TeOOH$

$HO-C_6H_4-TeOOH \xrightarrow{\text{excess } HNO_3} HO-C_6H_2(NO_2)_2-Te(O)ONO_2$ *(21)*

g) Hydrolysis of tellurinic acid chlorides and nitrates: Tellurinic acid chlorides were hydrolyzed by hot 70-80 per cent sodium hydroxide solution. Upon cooling white crystalline solids precipitated which were converted to tellurinic acids by treatment with sulfuric acid[358]. Tellurinic acid nitrates yielded the tellurinic acids upon treatment with sodium hydroxide[219] or with 20 per cent sulfuric acid[358].

h) Acetonyl and phenacyl aryl tellurium dichlorides *(22)* are cleaved by cold water and hot water, respectively, to arenetellurinic acid halides and halogenated ketones.

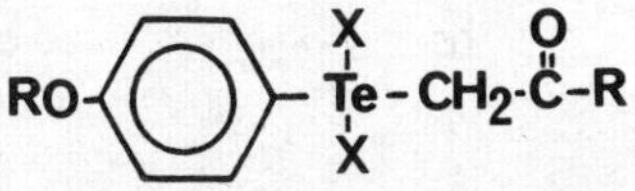

$$RO-C_6H_4-TeX_2-CH_2-C(=O)-R$$

(22)

R, R', X: CH_3, CH_3, Cl; C_2H_5, CH_3, Cl; CH_3, C_6H_5, Cl; CH_3, C_6H_5, Br: C_2H_5, C_6H_5, Cl;

Rohrbaech[371] claimed to have prepared thio derivatives of 4-methoxy- and 4-ethoxybenzenetellurinic acid anhydride. His results, however, are questionable, since the starting materials employed in these syntheses were probably mixtures of organyl tellurium trichlorides and diorganyl tellurium dichlorides[288].

Our knowledge of the reactivity of the few reported tellurinic acids is rather limited. The synthetic methods leading to tellurinic acids and their derivatives and the reactions given by these compounds are summarized in Fig. VI-6. It should be pointed out that tellurinic acids and tellurinic acid anhydrides can be reduced to ditellurides by potassium disulfite[348c,358] or sodium sulfide[427] and that treatment with hydriodic acid produces organyl tellurium triiodides[107]. These general statements, however, are based on only a few isolated experiments. Since organyl tellurium trichlorides hydrolyze in aqueous solution to tellurinic acid chlorides, the reduction of the trichlorides by aqueous potassium disulfite solution might actually proceed *via* the reduction of the tellurinic acid chlorides. Recuction experiments with acid chlorides have not been carried out. Drew[107] reported that methanetellurinic acid seemed to disproportionate on standing to dimethyl tellurium dihydroxide and tellurous acid.

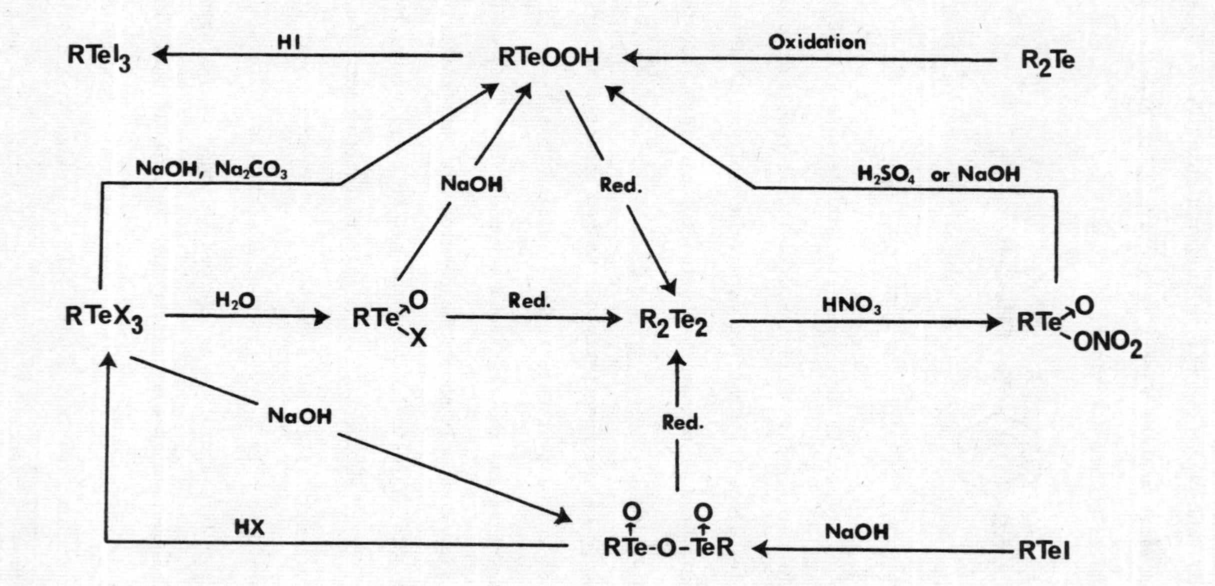

Fig. VI-6: Tellurinic Acids and Tellurinic Acid Derivatives

E) Diorganyl Ditellurides

Diorganyl ditellurides, R_2Te_2, are orange to red solids. Only the methyl and trifluoromethyl derivatives are liquids at room temperature. The known ditellurides are listed in Table VI-5. While the aromatic compounds are well represented, only a few aliphatic derivatives have been described. The aromatic derivatives are more stable and are much easier to handle than the aliphatic members which in addition to their relative instability possess an obnoxious, persistent odor. Liquid dimethyl ditelluride solidified upon standing for a few days at room temperature. Vacuum distillation of the solid regenerated the liquid[76]. It remained, however, liquid when stored in a sealed tube at -20°[447a]. Bis(4-chlorophenyl) ditelluride is the only ditelluride whose structure has been determined by single crystal X-ray analysis. The two organic groups bonded to the tellurium atoms form a dihedral angle of 72°. The structure of ditellurides thus resembles that of hydrogen peroxide[208]. Sink and Harvey[401a] assumed a similar structure for dimethyl ditelluride with a dihedral angle of 82° in their vibrational analysis of this molecule. More details about the structures of organic tellurium compounds are given in section XIV-C.

1) Synthesis of Diorganyl Ditellurides

The direct synthesis of ditellurides from sodium ditelluride and alkyl halides has been used very little. Most of the ditellurides were prepared by reduction of organyl tellurium trichlorides. A few other modes of formation of ditellurides have been found. Most of them, however, are of only very limited interest for preparative work. The following reactions are known to produce diorganyl ditellurides. Table VI-5 should be consulted for specific compounds.

Table VI-5

Diorganyl Ditellurides

R_2Te_2 R	Method	Yield %	mp. °C bp. (torr) °C	Reference
methyl	b	38	97(9)	37a, 76, 401a
	b	50	–	447a
	c	–	–	23, 365, 367
	c	94	-19.5[1)]	366
trideuteriomethyl	b	–	–	401a
trifluoromethyl	c	–	73	24
methylene	d	100	50-90	286
ethyl	*	–	–	258, 444
	b	71.4	92(10)	37a
carboxymethyl	d	–	140-4 (dec)	286
1-carboxyethyl	d	–	75	293
butyl	a	23	112-5(1.5)	26
benzyl	b	82	79 (dec)	37a, 397

Table VI-5 (cont'd)

1,2-diphenyl-2-chloroethenyl	e	100	151-2	305
phenyl	c	–	–	150
	d	76	66-7	122
	g	85	66	164a
	g	60	61-2	323d
	g'	43	66-7	348b
	h	–	–	150
	h	–	53-4	219
3-fluorophenyl	g	29	44	164a
4-fluorophenyl	g	58	77	164a
2-chlorophenyl	g	25	oil	164a
4-chlorophenyl	–	–	117	208
4-bromophenyl	g	20	153	164a
	i	44	153-5	348b
2-methylphenyl	g	12	oil	164a
	g'	51	oil	348b
	h	–	not pure(tar)	222

Table VI-5 (cont'd)

4-methylphenyl	d	-	52.5	122
	g	49	51	164a
	g'	58	52	348b
	h	-	not pure(tar)	222
	i	38	52	348b
4-biphenylyl	g	15	216	164a
4-methoxyphenyl	d	100	56-8	288,293
	e	100	60	339, 358, 359
	i	-	59-60	340, 426[2)]
4-ethoxyphenyl	d	100	108-9	50, 288, 358
	e	100	107,109	339, 359
	i	-	107	304, 340
2-phenoxyphenyl[3)]	d	-	-	103
4-phenoxyphenyl	d	100	87-8	103
	e	100	88	339, 359
	i	-	88	340
2-thiophenoxyphenyl	e	100	130.5-1.0	339

Table VI-5 (cont'd)

4-thiophenoxyphenyl	e	100	89-90	339
4-aminophenyl	**	-	-	155
4-acetamidophenyl	d	7	172 (dec)	358
	d	-	196-8 (dec)	155
2,4-dimethoxyphenyl	d	100	134-5	288
3-methyl-4-methoxyphenyl	d	-	77-8	293
5-methyl-2-methoxyphenyl	d	-	-	293
2-methyl-4-methoxyphenyl[3)]	d	-	-	293
3-hydroxy-4-methoxyphenyl	d	-	117-8 (dec)	288
2-bromo-5-methoxyphenyl[4)]	d	-	oil	293
2-bromo-6-methoxyphenyl	i	30	180-3	348b
4-hydroxy-3-nitrophenyl	d	100	150 (dec)	358
4-hydroxy-3,5-dinitrophenyl	d	100	153 (dec)	358
pentafluorophenyl	f	-	43-5, 42-3	197, 357
2-formylphenyl	g[5)]	40	147-52	348c
	i[6)]	45	147-52	348c
	j[7)]	30	147-52	348c

Table VI-5 (cont'd)

2-carboxyphenyl	d	55	216-20 (dec)	348c
	k	60	216-20 (dec)	348c
2-chloroformylphenyl	k	85	165-70	348c
1-naphthyl	e	100	123.5-4.5	339
	g'	66	119-122	348b
2-naphthyl	e	100	123	427
	i	-	123	426
	j	64	120-2	346
Bz-1,Bz-1'-dibenzanthronyl	-	-	-	P-27

* as by-product in telluride synthesis

** from 4-acetamidophenyl compound by hydrolysis with 4N ethanolic NaOH

1) boiling at 196° at 1 atm. with decomposition

2) reported mp. of 88° in ref. 426 is in error

3) not isolated in pure state

4) positions of substituents uncertain

5) from 2-(diethoxymethyl)phenyl lithium

6) from 2-formylphenyl methyl tellurium dichloride

7) from 2-formylphenyl butyl tellurium dichloride

a) Sodium ditelluride and alkyl halides in aqueous medium: Sodium ditelluride was prepared in basic aqueous medium according to a procedure widely used for the synthesis of sodium telluride[420] (see section III). The ditelluride reacted with butyl bromide to give a 23 per cent yield of dibutyl ditelluride. Chloroacetic acid, however, did not produce the expected ditelluride[26].

b) Sodium ditelluride and alkyl halides in liquid ammonia: Sodium ditelluride prepared from the elements in liquid ammonia (see section III) combined with alkyl halides to give the organic ditellurides in yields up to 80 per cent. Evaporation of the ammonia left solid residues from which the ditellurides were recovered by extraction[37a,76,397]. This method should be capable of further application in the synthesis of aliphatic and perhaps aromatic ditellurides.

c) Tellurium mirrors and organic radicals: Methyl[23,365-367], trifluoromethyl[24] and phenyl[150] radicals attacked mirrors of elemental tellurium and formed the respective ditellurides. These reactions were employed in the study of organic radicals generated by thermal decomposition of various organic compounds.

d) Reduction of organyl tellurium trichlorides by $K_2S_2O_5$: Ditellurides are conveniently synthesized by reduction of organyl tellurium trichlorides with potassium disulfite or hydrogen sulfite. The reductions proceed quantitatively in an aqueous medium at 0°. For literature references see Table VI-5. 2-Carboxybenzenetellurinic acid was similarly reduced to bis(2-carboxyphenyl) ditelluride[348c]. However, 4-hydroxyphenyl[358], 2-chlorocyclohexyl[304], 2-chlorostyryl[305], 4-(4'-carboxy-2'-quinolyl)phenyl[360] tellurium trichlorides and derivatives obtained from 1,3-diketones[276,282] were decomposed by potassium disulfite.

e) Reduction of organyl tellurium trichlorides by $Na_2S \cdot 9H_2O$: In the reduction of tellurium trichlorides with sodium sulfide, $Na_2S \cdot 9H_2O$, the mixture of the reactants with a large sulfide excess is heated to 95–100° and kept in the molten state at this temperature for about 15 minutes. The ditelluride separated quantitatively upon treatment of the cooled melt with water. The trichlorides listed above were also decomposed by sodium sulfide.

f) Reduction of organyl tellurium trichlorides by zinc: The reduction with zinc in ethanol was employed only in the synthesis of bis(pentafluorophenyl) ditelluride from the corresponding trichloride[197,357]. The yield was not reported.

g) Elemental tellurium and Grignard reagents or oganic lithium compounds: Aromatic Grignard reagents react with commercial elemental tellurium in tetrahydrofuran in the presence of oxygen (method g). The ditellurides were obtained in yields ranging from 12 to 85 per cent. However, not all aromatic Grignard reagents combine with tellurium. Alkyl magnesium halides were found to be unreactive[164a].

Aryl lithium compounds and elemental tellurium produce diaryl ditellurides (method g') according to equation (74)

(74)
$$\text{RLi} + \text{Te} \longrightarrow \text{R Te Li}$$
$$2\,\text{RTeLi} \xrightarrow{H_2O,\ O_2} \text{R TeTe R} + 2\,\text{LiOH}$$

h) "Tellurium dihalides" and Grignard reagents: The reaction of Grignard reagents with "tellurium dihalides" produced diorganyl ditellurides as by-products[150,219,222]. It was pointed out in section IV-B-3a, that the dihalides very easily disproportionate into tellurium and tellurium tetrahalides. The ditellurides were therefore probably formed from elemental tellurium and Grignard reagents[343].

i) Ditellurides from certain unsymmetric diorganyl tellurium dihalides: The reduction of certain unsymmetric organic tellurium dihalides $RR'TeX_2$, by sodium sulfide and the action of atmospheric agents on benzyl aryl tellurides gave diorganyl ditellurides. Excess sodium sulfide, zinc in chloroform, sodium hydrogen sulfite and hydrazine sulfate, reduced acetonyl, phenacyl and 2,4-dihydroxyphenyl aryl tellurium dichlorides to diaryl ditellurides[340] (eq. 75).

(75) 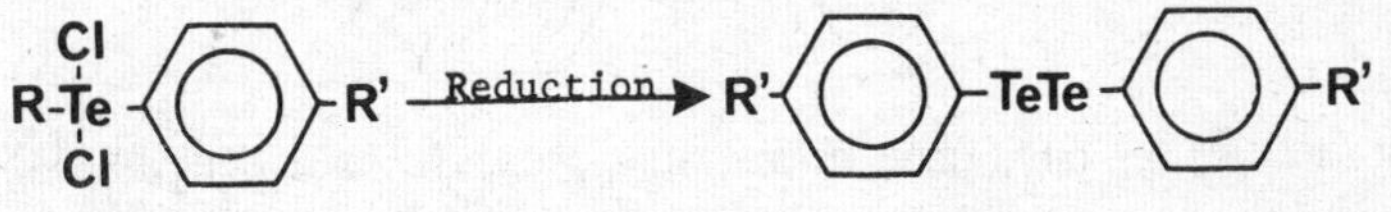 $R\text{-}TeCl_2\text{-}C_6H_4\text{-}R' \xrightarrow{\text{Reduction}} R'\text{-}C_6H_4\text{-}TeTe\text{-}C_6H_4\text{-}R'$

$R' = CH_3O,\ C_2H_5O,\ C_6H_5O$

The fate of the organic group cleaved from the tellurium atom is not known. The tellurium diiodide *(23)* treated with $Na_2S\cdot 9H_2O$ at 100° formed bis(4-ethoxyphenyl) ditelluride[304].

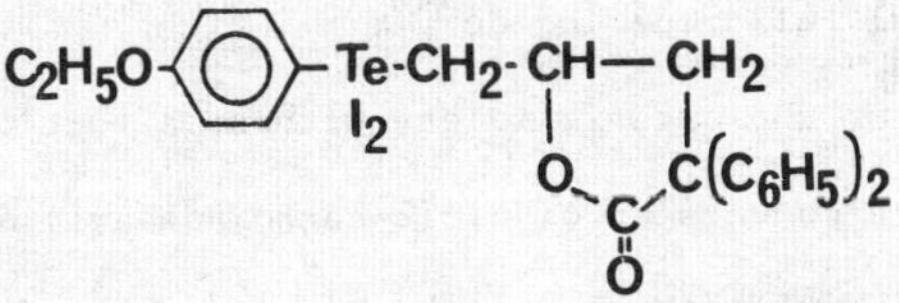

$C_2H_5O\text{-}C_6H_4\text{-}TeI_2\text{-}CH_2\text{-}CH(O\text{-}C(=O)\text{-})\text{-}CH_2\text{-}C(C_6H_5)_2$ (ring: O–C(=O)–C(C_6H_5)$_2$)

(23)

4-Methoxyphenyl benzyl and 2-naphthyl benzyl telluride were converted to the corresponding diaryl ditellurides upon exposure to the atmosphere. This conversion is facilitated by the presence of moisture. Benzaldehyde was detected in the reaction mixture[426].

Refluxing solutions of the alkyl aryl tellurium dihalides *(24)* in pyridine produced diaryl ditellurides according to equation (75a)[348b,c]. The thermal dissociation of *(24)* into alkyl halide and the unstable tellurenyl compound, R-Te-X, is followed by the loss of halogen from two molecules of the tellurenyl compound and concomitant formation of the diaryl ditelluride. The reaction of the cleavage products RX and X_2 with pyridine favors the formation of the ditellurides.

(75a) $2\ \text{R'},\text{Y-C}_6\text{H}_3\text{-TeR}X_2$ (24) $\xrightarrow{\text{pyridine}}$ $[2\ \text{R'},\text{Y-C}_6\text{H}_3\text{-TeX}] + 2RX$ → R',Y-C₆H₃-TeTe-C₆H₃-R',Y

R, R', Y, X: CH_3, CHO, H, Br; C_4H_9, CHO, H, Br; C_4H_9, Br, 2-CH_3O, Cl; C_4H_9, H, 4-CH_3, Cl; C_4H_9, H, 4-Br, Cl

j) Alkaline hydrolysis of 2-naphthyl tellurium iodide: Di(2-naphthyl) ditelluride was obtained by alkaline hydrolysis of 2-naphthyl tellurium iodide, by decomposition of the triphenylphosphine adduct of the iodide, and by action of triethylamine or triisopropyl phosphite on 2-naphthyl tellurium iodide[346].

k) Modification of the organic moiety without affecting the ditelluride group: The conversion of bis(2-carboxyphenyl) ditelluride to bis(2-chloroformylphenyl) ditelluride by butyl dichloroformate, $HCCl_2(OC_4H_9)$, in the presence of anhydrous zinc chloride is the only known reaction of this type[348c].

Bis(2-carboxyphenyl) ditelluride was obtained by treating 2-formylphenyl methyl telluride in aqueous ethanolic sodium hydroxide with silver nitrate. It is not known, whether the ditelluride link was formed in this reaction prior to the oxidation of the aldehyde group[348c].

2) Reactions of diorganyl ditellurides

Most of the reactions of diorganyl ditellurides proceed either with elimination of one tellurium atom or with fission of the tellurium-tellurium bond. Halogenolysis of ditellurides produces organyl tellurium trihalides as discussed in section VI-C-1g. Bis(thiocarbamoyl) disulfides react in the same way as do halogens to give organyl tellurium tris(dithiocarbamates)[126]. Controlled halogenolysis of bis(2-naphthyl) ditelluride with iodine yields 2-naphthyl tellurium iodide (see section VI-B)[427]. Chen and George[76] reported that dimethyl ditelluride in ether solution was partially decomposed to tellurium tetrabromide upon treatment with bromine while in benzene solution the expected tribromide was obtained in quantitative yield.

Diorganyl ditellurides upon heating eliminate tellurium and form diorganyl tellurides[122,164a,341]. The reaction commences in the absence of any solvent at 250° but proceeds in good yields only above 300°. Heating bis(4-methoxyphenyl) ditelluride and the corresponding ethoxy derivative in xylene for four hours was without effect[341].

Measurements of the molecular mass of diphenyl ditelluride in camphor[122] indicated that 30 per cent of the compound was dissociated into radicals, RTe . Upon raising the temperature of an ethanolic ditelluride solution from -80° to +80° the color intensified. Oxygen was absorbed by the solution. These observations and the low molecular masses of other ditellurides found by earlier investigators[288,293,358] can be explained by radical formation. Diorganyl disulfides are known to dissociate into radicals[382].

Ditellurides in ethereal solution treated with a three- to five-fold molar excess of an aromatic Grignard reagent gave in quantitative yield the unsymmetric tellurides (see section VII-A) according to equation (76)[164a,342].

(76) $$\text{R-TeTe-R} + \text{R'MgX} \longrightarrow \text{R-Te-R'} + \text{RTeMgX}$$

Aliphatic Grignard reagents did not react[342]. Phenyl lithium gave a complex mixture of tellurides from which definite compounds could not be isolated[342]. Diphenyl ditelluride and dialkyl mercury compounds refluxed in dioxane formed alkyl phenyl tellurides[323d].

Methyl iodide cleaved the tellurium-tellurium bond in bis(4-methoxyphenyl) ditelluride[288,358] (eq. 77).

$$(77)\quad X\text{-}C_6H_4\text{-}TeTe\text{-}C_6H_4\text{-}X + 3\,CH_3I \longrightarrow X\text{-}C_6H_4\text{-}TeI_2\text{-}CH_3 + [X\text{-}C_6H_4\text{-}Te(CH_3)_2]^+ I^-$$

$X = CH_3O$

The 4-ethoxyphenyl derivative, however, was decomposed by methyl iodide[288].

Hydrogen peroxide[293] or air[286,288,293,103] oxidized ditellurides to white substances which were not further characterized. Concentrated nitric acid decomposed bis(4-phenoxyphenyl)[103] and bis(4-methoxyphenyl) ditelluride[288] into 4,4'-dinitrodiphenyl ether, while diphenyl ditelluride gave benzenetellurinic acid nitrate[147,219] (see section VI-D-e).

Lederer[219] made an attempt to reduce diphenyl ditelluride to benzenetellurol with sodium in ethanol. He treated the reaction product with mercuric chloride and isolated $C_6H_5Te\cdot HgCl$ (see section VI-A).

A number of aromatic ditellurides were reduced to aryl sodium telluride by sodium borohydride in ethanol-benzene in the presence of sodium hydroxide[348b,c,d]. These tellurides (see section XI-A) were not isolated but were reacted immediately with organic halides to produce aryl alkyl tellurides (see section VII-A-2). Diethyl ditelluride in ether solution when treated with lithium at room temperature gave C_2H_5-Te-Li[117a]. Diethyl ditelluride was also cleaved by $(C_2H_5)_3GeLi$ in hexane[117a]. Diphenyl ditelluride in benzene when shaken with mercury formed $(C_6H_5Te)_2Hg$[323d].

Ditellurides can serve as dehalogenating agents. Diorganyl selenium dibromides, organyl tellurium trichlorides, tellurium tetrahalides, diorganyl tellurium dibromides and *vic*-dibromides transformed ditellurides into diorganyl tellurium dihalides with elimination of one tellurium atom per ditelluride molecule[341]. These reactions are discussed in more detail in section VII-B.

A methylene group was inserted into the Te-Te bond when bis(4-methoxyphenyl) or bis(4-ethoxyphenyl) ditelluride was treated with diazomethane in ether solution. With dichlorocarbene, generated from phenyl trichloromethyl mercury, definite products could not be obtained[347b]. Bis(4-ethoxyphenyl) ditelluride reacted with benzyne, obtained from diphenyliodonium 2-carboxylate, to give 1,2-bis(phenyltelluro)benzene[347a].

Ditellurides react with transition metal salts and carbonyl complexes to form compounds in which the organyltelluro group serves as a bridging group in binuclear complexes[176,177,293,378a,378b,419]. The ditelluride linkage seems to remain intact upon complexation with mercuric halides[150] and uranium pentachloride[392,392a]. Shlyk[396] reported that diphosphines and ditellurides yield organyltellurophosphines. For a detailed discussion of these compounds consult chapter XI. The reactions given by ditellurides and the synthetic routes to ditellurides are presented in Fig. VI-7.

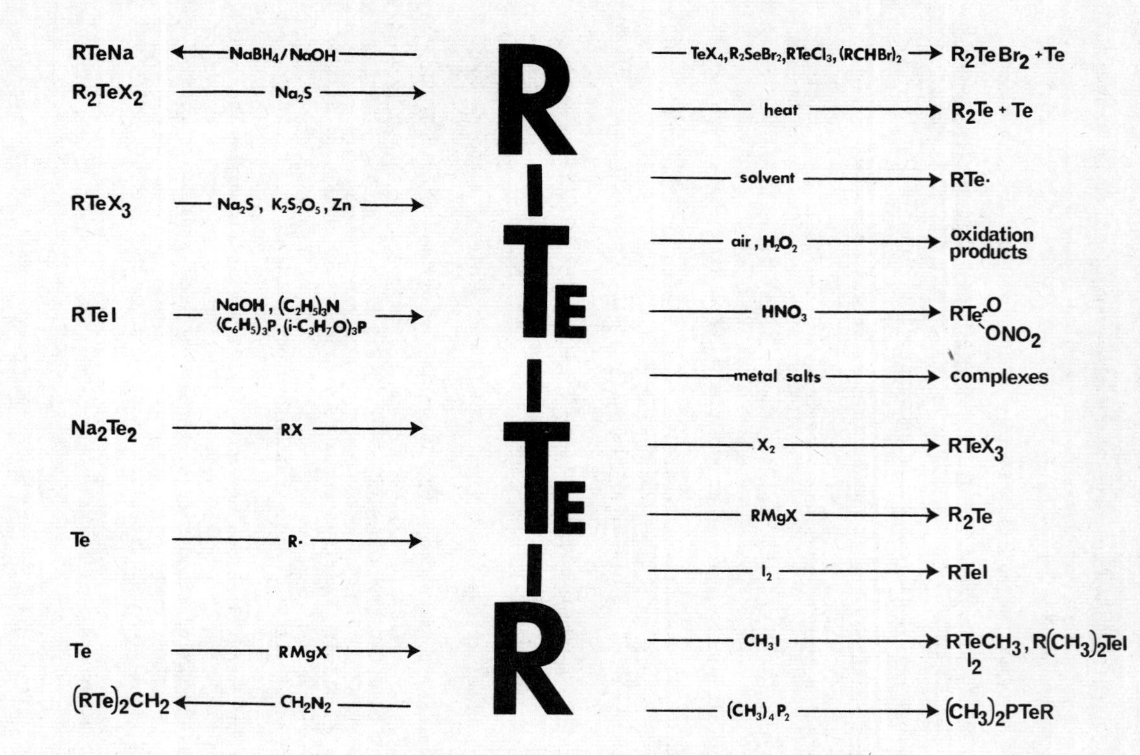

Fig. VI-7: Syntheses and Reactions of Diorganyl Ditellurides

VII. COMPOUNDS CONTAINING A CARBON-TELLURIUM-CARBON MOIETY

Diorganyl tellurides, diorganyl tellurium dihalides, diorganyl telluroxides and diorganyl tellurones comprise the compounds in which the tellurium atom is bonded to two organic groups. Compounds of the general formula R_2TeX_2, with X representing anionic groups like NO_3^-, ClO_4^-, CrO_4^{--} and others will be discussed together with the dihalides. The aromatic derivatives far outnumber the aliphatic compounds. Although methods for the preparation of aliphatic tellurides and their derivatives have been known for a long time, their sensitivity towards oxygen and their penetrating, obnoxious odor have discouraged more extensive investigations. Telluroxides and tellurones are not well characterized and much more work is required to establish with some confidence the chemical behavior of these compounds.

A) Diorganyl Tellurides

The section on diorganyl tellurides will be divided into three parts. Symmetric diorganyl tellurides will be discussed first, followed by unsymmetric compounds. The third part will deal in general terms with the reactions of these tellurides.

1) Symmetric diorganyl tellurides

The first organic tellurium compound, diethyl telluride, was prepared by Woehler[443] in 1840. His method of reacting potassium telluride with an alkyl sulfate was later employed in the synthesis of dimethyl[170,445], diethyl[21,170,258,444] and dipentyl telluride[446]. Potassium telluride was prepared by keeping tellurium and the carbonaceous residue obtained by thermal decomposition of potassium hydrogen D-tartrate at red heat until the carbon monoxide evolution had ceased[21]. These reactions are only of historic interest, since better ways of synthesizing these tellurides are now available. Natta[310] reported in 1926, that aliphatic alcohols and ethers, reacted with aluminum telluride, Al_2Te_3, at 250-300°, gave good yields of alkyl tellurides. This method has not been used again, although it seems to be the reaction of choice to produce large quantities of tellurides from cheap starting materials. Dimethyl telluride was formed when sodium tellurite, Na_2TeO_3, was heated with betaine or choline chloride in the presence of sodium formate[62a,368a].

Dialkyl tellurides are prepared most conveniently by the alkylation of sodium telluride or by reduction of dialkyl tellurium dihalides, which can be obtained by alkylation of tellurium tetrachloride (see section VII-B). Most of the other methods, with the exception of the reaction between diorganyl mercury and tellurium, are of limited value in the preparation of tellurides. Aromatic tellurides can be synthesized employing the methods developed for aliphatic derivatives. Sodium telluride, however, does not react with aromatic halides. The reaction of diaryl mercury with tellurium has produced several tellurides. This route requires the synthesis of the organic mercury compound. The more direct preparation from tellurium tetrachloride and a large excess of a Grignard reagent gives tellurides in acceptable yields.

Rohrbaech[371] claimed to have prepared bis(4-methoxyphenyl) and bis(4-ethoxyphenyl) telluride by reduction of the condensation product obtained from tellurium tetrachloride and the respective alkoxybenzenes. Morgan and Drew[288], however, showed that under Rohrbaech's conditions, only organyl tellurium trichlorides are formed. The reduced compounds, therefore, were not tellurides but ditellurides[222,37

Anderson[11] calculated the boiling points of dimethyl and diethyl telluride using the Kinney equation. The refractive index of diphenyl telluride[329] and the thermal characteristics of mixtures of this telluride with the corresponding sulfide and selenide[328] were determined by Pascal. He found that the telluride is isomorphous with the selenide and sulfide but not with the oxide.

The following methods have been developed for the synthesis of symmetric diorganyl tellurides. In order to learn whether a particular reaction has been used for the synthesis of a specific telluride, Table VII-1 should be consulted.

a) Alkylation of sodium telluride in an aqueous medium: Sodium telluride, prepared by the method of Tschugaeff and Chlopin[420] from tellurium and sodium formaldehyde sulfoxylate (Rongalite) in basic aqueous medium, combines with alkyl halides to give dialkyl tellurides. The insolubility of the alkyl halides in water necessitates the addition of ethanol to the reaction mixture. Water soluble alkylating agents react satisfactorily without the presence of an organic solvent. Thus, benzyl dimethyl phenyl ammonium chloride gave dibenzyl telluride[420]. *tert*-Butyl chloride did not produce the telluride in an aqueous ethanolic reaction medium[18].

b) Alkylation of sodium telluride in liquid ammonia: Elemental sodium and tellurium in liquid ammonia form sodium telluride, which upon addition of an alkyl halide, gives the dialkyl telluride. The utility of this method for the synthesis of long chain aliphatic tellurides is shown by the successful preparation of the hexadecyl derivative[P-50]. Experiments with aromatic halides have not yet been carried out.

c) Reduction of diorganyl tellurium dihalides by potassium disulfite, sodium hydrogen sulfite or sodium sulfite:

The reactions of diorganyl tellurium dihalides with $K_2S_2O_5$,

Table VII-1

Symmetric Diorganyl Tellurides, R_2Te

R	Method	Yield	mp. °C	bp. °C(torr)	Reference
methyl	a	-	-	93.5 (749)	35
	c	-	-	94 (770)	424
	i	-	-	-	23,149,150
	m	-	-	80-3	359
ethyl	b	38	-	137-8	397
	b	>80	-	38(14)	42
	g	-	-	137	449
	j	-	-	137-8	261
propyl	-	-	-	-	68
i-propyl	b	>80	-	49 (18)	42
butyl	a	57	-	109-12(12)	17
	e	-	-	132-5(99)	50
	j	-	-	100-10(15)	172
pentyl	a	55	-	138-40(18)	18

Table VII-1 (cont'd)

hexadecyl	b	-	43-5	-	P-50
benzyl	a	-	53	-	138,420
	a	90	50-2	-	405
iodomethyl	c	100	180-5	-	121,153
carboxymethyl	c	100	140-1	-	286
2-carboxyethyl	n	-	159	-	191
2-cyanoethyl	o	-	58-9	-	191
2-chloro-2-phenylethenyl	d	100	59-61	-	304
2,2-diphenylethenyl	c	56	173-5	-	119a
2,2-bis(4-methylphenyl)ethenyl	c	80	119-20	-	119a
	f	78	119-20	-	119a
carbo-ℓ-menthoxymethyl	m	-	58	-	17
phenyl	c	-	4.2	182-3(16.5)	3,219,328
	c[1)]	-	-	182(14)	227
	d	100	-	182-3(16)	359
	e[2)]	-	-	-	211
	f	-	-	-	238

Table VII-1 (cont'd)

phenyl	g	–	–	–	3,246,336,337,429
	g	70	–	174(10),312-20(760)	199,407
	g	77	–	178(12)	25
	h	–	–	182(14)	150,219
	i	–	–	–	150
	j	–	–	–	120,212,213,314,442
	k	100	–	–	122,343
	l	–	–	–	147
	m	25	–	–	213,215,359
	p	58	–	174(10)	379
	p[3)]	–	–	180(16.5)	35a
	q	–	–	–	376
2-methylphenyl	c	60	37-8	202-3(15)	222
	g	–	37-8	202.5(16)	429,449
	h	–	37-8	202-3(15)	222
3-methylphenyl	f	89	–	205-6(18)	225
	g	–	–	–	429

Table VII-I (cont'd)

3-methylphenyl	h	56	-	205-6(18)	225
4-methylphenyl	c	-	69-70	211-2(18)	222
	d	100	70	210(16)	359
	d[3)]	-	65-6	-	306
	g	-	63-4	210(16)	429,449
	h	55	69-70	211-2(18)	222
	j	-	-	-	318
4-biphenylyl	k	-	167-8	-	164a
2,4-dimethylphenyl	f	92	-	202-3(10)	223
	h	-	-	-	223
2,5-dimethylphenyl	g	-	-	-	429
	h	52	72	250-75(34)	223
2,4,6-trimethylphenyl	h	70	129	233-7(16)	224
2-methoxyphenyl	f	-	73-4	235-41(18) 248-51(30)	236
3-methoxyphenyl	f	-	-	247-52(35)	235
4-methoxyphenyl	d	93	56	-	415
	e[5)]	90	53-4	-	50

Table VII-1 (cont'd)

4-methoxyphenyl	e[6)]	37	53-4	-	293
	f	-	56-7	-	226
	g	-	-	-	429
	h	-	56-7	-	226
	j	-	-	-	317
	r	67.4	52-3	-	341
2-ethoxyphenyl	f	60	-	245(18)	232
	h	-	-	-	232
4-ethoxyphenyl	e[5)]	-	-	-	297
	f	-	63	235-40(18)	234
	h	39	64	-	297
	r	67.7	63-4	-	341
4-bromophenyl	d	-	121	-	415
2-carboxyphenyl	f	20	278	-	122,267
4-dimethylaminophenyl	c	-	128-30	-	295
4-hydroxy-2-methylphenyl	c	4	143-4	-	296
pentafluorophenyl	g	100	50-1	-	83

Table VII-1 (cont'd)

4-methoxyphenyl (cont'd)	j	30	50-1	-	83
	e	18	54-5	-	95a
1-naphthyl	f	-	-	-	233
	g	53	126.5	-	254,429
	h	62	-	-	233
	j	75	-	-	316
2-naphthyl	d	92	144-5	-	363
2-thienyl	e[7)]	-	50.5	-	205
	e[8)]	-	-	-	204
	f	-	-	-	205

1) reduction of $(R_2TeNO_3)_2O$ with SO_2

2) reducing agent: $LiAlH_4$

3) $(C_6H_5)_2Te$ containing ^{132}Te, ^{133}Te, ^{134}Te

4) reduction of $(R_2TeCl)_2O$ and $R_2Te(O_2CC_6H_5)_2$ by Na_2S

5) reducing agent: Zn/glac.CH_3COOH

6) reducing agent: Zn/benzene

7) reducing agent: $Na_2S_2O_3$

8) reducing agent: $SnCl_2$/HCl

$NaHSO_3$ or Na_2SO_3 proceed in aqueous solution containing sodium hydroxide or sodium carbonate. The few reported yields indicate that these reductions are almost quantitative. However, the less stable tellurium dichlorides like bis(4-ethoxyphenyl) and bis(2-chlorocyclohexyl) tellurium dichloride and those derived from acetylacetone, acetone, 1,1,1-trimethylacetone and acetophenone deposited elemental tellurium when treated with these reducing agents.

In order to facilitate purification diorganyl tellurides are often converted to their crystalline dihalides, which are easier to isolate and recrystallize. The reduction of the pure dihalides with reducing agents as discussed in sections VII-A-1c to VII-A-1f give the pure tellurides.

d) Reduction of diorganyl tellurium dihalides with sodium sulfide nonahydrate: The reduction with sodium sulfide is applicable to aliphatic and aromatic tellurium dihalides and gives almost always quantitative yields. The dihalide and a 15-fold molar excess of the sulfide is kept in the molten state at 100° for about 15 minutes. Equation (78) describes the process.

(78) $$R_2TeX_2 + Na_2S \xrightarrow{100°} R_2Te + 2NaX + S$$

e) Other reducing agents for diorganyl tellurium dihalides: Lithium aluminum hydride[211], zinc dust in glacial acetic acid[50] or benzene[293], tin dichloride in hydrochloric acid[204] and sodium thiosulfate[205] have been employed as reducing agents in a few isolated cases. There seems to be no advantage in using these reagents. Bis(4-methylphenyl) polonide dehalogenated the corresponding tellurium dichloride forming the polonium dichloride[318].

f) Reduction of diorganyl tellurium dihalides with methyl magnesium iodide: Aromatic tellurium dihalides which are insoluble in water can be reduced with a three-fold molar excess of methyl magnesium iodide in ethereal solution[233-236]. The yields vary between 60 and 92 per cent (eq. 79).

$$R_2TeX_2 + 2CH_3MgI \longrightarrow R_2Te + C_2H_6 + 2MgXI \tag{79}$$

Tetraorganyl tellurium compounds of the type $R_2Te(CH_3)_2$ have been shown to be unstable eliminating ethane and forming the telluride (see section IX). It is therefore likely that such a compound is formed as intermediate in these reductions. Ethyl magnesium iodide converted bis(1-naphthyl) tellurium dibromide into the telluride[233]. When phenyl or 2-methylphenyl magnesium bromide were reacted with aromatic tellurium dihalides, symmetric and unsymmetric tellurides were isolated[229,238].

g) Diaryl mercury and elemental tellurium: Diaryl mercury compounds transfer the organic groups to tellurium upon heating the reactants under an inert atmosphere in a sealed tube to 230° for several hours[3,25,83,199,246,254,407,429,449] (eq. 80).

$$R_2Hg + 2Te \longrightarrow R_2Te + HgTe \tag{80}$$

The yields obtained range from 50-100 per cent. The reaction temperature must be kept below the decomposition temperature of the mercury compound.

h) "Tellurium dihalides" and Grignard reagents: Lederer[219,222-226, 233,236] introduced the reaction of aromatic Grignard reagents with "tellurium dihalides" into organic tellurium chemistry. It is almost certain, that Lederer unknowingly handled a mixture of tellurium and tellurium tetrahalide, into which dihalides easily disproportionate (see section VI-B-3a). In most cases, a four-fold molar excess of Grignard reagent was employed. The complexity of the reaction is manifested by the isolation of tellurides, ditellurides, aromatic hydrocarbons and elemental tellurium from the reaction mixture. The yields of tellurides range from 30 to 70 per cent. The tellurides were usually converted to the dihalides to facilitate their isolation. Although the exact reaction mechanism accounting for all these products is still unknown, sections VI-B-3a and VI-A-2 present possible reaction sequences, which are applicable to the "tellurium dihalide" problem. Glazebrook and Pearson[150] obtained diphenyl telluride together with the other possible tellurides from a fused mixture of tellurium and iodine, stoichiometrically corresponding to TeI_2, and a Grignard solution containing methyl and phenyl magnesium bromide.

i) Organic radicals and tellurium mirrors: The only tellurides obtained during investigations of the attack of organic radicals on tellurium mirrors were dimethyl[23,150,249] and diphenyl tellurides[150]. The methyl and phenyl radicals were generated by thermal decomposition of various organic compounds. These reactions are unimportant for preparative purposes.

j) Diorganyl tellurides *via* tetraorganyl tellurium compounds: Diorganyl tellurides were the products of reactions between tellurium tetrachloride and butyl lithium[172], pentafluorophenyl lithium[83], phenyl magnesium bromide[120,212,213,314], 4-methoxyphenyl[317] and 4-methylphenyl[318] magnesium bromide. The amount of reagents present was sufficient to form a tetraorganyl tellurium compound as intermediate, which then probably decomposed to the diorganyl telluride. The following equations (81)-(82) present the likely reaction path.

(81) $$TeCl_4 + 4RM \longrightarrow R_4Te + 4MCl$$

M = Li, MgBr

(82) $$R_4Te \longrightarrow R_2Te + R_2$$

Wittig and Fritz[442] and Cohen and coworkers[83] have shown that tetraphenyl and tetrakis(pentafluorophenyl) tellurium decompose at their melting temperature into the respective tellurides.

k) Pyrolysis of diaryl ditellurides: Bis(4-biphenylyl) ditelluride heated with copper powder at 210° produced the telluride[164a]. Diphenyl telluride[122] was prepared by heating the ditelluride to about 300°. The yield, reported to be quantitative, makes this method very interesting. Starting with organyl tellurium trihalides, obtainable by reactions described in section VI-C-1, the ditellurides can be prepared quantitatively by reduction with sodium sulfide. The overall yields of tellurides would depend only on the efficiency of the organyl tellurium trichloride syntheses.

l) Elemental tellurium and Grignard reagents or organic thallium compounds: The reactions between elemental tellurium and Grignard reagents are not of importance for the preparation of tellurides except in connection with "tellurium dihalide" reactions (see subsection h above). Only diphenyl telluride was obtained by this method[147]. Petragnani and de Moura Campos further investigated this reaction and proposed two mechanisms[343] (see section IV-A-2), which account for the various observed products.

Bis(pentafluorophenyl) thallium bromide and tellurium heated in a sealed tube for 3 days at 190° produced bis(pentafluorophenyl) telluride[95a].

m) Diorganyl tellurides from triorganyl telluronium compounds: Five reactions shown in equations (83-87) are reported in the literature, which convert triorganyl telluronium halides into diorganyl tellurides. Reaction (83)[213] probably proceeds *via* tetraphenyl tellurium as discussed in subsection j above for similar reactions. Diphenyl telluride[215] was formed by refluxing the telluronium salt in dimethylaniline (eq. 84). Heating the telluronium salt of equation (87) in vacuum produced bis(carbo-ℓ-menthoxymethyl) telluride[17]. These two reactions probably proceed *via* the thermal dissociation of the respective telluronium salts. Telluronium salts seem to be reduced by hydrated sodium sulfide

(83) $\left[(C_6H_5)_3Te\right]^+Cl^- + C_6H_5MgBr \longrightarrow (C_6H_5)_2Te + (C_6H_5)_2 + MgClBr$

(84) $\left[(C_6H_5)_2CH_3Te\right]^+I^- + (CH_3)_2NC_6H_5 \longrightarrow (C_6H_5)_2Te + \left[(CH_3)_3NC_6H_5\right]^+I^-$

(85) $\left[CH_3O\text{-}C_6H_4\text{-}Te(CH_3)_2\right]^+I^- + NaSH \longrightarrow CH_3OC_6H_5 + (CH_3)_2Te + NaI + S$

(86) $\left[(C_6H_5)_3Te\right]^+I^- + NaSH \longrightarrow C_6H_6 + (C_6H_5)_2Te + NaI + S$

(87) $\left[(C_4H_9)_2TeCH_2COOR\right]^+Br^- \longrightarrow Te(CH_2COOR)_2$

R = ℓ-(2-i-C_3H_7-5-CH_3-cyclohexyl)

according to equations (85) and (86)[359]. Benzene and anisole were detected as reduction products.

<u>n)</u> Reactions affecting the organic part of diorganyl tellurides: The only reaction known to change the organic part of a symmetric telluride is the hydrolysis of bis(2-cyanoethyl) telluride by hydrochloric acid[191] (eq. 88).

(88) $(NC\text{-}CH_2CH_2)_2Te \xrightarrow{HCl/H_2O} (HOOC\text{-}CH_2CH_2)_2Te$

<u>o)</u> The electrolysis of a mixture of tellurium and acrylonitrile in 1<u>N</u> sodium sulfate yielded bis(2-cyanoethyl) telluride[191]. It is likely that the telluride is formed by addition of electrolytically generated hydrogen telluride to the carbon-carbon double bond. No further application of this reaction could be located in the literature.

<u>p)</u> Elemental tellurium and tetraphenyl tin: Schmidt and Schumann[379] heated tetraphenyl tin and tellurium for 80 hours at 240° in an evacuated tube and isolated diphenyl telluride. Telluranthrene (see section

XII-C-5) was also formed.

q) Diazonium and iodonium salts in the synthesis of diaryl tellurides: Farrar[122] obtained bis(2-carboxyphenyl) telluride from sodium telluride and 2-carboxybenzenediazonium chloride. The physical properties of Farrar's telluride differed from those reported for the same telluride by Mazza and Melchionna[267]. Diphenyl iodonium chloride in the presence of thioglycolic acid, sodium carbonate and tellurium produced diphenyl telluride[376].

r) Bis(4-methoxyphenyl) and bis(4-ethoxyphenyl) ditelluride were converted into tellurides[341] by the respective diaryl tellurium dibromides upon refluxing of the reactants in toluene according to equation (89).

$$R_2Te_2 + R_2TeBr_2 \longrightarrow R_2Te + R_2TeBr_2 + Te \quad (89)$$

Since the reactions were performed with starting materials containing the same organic groups, conclusions as to the reaction sequence cannot be drawn.

s) Miscellaneous methods: Nefedov and coworkers[312] obtained diphenyl and bis(4-methylphenyl) telluride containing the isotope ^{125m}Te by β-decay of $R_3{}^{125}Sb$. Diphenyl and dibutyl telluride with ^{132}Te were prepared from ^{132}Te and diphenyl mercury[3,246] and dibutyl mercury, respectively. Syntheses, which are successful in sulfur and selenium chemistry, do not necessarily work with tellurium. For example, the telluride $[CCl_3CH(OH)]_2Te$ could not be obtained by the action of hydrogen telluride on an ethereal solution of chloral[244], whereas hydrogen sulfide and selenide gave the expected products.

Figure VII-1 summarizes the reactions which yield diorganyl ditellurides.

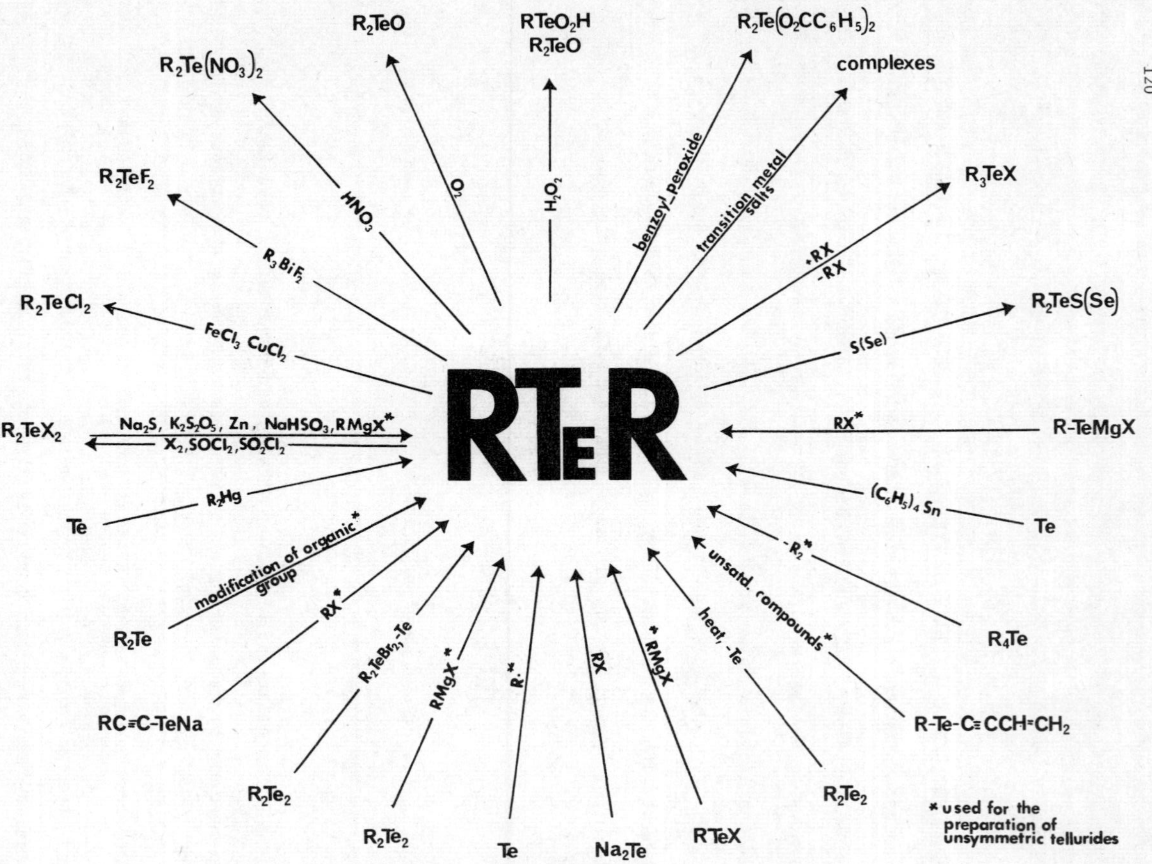

Fig. VII-1: Syntheses and Reactions of Diorganyl Tellurides

2. Unsymmetric Diorganyl Tellurides

Many of the methods, which have been employed for the preparation of specific unsymmetric diorganyl tellurides as described below, might have general applicability. Extensive investigation in this field have, however, not yet been carried out.

Alkyl aryl tellurides seem to be less stable than diaryl tellurides. Bowden and Braude[39] reported, that ethyl phenyl telluride in solution formed the symmetric tellurides. The tellurium-carbon(benzyl) bond in aryl benzyl tellurides is cleaved by bromine, iodine and sulfuryl and thionyl chloride[426]. Such tellurides are also easily oxidized to diaryl ditellurides[426].

The unsymmetric tellurides, which are listed in Table VII-2, have been prepared according to the following methods:

a) The reaction of organyl alkali metal tellurides with aliphatic halides: Organyl alkali metal tellurides, RTeM (M = Li,Na), (chapter XI-A), have been synthesized from sodium acetylides and tellurium in liquid ammonia[38,38a,42,43,348,355], from organyl lithium compounds and tellurium in diethyl ether[348b,c] and by reduction of diaryl ditellurides in ethanol/benzene solution with sodium borohydride in aqueous sodium hydroxide[348b,c].

Acetylenic hydrocarbons with a hydrogen atom on at least one of the triply bonded carbon atoms form sodium acetylides upon treatment with sodium or sodium amide in liquid ammonia. Tellurium, added to such a reaction mixture, inserts into the carbon-sodium bond. The organyl sodium telluride can then react with an organyl halide to produce the telluride in yields up to 65 per cent[38,38a,42,43,348,355] (eq. 90).

Table VII-2

Unsymmetric Diorganyl Tellurides, RTeR'

R	R'	Method	Yield %	mp.°C	bp.°C(torr)	Reference
methyl	$CH_3-C{\equiv}C-$	a	48	-	71-2(50)	38a
	$CH_2{=}CH-C{\equiv}C-$	a	50	-	83-4(20)	348
	$CH_2{=}C(CH_3)-C{\equiv}C-$	a	49	-	92-3(20)	355
	$C_6H_5-C{\equiv}C-$	a	46	-	122-4(2)	38
					92(0.05)	38
	3-C_6H_5-4,5-dihydroisoxazol-5-yl-$C{\equiv}C-$ ($C_6H_5-C{=}N-O-CH(C{\equiv}C-)-CH_2$, ring)	b	89	66-7	--	353
	1,3-$(C_6H_5)_2$-2-pyrazolin-5-yl-$C{\equiv}C-$ ($CH_2-CH(C{\equiv}C-)-N(C_6H_5)-N{=}C(C_6H_5)$, ring)	b	-	127-8	--	354
ethyl	$CH_3-C{\equiv}C-$	a	65	-	71(11)	42,43
	$CH_2{=}CH-C{\equiv}C-$	a	58	-	80-1(10)	348
	2,3-$(C_6H_5)_2$-isoxazolidin-5-yl-$C{\equiv}C-$ ($C_6H_5-CH-CH_2-CH(C{\equiv}C-)-O-N(C_6H_5)$, ring)	b	20	90	-	353

Table VII-2 (cont'd)

ethyl (cont'd)	$C_6H_5-C(=N-N(C_6H_5)-CH(-C\equiv C-)-CH_2-)$	b	-	116	-	354
butyl	$-CH_2COOC_2H_5$	c	80	-	135-8(21) 85(<0.1)	17
	$C_6H_5C(O)-$	a	20	-	100-115(0.3)	348d
	$-CH_2-COO-$(ring) $i-C_3H_7$, CH_3	c	61	-	120-5(0.1)	17
pentyl	$-CH_2COOC_2H_5$	c	-	-	140-50(17)	18
	$-CH_2COO-$(ring) CH_3, $i-C_3H_7$	c	-	-	78-85(17)	18
phenyl	methyl	c	45	-	105-9(3)	348b
		k	40	-	118-22(22)	150,370
		m	-	-	-	150
	ethyl	h	43	-	107-8(22)	39
	i-propyl	f	79	-	83(7)	323d
	butyl	f	75	-	115(7)	323d
	benzyl*	f	-	-	-	323d

Table VII-2 (cont'd)

phenyl	$HC\equiv C-CH_2-$	h	-	-	-	352
	$H_2C=C=CH-$	b	-	-	-	352
	$HC\equiv C-CH(CH_3)-$	h	-	-	-	352
	$H_2C=C=C(CH_3)-$	b	27	-	65-75(0.001)	352
	2-carboxymethyl	a	85	65-7	-	348b
	$CH_3C(O)-$	a	80	-	105-10(0.3)	348d
	$C_2H_5C(O)-$	a	65	-	108-10(0.3)	348d
	$C_6H_5C(O)-$	a	60	66-70	-	348d
	2-$CH_3C_6H_4C(O)-$	a	50	30-4	-	348d
	2-$FC_6H_4C(O)-$	a	86	44-6	-	348d
	2-$ClC_6H_4C(O)-$	a	53	42-4	-	348d
	2-$BrC_6H_4C(O)-$	a	86	65-8	-	348d
	2-$IC_6H_4C(O)-$	a	75	70-2	-	348d
	2-$CH_3OC_6H_4C(O)-$	a	56	75-7	-	348d
	4-$CH_3OC_6H_4C(O)-$	a	78	100-3	-	348d
	2-$CH_3SC_6H_4C(O)-$	a	65	52-3	-	348d
	2-$CH_3SeC_6H_4C(O)-$	a	80	78-9	-	348d

Table VII-2 (cont'd)

phenyl	2-methylphenyl	i	77	-	-	238
		j	24	-	212-3(22)	238
	4-methylphenyl	i	84	-	-	229
		j	79	63-4	207-8(16)	229
	4-methoxyphenyl	d	87	61	-	363
		e	100	61	-	342
	4-phenoxyphenyl	d	70	-	177(0.03)	363
		e	100	oil	-	342
	1-naphthyl	d	80	-	147-8(0.01)	363
	2-naphthyl	d	75	49-50	-	363
		g	100	49-50	-	427
2-methylphenyl	methyl	a	80	-	130-1(5)	348c
	$CH_3C(O)$-	a	35	-	100-5(0.3)	348d
	$C_6H_5C(O)$-	a	60	69-70	-	348d
	2-$CH_3C_6H_4C(O)$-	a	37	25-30	-	348d
	2-$CH_3OC_6H_4C(O)$-	a	32	58-63	-	348d
4-methylphenyl	methyl	a	75	-	150-2(18)	348b

Table VII-2 (cont'd)

4-methylphenyl	butyl	a	–	–	–	348b
	2-I-$C_6H_4C(O)$-	a	90	95-100	–	348d
4-biphenylyl	methyl	k	28	84	–	370
2(hydroxymethyl)phenyl	methyl	b	90	–	135-6(0.25)	348c
2-(1-hydroxyethyl)phenyl	methyl	b	53	–	130-5(0.1)	348c
4-methoxyphenyl	methyl	i	67	–	150-2(20)	358
		i	–	–	–	288
		c	–	–	150-3(20)	359
	ethyl	d	30	–	120(0.1)	426
	benzyl	d	100	36-8	–	426
	4-ethoxyphenyl	i	–	45	–	297
	4-dimethylamino	i	100	96-7	–	340
	1-naphthyl	d	83	87-8	–	363
	2-naphthyl	d	95	72	–	363
4-ethoxyphenyl	4-dimethylaminophenyl	i	100	126-7	–	340
	1-naphthyl	e	100	65-6	–	342
4-phenoxyphenyl	4-dimethylaminophenyl	i	100	76	–	340

Table VII-2 (cont'd)

4-phenoxyphenyl	1-naphthyl	d	76	71-2	-	363
	2-naphthyl	d	87	69-70	-	363
2-(diethoxymethyl)phenyl	methyl	c	100	-	118-22(0.1)	348c
	butyl	a	68	-	148-50(1.0)	348c
4-bromophenyl	methyl	a	60	49.5	-	348b
	butyl	a	30	-	-	348b
2-bromo-6-methoxyphenyl	butyl	a	60	-	155-60(1.5)	348b
	2-carboxyethyl	a	66	123-5	-	348b
2-formylphenyl	methyl	c	77	38-9	120-30(0.1)	348c
2,4-dinitrophenylhydrazone		b	-	204-8	-	348c
oxime		b	85	95-8	-	348c
2-formylphenyl	butyl	b	80	-	140-2(1.0)	348c
2,4-dinitrophenylhydrazone		b	-	202-4	-	348c
2-formylphenyl	$-CH_2COOH$	c	-	140-3	-	348e
2-carboxyphenyl	methyl	a	60	195-200	-	348c
	butyl	b	85	-	-	348c
	$-CH_2COOH$ †	-	-	195	-	267

Table VII-2 (cont'd)

2-chloroformylphenyl	methyl	b	70	56-62	-	348c
	butyl	b	-	-	-	348c
2-carbethoxyphenyl	methyl	b	80	61-5	-	348c
	butyl	b	80	-	155-60(0.1)	348c
2-cyanophenyl	methyl	b	55	-	145-50(0.1)	348c
2-naphthyl	ethyl	g	75	-	143(0.03)	427
	benzyl	d	-	58-9	-	426
	cyclohexyl	d	59	-	190(0.024)	426
		g	91	-	190(0.024)	427
	$(C_6H_5)_2C$—C═O, O, H_2C—CH, CH_2- (ring)	l	61	174-6	-	302
		i	89	174-6	-	302
	C_6H_5-C≡C-	g	95	88-9	-	305
	1-naphthyl	d	91	93-4	-	363

* unstable at room temperature

† structure of product doubtful (see section VII-A-1g)

(90)

$$R\text{-}C{\equiv}CH + Na \xrightarrow{NH_{3}(\ell)} R\text{-}C{\equiv}C\text{-}Na \xrightarrow{Te} R\text{-}C{\equiv}C\text{-}Te\text{-}Na \xrightarrow[-NaX]{+R'X} R\text{-}C{\equiv}C\text{-}Te\text{-}R'$$

Aryl lithium compounds, prepared from aryl bromides and butyl lithium in diethyl ether solution, react with elemental tellurium. The thus formed aryl lithium tellurides combine with the butyl bromide generated in the lithium-bromine exchange reaction (eq. 90a). The aryl butyl tellurides were isolated in yields ranging from 30-68 per cent[348b,c]. Butyl lithium telluride, obtained from butyl lithium and tellurium in diethyl ether, reacted with benzoyl chloride to give benzoyl butyl telluride in 20 per cent yield[348d].

(90a)

$$R,R'\text{-}C_6H_3\text{-}Br + C_4H_9Li \longrightarrow C_4H_9Br + R,R'\text{-}C_6H_3\text{-}Li \xrightarrow{Te} R,R'\text{-}C_6H_3\text{-}Te\text{-}Li \xrightarrow[-LiBr]{+C_4H_9Br} R,R'\text{-}C_6H_3\text{-}Te\text{-}C_4H_9$$

R, R': 4-CH_3, H; 4-Br, H; H, $CH(OC_2H_5)_2$; 6-CH_3O, Br

Diaryl ditellurides in ethanol/benzene solution are converted to aryl sodium tellurides upon treatment with <u>1N</u> sodium hydroxide solution containing sodium borohydride. The addition of carboxylic acid chlorides[348d], β-chloropropionic acid[348b] or dimethyl sulfate[348b,c] to this reaction mixture produced aryl organyl tellurides in yields as high as 90 per cent.

<u>b)</u> Reactions affecting the organic moiety in diorganyl tellurides: The reactions, which modify the organic moiety in unsymmetric diorganyl tellurides comprise addition to the vinyl group in alkyl vinylacetylenyl tellurides[353,354], the isomerization of phenyl propargyl tellurides to allenic derivatives[352], the introduction

of a substituent into an aromatic ring in methyl aryl tellurides[370] and the modification of a substituent already present in an aryl group[348c,370].

Radchenko and coworkers[353,354] prepared *via* a 1,3-dipolar addition of benzonitrile N-oxide *(25)*[353], N-α-diphenylnitrone *(26)*[353], and 1,3-diphenylnitrilimine *(27)*[354] to alkyl vinylacetylenyl tellurides in benzene the unsymmetric tellurides presented in scheme (91).

(91)

CH_3-Te-C≡C-CH=CH_2 + *(25)* → 5-(methyltelluroethynyl)-3-phenyl-2-isoxazoline

C_2H_5-Te-C≡C-CH=CH_2 + *(26)* → 5-(ethyltelluroethynyl)-2,3-diphenylisoxazolidine

R-Te-C≡C-CH=CH_2 + *(27)* → 5-(alkyltelluroethynyl)-1,3-diphenyl-2-pyrazoline

R = CH_3, C_2H_5

Pourcelot[352] found that phenyl propargyl tellurides *(28)* isomerize to the allenic derivatives *(29)*. The isomerization is facilitated by strong base. The propargyl derivatives have been prepared from

$$C_6H_5\text{-Te-}\underset{}{\overset{R}{C}}H\text{-C}\equiv CH \rightleftarrows C_6H_5\text{-Te-}\overset{R}{C}\text{=C=}CH_2$$

(28) *(29)*

propargyl bromide and phenyltelluro magnesium bromide. The isomerization cannot have taken place prior to the formation of the telluride, since allenyl bromide was unreactive under these conditions.

Methyl phenyl and methyl biphenylyl tellurides are susceptible to substitution in the 4-position of the aromatic group[370]. Rogoz[370] introduced the bromoacetyl group by reacting the tellurides with bromoacetyl chloride in presence of aluminum chloride in chloroform. The bromoacetyl group was then modified in a series of reactions given in scheme (92). The compounds obtained, together with their melting points and yields, are listed in Table VII-3.

The reactive aldehyde group in 2-formylphenyl alkyl tellurides can be easily converted into other functional groups without affecting the carbon-tellurium bonds[348c]. These reactions are summarized in scheme (93).

c) The conversion of triorganyl telluronium salts to diorganyl tellurides: The thermal decomposition of triorganyl telluronium halides of the type $[R_2R'Te]^+X^-$ seems to be a convenient way to prepare unsymmetric tellurides, RTeR', according to equation (94).

(92)

$$CH_3\text{-}Te\text{-}C_6H_5 \xrightarrow{-5° \text{ to } -10°}$$

$$CH_3\text{-}Te\text{-}C_6H_4\text{-}C_6H_5 \xrightarrow{+15°}$$

$$\xrightarrow[-HCl]{BrCH_2COCl} R\text{-}\overset{O}{\overset{\|}{C}}\text{-}CH_2Br$$

$$\xrightarrow{(CH_2)_6N_4} R\text{-}\overset{O}{\overset{\|}{C}}\text{-}CH_2Br\cdot\text{hexamethylenetetramine} \xrightarrow{\text{HCl conc.}} R\text{-}\overset{O}{\overset{\|}{C}}\text{-}CH_2NH_2$$

$$\xrightarrow{(CH_3CO)_2O} R\text{-}\overset{O}{\overset{\|}{C}}\text{-}CH_2NH\text{-}\overset{O}{\overset{\|}{C}}\text{-}CH_3 \xrightarrow{CH_2O} R\text{-}\overset{O}{\overset{\|}{C}}\text{-}\underset{CH_2OH}{CH}\text{-}NH\text{-}\overset{O}{\overset{\|}{C}}\text{-}CH_3$$

$$\xrightarrow[i\text{-}C_3H_7OH]{(i\text{-}C_3H_7O)_3Al} R\text{-}\overset{OH}{CH}\text{-}\underset{CH_2OH}{CH}\text{-}NH\text{-}\overset{O}{\overset{\|}{C}}\text{-}CH_3 \xrightarrow{HCl} R\text{-}\overset{OH}{CH}\text{-}\underset{CH_2OH}{CH}\text{-}NH_2$$

$$\xrightarrow[CaCO_3/NaCN]{CCl_3CH(OH)_2} R\text{-}\overset{OH}{CH}\text{-}\underset{CH_2OH}{CH}\text{-}NH\text{-}\overset{O}{\overset{\|}{C}}\text{-}CHCl_2$$

$R = CH_3Te\text{-}C_6H_4\text{-}$, $CH_3Te\text{-}C_6H_4\text{-}C_6H_4\text{-}$

Table VII-3

Alkyl Aryl Tellurides Prepared by Rogoz[370]

	CH_3-Te-C_6H_4-		CH_3-Te-C_6H_4-C_6H_4-	
	Yield %	mp. °C	Yield %	mp. °C
$-C(=O)-CH_2Br$	52	55-6	22.3	118
$-C(=O)-CH_2Br \cdot$ HMTA*	88.1	107(dec)	--	--
$-C(=O)-CH_2-NH_2 \cdot HCl$	30.2	233	57.3	243
$-C(=O)-CH_2-NH-C(=O)-CH_3$	78.9	163	86	190
$-C(=O)-CH(CH_2OH)-NH-C(=O)-CH_3$	45.8	155	52	161-3
$-CH(OH)-CH(CH_2OH)-NH-C(=O)-CH_3$	79.5	128	89	147-8
$-CH(OH)-CH(CH_2OH)-NH_2$	83.3	108	91.6	131
$-CH(OH)-CH(CH_2OH)-NH-C(=O)-CHCl_2$	92.9	114	71.6	146

*HMTA = Hexamethylenetetramine

(93)

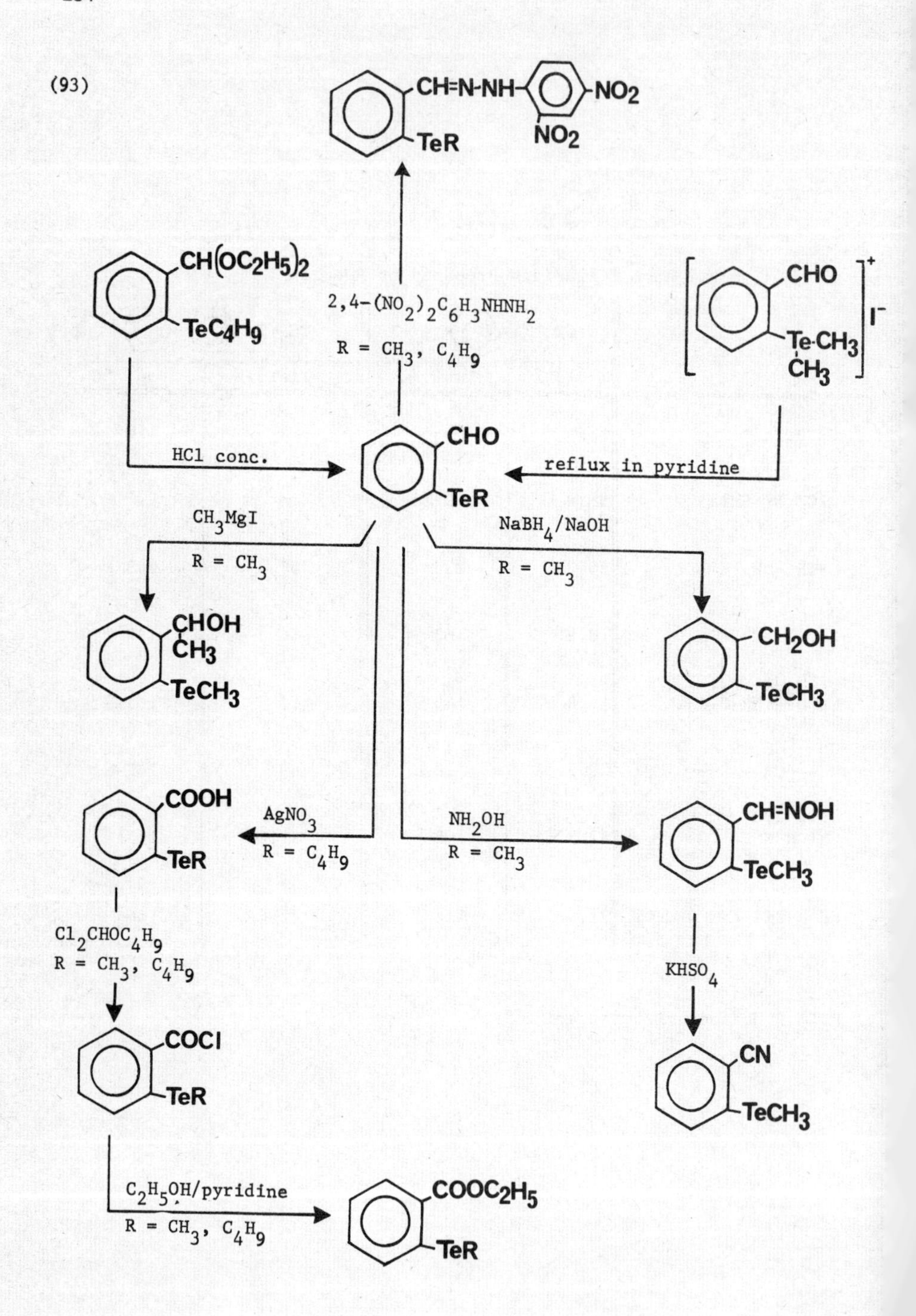
CH=N-NH
NO2
NO2
TeR
CH(OC2H5)2
TeC4H9
2,4-(NO2)2C6H3NHNH2
R = CH3, C4H9
CHO
Te-CH3
CH3
I−
HCl conc.
CHO
TeR
reflux in pyridine
CH3MgI
R = CH3
NaBH4/NaOH
R = CH3
CHOH
CH3
TeCH3
CH2OH
TeCH3
COOH
TeR
AgNO3
R = C4H9
NH2OH
R = CH3
CH=NOH
TeCH3
Cl2CHOC4H9
R = CH3, C4H9
KHSO4
COCl
TeR
CN
TeCH3
C2H5OH/pyridine
R = CH3, C4H9
COOC2H5
TeR

(94) $$[R_2R'Te]^+X^- \xrightarrow{heat} RTeR' + RX$$

Balfe and coworkers[17,18] applied this method in the synthesis of compounds of the type, $R\text{-}Te\text{-}CH_2COOR'$ ($R = C_4H_9, C_5H_{11}$; $R' = C_2H_5$, ℓ-menthyl). They heated the telluronium salts $[R_2TeCH_2COOR']^+Br^-$ in vacuum to 100°. 2-Formylphenyl carbethoxymethyl methyl telluronium bromide lost methyl bromide at 170° at 0.2 mm Hg[348e]. The telluronium salts $[2\text{-}RC_6H_4\text{-}Te(CH_3)_2]^+I^-$ [$R = -CHO, -CH(OC_2H_5)_2, H$], when refluxed in pyridine, were converted to the corresponding aryl methyl tellurides[348b,c].

Reichel and Kirschbaum[359] treated 4-methoxyphenyl dimethyl telluronium iodide with molten sodium sulfide nonahydrate and isolated among other products 4-methoxyphenyl methyl telluride.

Further investigations of the thermal properties of telluronium salts should make it possible to arrange the organic groups bonded to the tellurium atom in a sequence of thermal "cleavage" stability. Such a sequence will be very helpful in developing methods for the preparation of unsymmetric tellurides from symmetric tellurides *via* telluronium salts.

d) The reaction of organyl tellurium trichlorides with organyl mercury chlorides: The reaction of organyl tellurium trichlorides with organyl mercury chlorides is a convenient method for the preparation of unsymmetric aryl tellurides, since the tellurium containing starting materials are easily accessible (see section VI-C-1). Rheinboldt and Vicentini[363] introduced this method in 1956. Later it was employed for the synthesis of benzyl, ethyl and cyclohexyl aryl tellurides[426]. The primary product in these reactions is the tellurium dichloride, which is subsequently reduced by sodium sulfide (eq. 95).

(95) $$RTeCl_3 + R'HgCl \xrightarrow{-HgCl_2} R\text{-}\underset{Cl}{\overset{Cl}{Te}}\text{-}R' \xrightarrow{Na_2S} R\text{-}Te\text{-}R'$$

<u>e)</u> The reaction of diaryl ditellurides with aromatic Grignard reagents: Aromatic ditellurides obtainable through reduction of aromatic trichlorides (see section VI-E-1d to 1f) are cleaved by aromatic but not by aliphatic Grignard reagents in ether solution to give the unsymmetric tellurides in quantitative yields based on equation (96).

(96) $$R_2Te_2 + R'MgBr \longrightarrow R\text{-}Te\text{-}R' + RTe\text{-}MgBr$$

The aryltelluro magnesium bromide which is formed in this reaction, precipitates upon addition of petroleum ether as a syrup. This unstable substance decomposes upon contact with the atmosphere to tellurium and hydrogen telluride[342].

<u>f)</u> The reaction of diphenyl ditelluride with dialkyl mercury compounds: It has been found, that refluxing equimolar amounts of diphenyl ditelluride and di-*i*-propyl, dibutyl or dibenzyl mercury in dioxane produces alkyl phenyl tellurides in yields as high as 79 per cent[323d]. Elemental mercury precipitates during the reaction. Other ditellurides, which should react similarly, have not yet been tested.

<u>g)</u> The reaction of 2-naphthyl tellurium iodide with Grignard reagents or phenyl mercury chloride: Aryl tellurium halides, RTeX, would be very desirable starting materials for the preparation of unsymmetric tellurides. However, only 2-naphthyl tellurium iodide is available (see section VI-B). It undergoes reaction with Grignard reagents in ether solution and with phenyl mercury chloride in dioxane solution to produce the tellurides in good yields[427,305].

h) The reaction of elemental tellurium with Grignard reagents: The addition of elemental tellurium to a Grignard solution produces organyltelluro magnesium bromide, which without being isolated, reacts with an organic halide and forms the unsymmetric telluride[39,352] (eq. 97). Petragnani and de Moura Campos[343] were unable to prepare ethyl phenyl telluride by this method, while Bowden and Braude[39] claimed a yield of 43 per cent.

$$\text{RMgBr} + \text{Te} \longrightarrow \text{RTeMgBr} \xrightarrow{\text{R'X}} \text{R-Te-R'} + \text{MgBrX} \qquad (97)$$

i) The reduction of unsymmetric diorganyl tellurium dihalides: The reduction of unsymmetric diorganyl tellurium dihalides has been carried out with sodium sulfide[302,340,363,426], sodium sulfite[229], potassium disulfite[288,358], zinc dust in glacial acetic acid[297] or methyl magnesium iodide in the manner described in section VII-A-1c to 1f. The unsymmetric diorganyl tellurium dichlorides obtained from aryl tellurium trichlorides and acetone, acetophenone and resorcinol gave, on reduction, the symmetric diaryl ditellurides and not the unsymmetric tellurides (see section VI-E-1h).

j) The reaction of symmetric diaryl tellurium dihalides with aromatic Grignard reagents: Symmetric diaryl tellurium dihalides are reduced to symmetric diaryl tellurides by methyl magnesium halides (see section VII-A-1f). However, when bis(4-methoxyphenyl) tellurium dibromide was reacted with phenyl magnesium bromide, 4-methylphenyl phenyl telluride was the main product[229]. Diphenyl tellurium dibromide and 2-methylphenyl magnesium bromide gave diphenyl telluride and 2-methylphenyl phenyl telluride[238]. A likely mechanism for these reactions has been presented in section VII-A-1f.

k) The reaction of tellurium dihalides with solutions containing methyl magnesium halide and phenyl or 4-biphenylyl magnesium halide. A fused 1:1 molar mixture of tellurium and iodine reacted with an ethereal solution containing methyl and phenyl magnesium iodide. Methyl phenyl telluride was isolated in 40 per cent yield in addition to diphenyl ditelluride and the symmetric tellurides[150]. Tellurium dibromide and the appropriate mixture of Grignard reagents produced similarly methyl phenyl telluride and methyl 4-biphenylyl telluride in moderate yields[370].

l) The reaction of 2-naphthyl tellurium iodide with 2,2-diphenyl-2-allylacetic acid: 2-Naphthyl tellurium iodide combined with diphenyl allylacetic acid to yield 2-naphthyl 2-oxo-3,3-diphenyl-5-tetrahydrofurylmethyl telluride.[302] Detailed equations describing this reaction are presented in scheme *(61)* in section VI-B.

m) Elemental tellurium and organic radicals: Methyl and phenyl radicals generated by the thermal decomposition of acetophenone attacked tellurium mirrors and produced among other products methyl phenyl telluride[150].

3) Tellurides with Two Tellurium Atoms in the Molecule

Petragnani obtained 4-methoxyphenyltelluromethyl 4-methoxyphenyl telluride and the corresponding ethoxy derivative, which melted at 98-99° and 40-41°, respectively, in quantitative yield from the ditellurides and diazomethane in ether solution[347b].

Benzyne, generated from diphenyliodonium 2-carboxylate in 1,2-dichlorobenzene at 190°, reacted with bis(4-ethoxyphenyl) ditelluride. 2-(4-Ethoxyphenyltelluro)phenyl 4-ethoxyphenyl telluride melting at 114-115° was produced in quantitative yield[347a].

The other known tellurides, which contain more than one tellurium atom in the molecule, are either cyclic or polymeric compounds derived from telluroformaldehyde, CH_2Te. They are discussed in chapter XII and XIII, respectively.

4) Reactions of Diorganyl Tellurides

The reactions of diorganyl tellurides can be divided into three groups:

a) Reactions, in which the valency of the tellurium atom is increased without cleavage of a tellurium-carbon bond

b) Reactions, which proceed with tellurium-carbon bond cleavage.

c) Reactions, in which the organic moiety is modified without affecting the tellurium carbon bonds.

A general outline of the reactive behavior of diorganyl tellurides is given in Fig. VII-1. Many of these reactions have been employed to synthesize other organic tellurium compounds. Cross-reference to the appropriate section, in which a particular reaction is treated in detail, are included in the following discussion.

a) Reactions, which increase the valency of the tellurium atom without cleavage of a carbon-tellurium bond: Organic halides add to aliphatic and aromatic tellurides to form triorganyl telluronium salts (eq. 98).

(98) $$R_2Te + RX \longrightarrow [R_3Te]^+X^-$$

Aromatic tellurides seem to be less reactive than aliphatic derivatives. Diazotetraphenylcyclopentadiene, heated with diphenyl telluride in diethyl ether, produced diphenyl telluronium tetraphenylcyclopentadienylide[132a]. For details on these reactions and a list of known compounds please turn to section VIII-1.

Diorganyl tellurides are easily and almost always quantitatively converted into diorganyl tellurium dihalides. Elemental halogens, sulfuryl halides, thionyl chloride, transition metal chlorides, *vic*-dibromides and tris(4-methylphenyl) bismuth difluoride were employed as halogenating agents. This reaction is especially important for the synthesis of diorganyl tellurium dibromides and diiodides, which are not obtainable through condensation reactions. Literature references are to be found in section VII-B-1.

Nefedov[319a,d] studied the reaction $R_2TeX_2 + R_2Te \rightleftharpoons R_2Te + R_2TeX_2$ (X = F, Cl, Br, I) in ehloroform solution and determined the entropy changes and activation energies. The reaction mixtures were analyzed employing thin-layer chromotography[319b].

Nitric acid oxidizes diorganyl tellurides to diorganyl tellurium dinitrates, whereas oxygen or hydrogen peroxide yield diorganyl telluroxides (section VII-C), tellurinic acids (section VI-D-a) or mixtures of oxides and acids, depending upon the conditions. Benzoyl peroxide gives diorganyl tellurium dibenzoates (section VII-B).

The complexes formed between diorganyl tellurides and transition metal salts, trimethyl aluminum, trimethyl gallium and boron tribromide are described in chapter XI. Lederer[212] reported that diphenyl telluride and hydrochloric acid gave the adduct $R_2Te \cdot HCl$, which melted at 233-4°. Peach[333] determined the specific conductance of a 0.32<u>M</u> dimethyl telluride solution in liquid hydrogen chloride at -95°. The value of 3123 $\mu\Omega^{-1}cm^{-1}$ indicates that the $R_2Te \cdot HCl$ complex, if formed, is only slightly ionized. Hellwinkel and Fahrbach[172] isolated the adduct *(30)* from an ethanolic solution containing the telluride and picric acid. Similarly, the picric

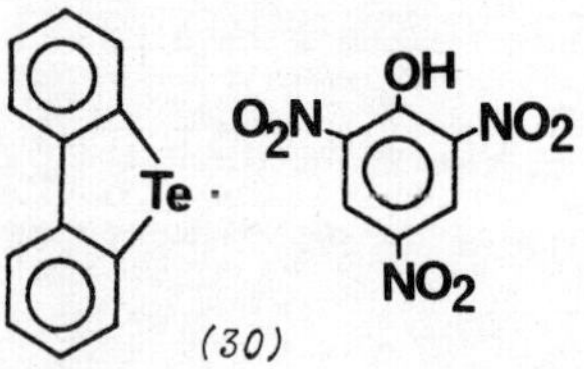

(30)

acid adduct of 4-methoxyphenyl methyl telluride was prepared[288]. The report by Hetnarski and Hofman[174], that dimethyl telluride produces a deep red complex with 1,3,5-trinitrobenzene raises the question whether the acidic hydrogen atom in picric acid is at all involved in the formation of the adducts.

b) Reactions, which proceed with tellurium-carbon bond cleavage: Atmospheric oxygen[17,18] or alkaline hydrogen peroxide[18] oxidize dialkyl tellurides to tellurinic acids (eq. 99) (see section VI-D-a).

$$(RCH_2)_2Te + 2O_2 \longrightarrow RCH_2TeOOH + RCOOH \quad (99)$$

Only two literature reports claim the exchange of organic groups in diorganyl tellurides. Mazza and Melchionna[267] converted bis(2-carboxyphenyl) telluride into 2-carboxyphenyl carboxymethyl telluride by adding sodium chloroacetate to a reaction mixture containing the symmetric telluride and zinc in a potassium hydroxide solution. Farrar[122] was unable to repeat the synthesis of Mazza's starting telluride.

Bowden and Braude[39] suggest that the rapid change in the absorption spectrum of methyl phenyl telluride in cyclohexane is caused by the conversion of the unsymmetric telluride to the symmetric tellurides.

Diphenyl telluride heated with sulfur at 220° for 10 hours under an atmosphere of carbon dioxide formed diphenyl sulfide and tellurium[200]. The tellurium atom in bis(pentafluorophenyl) telluride was replaced by a selenium atom upon keeping the reagents together at 320°[83].

Tellurides containing benzyl groups are rather unstable towards atmospheric oxidation. Dibenzyl telluride decomposes quickly when exposed to air[405,420]. 2-Naphthyl benzyl and 4-methoxyphenyl benzyl telluride heated with sulfuryl or thionyl chloride, with bromine or iodine produced aryl tellurium trihalides and benzyl halide. Upon exposure of these tellurides to the atmosphere ditellurides and benzaldehyde were formed[426]. 2-Formylphenyl methyl telluride and 2-formylphenyl butyl telluride were converted to bis(2-carboxyphenyl) ditelluride upon refluxing their aqueous-ethanolic solutions, which contained sodium hydroxide and silver oxide[348c]. Treatment of $R-C\equiv C-Te-CH_3$ [$R = CH_3, CH_2=C(CH_3)$] with mercuric acetate in water liberated the respective acetylene[38a].

The flash photolysis of dimethyl telluride generated the $CH_3Te\cdot$ radical[83a,b]. The irradiation of diphenyl telluride with neutrous for 15 seconds decomposed some of the diphenyl ditelluride. Organic tellurium compounds containing the isotope ^{131}Te were formed[439b]. Llabador and Adloff showed that ^{132}Te-diphenyl telluride[246,247,247a] and ^{132}Te-dibutyl telluride[247,247a] formed the corresponding ^{132}I-organic halides, RI, upon disintegration of the ^{132}Te isotope.

Potassium amide in liquid ammonia cleaved both carbon-tellurium bonds in diphenyl telluride[380] according to equation (100). Diphenyl selenide under the same conditions produced selenanthrene.

$$(100) \quad 7(C_6H_5)_2Te + 10KNH_2 + 2NH_3 \longrightarrow 4K_2Te(NH)_3 + K_2Te_3 + 14C_6H_6$$

Diethyl telluride when kept in contact with the hydrides of the elements silicon, germanium and tin of the general formula R_3MH (R = cyclohexyl, ethyl) at 20-200° gave compounds of the composition R_3M-Te-C_2H_5 or R_3M-Te-MR_3 depending upon the reaction conditions and the hydride employed. Ethane is liberated during these reactions. A detailed discussion of these compounds is given in section XI-B.

c) Reactions modifying the organic moiety: 2-Formylphenyl carboxymethyl telluride was cyclized to 2-carboxybenzotellurophene in refluxing pyridine in the presence of acetic anhydride[348e].

The other reactions, which modify organic groups in diorganyl tellurides without a tellurium-carbon bond cleavage have already been discussed in sections VII-A-1n and VII-A-2b.

B) Diorganyl Tellurium Compounds, R_2TeX_2

Diorganyl tellurium dihalides, R_2TeX_2 (X = F, Cl, Br, I), are crystalline solids, which can be easily purified by recrystallization. A large number of chlorides, bromides and iodides have been prepared. When diorganyl tellurides are the primary reaction products, their conversion to the dichlorides or dibromides facilitates the isolation and purification of the organic tellurium compounds. The reactions represented by equation (101) generally proceed in quantitative yield

$$R_2Te \underset{\text{Reduction}}{\overset{X_2}{\rightleftarrows}} R_2TeX_2 \qquad (101)$$

in both directions. The difluorides are prepared by exchanging halogen for fluorine with silver fluoride[120] or by fluorinating the telluride with tris(4-methylphenyl) bismuth difluoride[317]. The methyl, phenyl[120] and 4-methoxyphenyl[317] derivatives are the only known difluorides.

The structures of these diorganyl tellurium dihalides have been determined by X-ray methods[66,78,79]. The groups bonded to the tellurium atom expand its electronic shell to ten electrons. With four bonding electron pairs and one stereochemically active lone electron pair these molecules possess as expected[145] a trigonal bipyramidal shape with the lone electron pair and two organic groups occupying the equatorial positions. The halogen atoms take up the axial positions[308]. (Fig. VII-2).

Bond lengths and bond angles for these compounds are tabulated and discussed in section XIV-C. The dipole moments of the diphenyl tellurium dihalides were found to be 1.80, 2.48, 2.63 and 3.84D in benzene at 25° for the fluoride chloride, bromide and iodide, respectively. The molecular moments are believed to be the sum of the equatorial bond

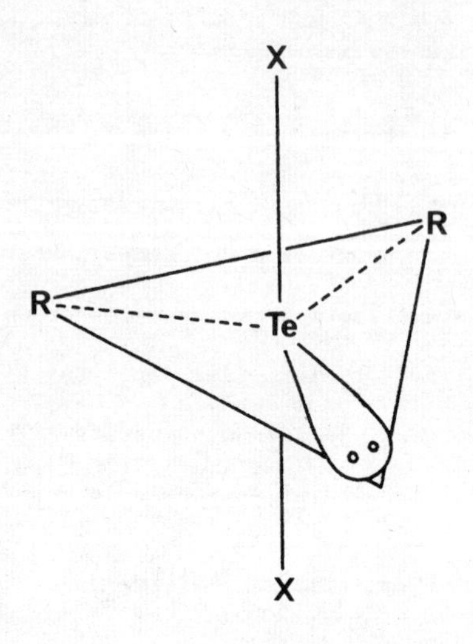

Fig. VII-2: The Molecular Structure of Diorganyl Tellurium Dihalides

moments[192a], since the halogen-tellurium-halogen angles are close to 180°[66,78]. The decrease in the dipole moment in going from the iodide to the fluoride has been attributed to a contraction in the size of the sp^3d hybrid orbital which contains the lone electron pair. This results because of the increasing electronegativity of the halogen atoms[192a]. The bis(4-methylphenyl) tellurium dichloride and dibromide have dipole moments of 2.98D and 3.21D in benzene at 25°, respectively[188].

The diaryl tellurium dihalides are stable substances. The carbon-tellurium bond is cleaved only under drastic conditions. Bis(4-ethoxyphenyl) tellurium dihalides, for instance, were decomposed by concentrated nitric acid to 2-halo-4-nitroethoxybenzenes[297]. Certain dihalides, however, containing labile organic groups undergo carbon-tellurium bond cleavage under rather mild conditions. More details concerning these reactions are presented in section VII-B-3.

Dialkyl tellurium dihalides, especially the dimethyl and diethyl derivatives, easily undergo cleavage of the carbon-tellurium bond. Dimethyl tellurium diiodide forms stable solutions in solvents of low polarity, but decomposes in acetone, 2-butanone, phenyl methyl ketone, cyclohexanone, methyl cyanide, pyridine, tris(dimethylamino)phosphine oxide and tetrahydrofuran. It has been suggested, that these decomposition reactions proceed through dissociation of $(CH_3)_2TeI_2$ to CH_3TeI and CH_3I[397 a].

Vernon[423,424] observed, that two substances, both of which analyzed for dimethyl tellurium diiodide, possessed entirely different properties. The following scheme (102) outlines Vernon's preparative route employing his designation of α- and β-compound for the two different forms.

(102) $$Te + 2CH_3I \underset{>180^\circ}{\overset{80^\circ}{\rightleftharpoons}} \alpha\text{-}(CH_3)_2TeI_2 \underset{HI}{\overset{Ag_2O/H_2O}{\rightleftharpoons}} \alpha\text{-}(CH_3)_2Te(OH)_2$$

α-base → evaporated to dryness, heated at 100°/15 torr →

$$\beta\text{-base} \underset{Ag_2O}{\overset{HX}{\rightleftharpoons}} \beta\text{-dihalides}$$

Gilbert and Lowry[144] extended these investigations to the ethyl derivatives. All the reactions, except the conversion of the α-base into the β-base, are reversible. The physical properties, like the conductivities in the molten state and in solution[142,249], the absorption spectra[144,249,423] and the parachors[50,249], of these two series of dihalides were determined. Vernon, unaware of the correct molecular structure of quadrivalent tellurium compounds, proposed a square-planar structure for the dialkyl tellurium dihalides and *cis-trans* isomerism as an explanation for the existence of the two series of dihalides.

Drew[107] disproved by chemical means the postulated *cis-trans* isomerism. The members of the α-series are non-polar compounds with a structure as shown in Fig. VII-2. The members of the β-series, however, have salt-like character. When the β-base was treated with hydrobromic or hydriodic acid, trialkyl telluronium halides and alkyl tellurium trihalides were recovered. The following reaction sequence (103) is consistent with the experimental observations.

(103) $$R_2TeI_2 \underset{HI}{\overset{Ag_2O/H_2O}{\rightleftharpoons}} R_2Te(OH)_2 \xrightarrow{heat} R_3Te\text{-}O\text{-}\overset{O}{\overset{\|}{Te}}R$$

"β-base"

$$R_3Te\text{-}O\text{-}TeOR \xrightarrow{HI} R_3TeI + RTeOOH$$

$$RTeOOH \xrightarrow{HI} RTeI_3$$

$$R_3TeI + RTeI_3 \rightarrow R_3TeI \cdot RTeI_3$$

"β-diiodide"

Drew[107] prepared a series of "β-dihalides" by mixing trimethyl telluronium halides with methyl tellurium trihalides (eq. 104). These compounds are listed in Table VII-4.

(104) $$RTeX_3 + R_3TeY \longrightarrow RTeX_3 \cdot R_3TeY$$

The properties of these "β-dihalides" suggested a salt-like structure of the type $[R_3Te]^+ [RTeX_3Y]^-$. The results of an X-ray structure analysis of $(CH_3)_3TeI \cdot CH_3TeI_3$[118] corroborated Drew's conclusion.

The observation of Morgan and Burgess[296], that bis(4-hydroxy-2-methylphenyl) tellurium dichloride formed the triaryl telluronium chloride upon heating in ethanol or upon dissolution in a hot aqueous sodium carbonate solution, might indicate the possibility of similar organic group migrations in aromatic compounds.

Table VII-4

Trialkyl Telluronium Alkyltetrahalotellurates(IV)[107]

Compound*	Color	mp. °C
$RTeI_3 \cdot R_3TeI$	purple-black with green luster	80-5 (dec)
$RTeI_3 \cdot R_3TeBr$	purple-red with golden luster	blackens at ~90
$RTeBr_3 \cdot R_3TeBr$	yellow	142
$RTeBr_3 \cdot R_3TeI$	orange-brown	120 (dec)
$2R_2TeI_2 \cdot R_3TeI$	purple	---
$R_2TeI_2 \cdot R_2TeI_4 \cdot R_3TeI$	steel-blue	~80
$R_2Te\text{-}O\text{-}\overset{O \atop \uparrow}{Te}R$	colorless	---
$R_3Te\text{-}O\text{-}TeI_3$	black	---

* $R = CH_3$

1) <u>Synthesis of symmetric and unsymmetric diorganyl tellurium compounds, R_2TeX_2 and $RR'TeX_2$</u>

The most economical methods for the synthesis of diorganyl tellurium dihalides are those which employ either elemental tellurium or tellurium tetrachloride and easily available organic halides, ketones, olefins or substituted aromatic compounds as starting materials. Many of these reactions can also be used to prepare organyl tellurium trihalides (see chapter VI-C). In order to obtain the dihalides the reactions are often carried out at higher temperatures with an excess of the organic component.

Of great importance for the preparation of diorganyl tellurium dichlorides are the condensation reactions of tellurium tetrachloride. Since tellurium tetrabromide and tetraiodide do not condense as easily as the tetrachloride, the heavier dihalides are prepared by halogenating the tellurides obtained by reduction of the diorganyl tellurium dichlorides.

A special method, which uses triaryl ^{125}Sb-antimony dichlorides, produces through β-decay of the antimony isotope diphenyl and bis(4-methylphenyl) ^{125m}Te-tellurium dichlorides[312]. Diphenyl ^{125}Sb-antimony trichloride forms similarly diphenyl ^{125m}Te-tellurium dichloride[439a].

The synthetic methods leading to diorganyl tellurium dihalides are presented schematically in Fig. VII-3. Symmetric and unsymmetric compounds are listed in Table VII-5 and VII-6, respectively.

a) Elemental tellurium and organic halides: Alkyl iodides react with elemental tellurium upon heating at 80° in a sealed tube. Thus, dimethyl[168,391,423], bis(iodomethyl)[121], diethyl[425] and dibenzyl[425] tellurium diiodide were prepared in yields of up to 50 per cent. Carillo and Nassiff[56] obtained dimethyl tellurium diiodide containing the tellurium isotopes ^{127}Te or ^{129}Te using this reaction. The only aromatic iodide investigated was pentafluorophenyl iodide which gave after seven days at 230° bis(pentafluorophenyl) telluride[83]. The expected diiodide probably dissociated thermally at this temperature. α,ω-Organic dihalides react with tellurium to give heterocyclic products, which are discussed in chapter XII.

b) Condensation of tellurium tetrachloride with 1,3-diketones: Tellurium tetrachloride condenses with 1,3-diketones to yield preferentially the cyclic product *(2)* under the conditions outlined in section IV-B-1a. Organyl tellurium trichlorides and diorganyl tellurium dichlorides were observed as side-products[277,282] (see also section VI-C-1a). The diorganyl tellurium dichlorides isolated from the reaction of tellurium tetrachloride with 1,3-diketones in a 1:2 molar ratio are listed in

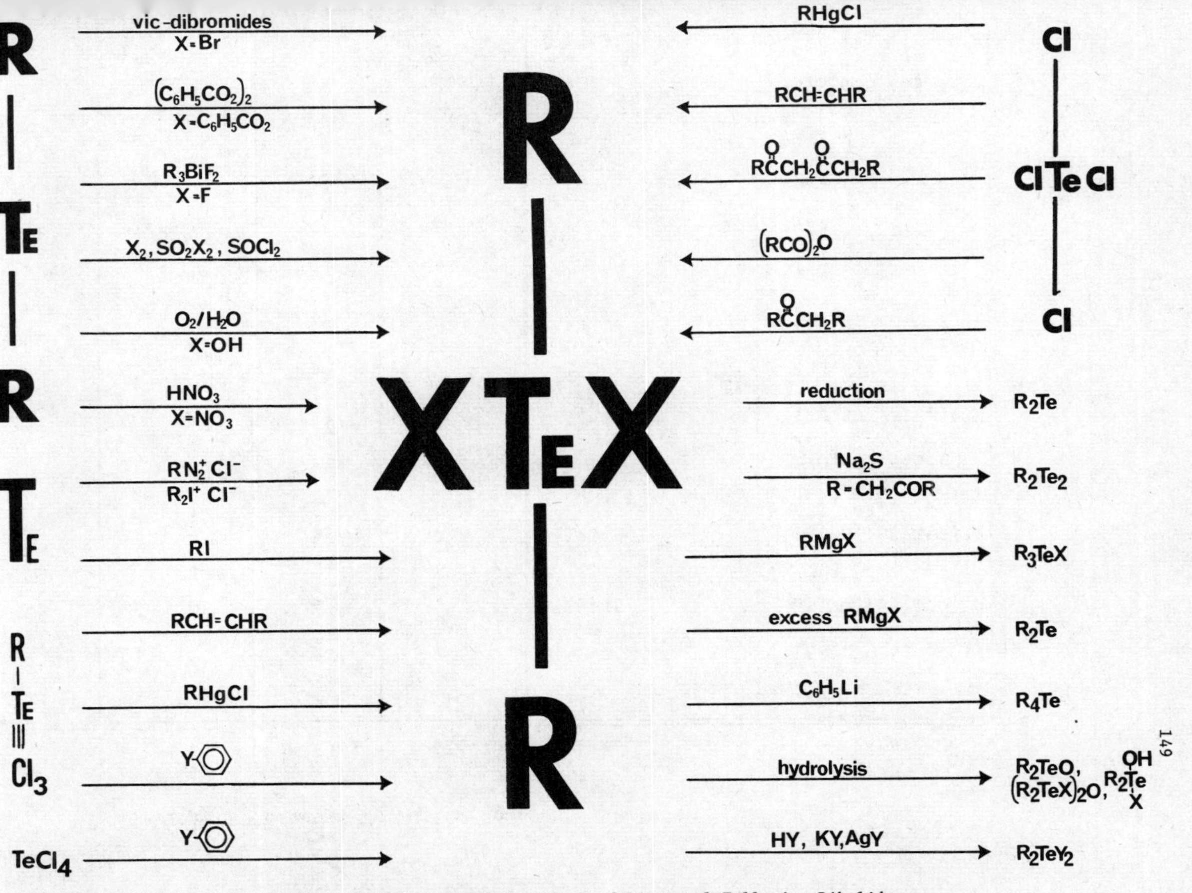

Fig. VII-3: Syntheses and Reactions of Diorganyl Tellurium Dihalides

Table VII-5

Symmetric Diorganyl Tellurium Compounds, R_2TeX_2

R	X	Method	Yield %	mp. °C	Reference
methyl	F	m	–	84	120
	Cl	m	–	92–7	168,170,423,425,445
	Br	m	–	89–95	35,77,168,170,423, 425,445
	I	a	50	127–30	56,96,168,359,391, 423,447
	I	m	–	–	170,249,445
	NO_3	l	–	142	170,307,423,445
	ClO_4	m	–	142	423,425
		m	–	37	249
	C_6H_5COO	m	–	154	423
	picrate	m	–	–	423
	CN	m	–	90	249
	OH	n	–	–	424
		m	–	–	423,425,445
iodomethyl	I	a	30	131,127	153,121

Table VII-5 (cont'd)

carboxymethyl	Cl	d	56	160-1	286
		i	-	160-1	286
ethyl	Cl	m	-	-5	144,170,258,444
	Br	m	-	-	170,444
	I	a	20	57	144,425
		m	-	50	170,444
	NO_3	l	-	-	170,258
	OH	m	-	-	144,249,425
2-chloroethyl	Cl	g	50	116 (dec)	12a, 140
2-cyanoethyl	Br	i	-	105-6	191
2-chloropropyl	Cl	g	20-30	100	12a,323b
butyl	I	i	-	61	17
	OH	n	-	175 (dec)	17
2-chlorobutyl	Cl	g	-	-	323,323b
4-chlorobut-2-enyl	Cl	g	20	83	12a
pentyl	Cl	m	-	-	446
	Br	m	-	-	446

Table VII-5 (cont'd)

pentyl (cont'd)	I	m	–	–	446
	OH	m	–	–	446
	NO_3	l	–	40	446
benzyl	I	a	–	134	425
2-chlorocyclohexyl	Cl	g	65–80	132–5	140,323,323b
CH_3COCH_2–	Cl	c	16.7	126–8	290
$C_3H_7COCH_2$–	Cl	c	14	92–3	290
i-$C_3H_7COCH_2$–	Cl	c	52	90	290
$C_4H_9COCH_2$–	Cl	c	28	62	290
i-$C_4H_9COCH_2$–	Cl	c	45.5	95	290
t-$C_4H_9COCH_2$–	Cl	c	26	191–2	290
$C_6H_5COCH_2$–	Cl	c	–	186–7	290,371,374
4-$CH_3OC_6H_4COCH_2$–	Cl	c	–	190	374
		m	–	–	374
	OH	m	–	–	374
4-$C_2H_5OC_6H_4COCH_2$–	Cl	c	–	212–3	371
4-$CH_3C_6H_4COCH_2$–	Cl	c	–	200	374

Table VII-5 (cont'd)

2,4-$(CH_3)_2C_6H_3COCH_2$-	Cl	c	-	180	371
2,4,5-$(CH_3)_3C_6H_2COCH_2$-	Cl	c	-	188	371
4-*i*-$C_3H_7C_6H_4COCH_2$-	Cl	c	-	183	371
1-$C_{10}H_7COCH_2$-	Cl	c	-	203-4	371
$CH_3COCH{=}C(OH)CH_2$-	Cl	b	18	115 (dec)	277
$CH_3COCCl{=}C(OH)CH_2$-	Cl	b	16	131-2 (dec)	277
i-$C_3H_7CH_2COCH{=}C(OH)CH_2$-	Cl	b	12	97-8	282
$(CH_3)_3CCOCH{=}C(OH)CH_2$-	Cl	b	23	133 (dec)	277
$C_6H_5COCH{=}C(OH)CH_2$-	Cl	b	13	148 (dec)	277
$C_6H_5CCl{=}CH$-	Cl	o	83	205-15 (dec)	305
	I	m	98	168-9 (dec)	305
$C_6H_5C{\equiv}C$-	I	i	33*	120-5	305
$(C_6H_5)_2C$–CH_2–CH(CH_2-)–O–C(=O)– (ring: 3,3-diphenyl-2-oxotetrahydrofuran-5-ylmethyl)	Cl	g	100	195-8	304
		p	100	195-8	304

Table VII-5 (cont'd)

phenyl	F	m	-	154	120
	Cl	i	-	161-2	3,246,254,314,429
		j	-	162-3	314,343,364
		m	100	160-1	214
		m	68	159-60	245a
		q	-	158	375,415,437
		q	21	160	374b
	Br	i	100	203-4	147,199,219,238, 359,375,442
		k	71-86	-	301
		k	-	-	221
		m	-	-	182-214
	I	i	-	237-8 (dec)	213,214,359
		k	-	-	221
		m	-	237-8 (dec)	182,214
	NO_3	l	-	160	227
		m	-	-	227

Table VII-5 (cont'd)

phenyl (cont'd)	OH	-	-	192-3	214
	CH_3COO	m	-	-	214
4-methoxyphenyl	F	j	-	-	317
	Cl	e	95	181-2	293
		i	-	183-4	226,317
		j	-	182-4	341
		q	-	182-3	415
		q	54	182-3	374b
	Br	i	-	190,198	226,293,415
	I	i	-	166-7	226
3-methoxyphenyl	Cl	i	-	162-3	235
		q	-	163	415
	Br	i	-	185-6	235
	I	i	-	167-8 (dec)	235
2-methoxyphenyl	Cl	i	-	184-5	236
		q	-	185	415
	Br	i	-	195-6	236

Table VII-5 (cont'd)

2-methoxyphenyl (cont'd)	I	i	-	198-9	236
2,4-dimethoxyphenyl	Cl	e	-	204-5	288
4-ethoxyphenyl	Cl	e	-	108	288,297
		f	-	-	297
		i	-	125,108	234,297
		j	-	110-1	341
	Cl	q	-	110	415
	Br	i	69	117,127	234,297
	I	i	-	135,144	234,297
3-ethoxyphenyl	Cl	q	-	144-5	415
2-ethoxyphenyl	Cl	i	-	163-4	232
		q	-	166	415
	Br	i	-	183-4	232
	I	i	-	214-5	232
4-phenoxyphenyl	Cl	f	-	157-8	103
4-dimethylaminophenyl	Cl	e	-	188-9	295
	[illegible]	m	-	158-9 (dec)	295

Table VII-5 (cont'd)

4-methyl(phenyl)aminophenyl	Cl	e	5	170-2	295
4-chlorophenyl	Cl	i	-	184-5	230
		q	-	182-3	415
		q	55	183-4	374b
3-chlorophenyl	Cl	q	-	119-20	415
2-chlorophenyl	Cl	q	-	192	415
4-bromophenyl	Cl	q	-	190	415
	Br	i	-	195	230,415
	I	i	-	231	415
3-bromophenyl	Cl	q	-	143-4	415
2-bromophenyl	Cl	q	-	244-6	415
2,4-dihydroxyphenyl	Cl	e	-	188-9	374
		m	-	-	374
	OH	m	-	-	374
4-methylphenyl	F	j	-	-	318
	Cl	i	-	166-7	188,214,318
		j	83	160-72	306,412,429

Table VII-5 (cont'd)

4-methylphenyl (cont'd)		m	35	166-9	306
		q	-	164	415
		q	36	162-3	374b
	Br	i	100	201-4	188,222,318,359, 361,375,449
		k	-	-	221
		m	-	-	214
	I	i	-	218-9	214,318
	OH	**	-	-	214
	$C_3H_7CO_2$	m	-	125-7	306
	$C_7H_{15}CO_2$	m	-	85-6	306
	$C_{11}H_{23}CO_2$	m	-	66-8	306
	$C_6H_5CO_2$	m	-	232-5	306,412
3-methylphenyl	Cl	i	-	131-2	225
		j	-	-	429
		q	-	128	415
	Br	i	-	165-6	225

Table VII-5 (cont'd)

3-methylphenyl (cont'd)	I	i	-	164	225
	ONa	m	-	-	225
2-methylphenyl	Cl	j	-	-	420
		m	-	183	214
		q	-	184	415
	Br	i	-	182	222,449
		m	-	-	214
	I	i	-	175-6	214
	NO_3	l	-	-	227
2,4-dimethylphenyl	Cl	i	-	187-8	223
	Br	i	-	200-1	223
	I	i	-	181-2	223
2,5-dimethylphenyl	Cl	i	100	197-8	223
	Br	i	-	189-90	223
	I	i	-	161-2	223
2,4,6-trimethylphenyl	Cl	i	-	178-9	224
	Br	i	-	205-6	224

Table VII-5 (cont'd)

2,4,6-trimethylphenyl (cont'd)	I	i	-	111	224
2-hydroxy-5-methylphenyl	Cl	e†	-	213-4	296
4-hydroxy-3-methylphenyl	Cl	e†	-	197-8 (dec)	296
4-hydroxy-2-methylphenyl	Cl	e†	-	-	296
1-naphthyl	F	j	-	-	316
	Cl	i	-	265	254
		j	80.3	205-6	316,364
	Br	i	-	244 (dec)	233,254,316
	I	i	-	184-6	233,316
2-naphthyl	Cl	j	100	245-7	363
	Br	i	100	253-5	363
	I	i	94.8	214-8 (dec)	363
2-thienyl	Cl	i	-	185.5	204
	Br	i	62	190.5	204
	I	i	-	125	204

* overall yield for reaction $TeCl_4 + 4RMgBr \rightarrow R_2Te \rightarrow R_2TeI_2$. ** from $R_2TeO + H_2O$

Table VII-6

Unsymmetric Diorganyl Compounds $RR'TeX_2$

R	R'	X	Method	Yield %	mp. °C	Reference
butyl	carbomenthoxymethyl	$C_6H_5CO_2$	1	87	133	17
pentyl	carbethoxymethyl	$C_6H_5CO_2$	1	-	77-8	18
2-chlorocyclohexyl	$-CH_2-CH-CH_2$ ring: $-O-C(=O)-C(C_6H_5)_2-$ (lactone)	Cl	h	57	95-120	304
phenyl	2-carboxyethyl	Cl	i	-	120-5	348b
	$-CH_2-CH-CH_2$ ring: $-O-C(=O)-C(C_6H_5)_2-$ (lactone)	Cl	h	90	188.9	304
	2-cyclohexyl	Cl	h	63	129-31 (dec)	304
	4-methylphenyl	Cl	i	100	135-6	229
		Br	i	-	175-6	229
			m	-	-	182

Table VII-6 (cont'd)

phenyl (cont'd)	4-methylphenyl (cont'd)	I	i	-	195,204	229
			m	-	-	182
		OH	m	-	148	251
	2-methylphenyl	Cl	i	-	179-80	238
		Br	i	-	154-5	238
		I	i	-	172-3	238
	4-methoxyphenyl	Cl	j	100	114-5	363
		Br	i,m	100	149	363
		I	i	100	166-7 (dec)	363
	4-phenoxyphenyl	Cl	j	100	129	342,363
		Br	i,m	100	156-8	363
		I	i	100	184.5 (dec)	363
	1-naphthyl	Cl	j	100	203	363
		Br	i,m	100	180-2	363
		I	i	100	212 (dec)	363
	2-naphthyl	Cl	j	100	172	363,427
		Br	i,m	100	186-7	363

Table VII-6 (cont'd)

phenyl (cont'd)	2-naphthyl (cont'd)	I	i	100	182 (dec)	363
4-bromophenyl	butyl	Cl	i	-	76-7	348b
	4-bromobiphenylyl	Br	*	-	>260	230
2-bromo-6-methoxyphenyl	butyl	Cl	i	90	138-41	348b
4-methoxyphenyl	methyl	I	r	-	109	288,358
			i	-	109	359
	ethyl	Cl	j	100	80-1	426
		Br	i	100	107-8	426
		I	i	100	135 (dec)	426
	benzyl	Cl	p	100	127	426
		Br	m	77.4	153 (dec)	426
		I	m	75.8	133 (dec)	426
	CH_3COCH_2-	Cl	f	65	137-8	340
	$C_6H_5COCH_2$-	Cl	f	60	136-7	340
		Br	f	29	151-2	340
	$-CH_2-CH-CH_2-C(C_6H_5)_2-C(=O)-O-$ (ring: O bonded to CH)	Cl	h	-	178-81	300
			p	100	178-81	304

Table VII-6 (cont'd)

4-methoxyphenyl (cont'd)	4-ethoxyphenyl	Cl	f	-	165-6	297
	4-dimethylaminophenyl	Cl	f	75	170-2	340
			j	-	-	340
			m	100	-	340
		Br	f	50	183-4 (dec)	340
			j	-	-	340
			m	100	-	340
		I	i	100	129-30	340
	2,4-dihydroxyphenyl	Cl	f	70	182-3	340
		Br	f	6	179-80	340
			m	100	179-80 (dec)	340
		I	m	100	130 (dec)	340
	2,4-bis(acetoxy)phenyl	Cl	s	88	166-7	340
	1-naphthyl	Cl	j	100	219-20	363
		Br	i,m	100	220-2	363
			k	70	-	301
		I	i	100	189 (dec)	363

Table VII-6 (cont'd)

4-methoxyphenyl (cont'd)	2-naphthyl	Cl	j	100	130	363
		Br	i,m	100	163-4	363
			k	66	-	301
		I	i	100	173 (dec)	363
4-ethoxyphenyl	CH_3COCH_2-	Cl	f	75	134-5	340
	$C_6H_5COCH_2$-	Cl	f	57	141-2	340
	2-chlorocyclohexyl	Cl	h	-	97-8	300
	-CH -CH——CH_2 O—C(=O)—$C(C_6H_5)_2$ (ring: O bonded to the second CH, C(C_6H_5)$_2$ bonded to CH_2)	Cl	h	-	193-6	300,304
		I	i	93	187-8 (dec)	304
			m	-	-	304
	-CH -CH——CH_2 O—C(=O)—C (ring: O bonded to the second CH, C bonded to CH_2; C also bonded to two benzene rings that are joined to each other)	Cl	h	58	206-9 (dec)	304
	4-dimethylaminophenyl	Cl	f	59	153-4	340
			j	-	-	340
			m	100	-	340

Table VII-6 (cont'd)

4-ethoxyphenyl (cont'd)	4-dimethylaminophenyl	Br	f	49	121-3 (dec)	340
			j	-	-	340
			m	100	-	340
		I	i	100	96-7	340
	2,4-dihydroxyphenyl	Cl	f	70	189-90 (dec)	340
		Br	f	28	190-1 (dec)	340
			m	100	-	340
		I	m	100	127 (dec)	340
	2,4-bis(acetoxy)phenyl	Cl	s	88	162-8	340
	1-naphthyl	Cl	j	100	193-4	342
4-phenoxyphenyl	$C_6H_5COCH_2$-	Cl	f	45	146-7	340
	2-chlorocyclohexyl	Cl	h	72	121-4	304
	$-CH_2-CH-CH_2-C(C_6H_5)_2-C(=O)-O-$ (ring: CH–O–C(=O)–C(C_6H_5)$_2$–CH_2)	Cl	h	80	122-5	304
			p	100	122.5	304

4-phenoxyphenyl (cont'd)	4-dimethylaminophenyl	Cl	f	41	194-5	340
			j	-	194-5	340
		Br	f	low	188-9	340
			j	-	188-9	340
			m	100	-	340
		I	i	100	135-6	340
	2,4-dihydroxyphenyl	Cl	f	61	181 (dec)	340
		Br	m	100	181-2 (dec)	340
		I	m	100	122-4 (dec)	340
	2,4-bis(acetoxy)phenyl	Cl	s	100	172-3	340
	1-naphthyl	Cl	j	100	175-6	363
		Br	i,m	100	189-90	363
		I	i	100	183-4 (dec)	363
	2-naphthyl	Cl	j	100	138-9	363
		Br	i,m	100	146-7	363
		I	i	100	204-6 (dec)	363
2-formylphenyl	methyl	Cl	i	100	154-9	348c

Table VII-6 (cont'd)

2-formylphenyl (cont'd)	butyl	Cl	i	85	105-10	348c
2-carbethoxyphenyl	methyl	Br	i	100	130-6	348c
	butyl	Cl	i	100	135-7	348c
2-methylphenyl	methyl	Cl	i	100	95-7	348c
4-methylphenyl	methyl	Cl	i	-	150-2	348c
	butyl	Cl	i	-	-	348c
1-naphthyl	-CH-CH(O-)-CH$_2$-C(C$_6$H$_5$)$_2$-C(=O)- (lactone ring)	Cl	h	75	209-11	304
	2-chlorocyclohexyl	Cl	h	87	158-60 (dec)	304
	2-naphthyl	Cl	j	100	232-3	363
		Br	i,m	100	215-7	363
		I	i	100	205 (dec)	363
2-naphthyl	ethyl	Cl	j	100	138-9	427
		Br	i	100	141-2	427
		I	i	100	102,131 (dec)**	427

Table VII-6 (cont'd)

2-naphthyl (cont'd)	benzyl	Cl	p	80	128-30	426
		Br	m	80	149 (dec)	426
		I	m	100	144 (dec)	426
	cyclohexyl	Cl	p	59	-	426
			j	100	140-1	426,427
		Br	i	100	157	426
		I	i	100	139 (dec)	426
	2-chlorocylohexyl	Cl	h	96	136-8 (dec)	304
	$C_6H_5CH{=}CH-$	I	i	59	153-4 (dec)	305
	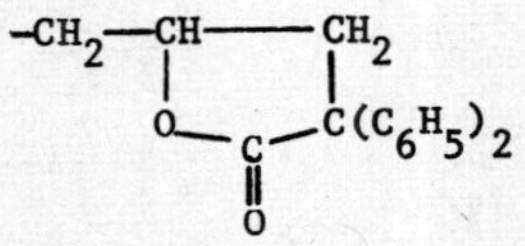	Cl	h	73	163-5	302
		I	i	61	174-6	302

* $2,4\text{-}BrC_6H_4MgBr \rightarrow 4\text{-}Br\text{-}C_6H_4\text{-}C_6H_4MgBr \xrightarrow[4\text{-}BrC_6H_4MgBr]{\text{"}TeBr_2\text{"}} RR'TeBr_2$

** dimorphic

Table VII-5.

c) Condensation of tellurium tetrachloride with monoketones: The reactions between tellurium tetrachloride and monoketones in ether or chloroform produce di- and trichlorides (see section VI-C-1b). Chloroform is preferred as a solvent, since ether forms an adduct with tellurium tetrachloride, which contaminates the final product[290,371,374]. Morgan and Elvins proposed that these reactions proceed *via* an addition of tellurium tetrachloride to the carbon-carbon double bond of the enolized ketone[290].

d) Condensation of tellurium tetrachloride with carboxylic acid anhydrides: Bis(carboxymethyl) tellurium dichloride was isolated from the reaction between tellurium tetrachloride and acetic acid anhydride[286]. The yield of the dichloride was increased when a six-fold molar excess of the anhydride was employed (see section IV-B-1c). Other carboxylic acid anhydrides gave only oily products[293].

e) Condensation of tellurium tetrachloride with aromatic compounds: Tellurium tetrachloride combines with aromatic compounds containing an activated ring position to form aryl tellurium trichlorides when equimolar quantities are refluxed in chloroform (see section VI-C-1d). Diorganyl tellurium dichlorides are formed - if at all - only as by-products in very small yields. Resorcinol, however, easily gave the dichloride[374]. A large excess of the aromatic compound and prolonged heating favors the formation of the diaryl tellurium dichlorides[288,293,297]. The insolubility of the trichlorides in the solvents employed in the condensation reactions largely prevents further condensation to the dichlorides. In section IV-B-1d are listed those aromatic compounds which only form adducts or do not react at all with tellurium tetrachloride. Morgan and Burgess[296] isolated diaryl tellurium dichlorides together with other products from reactions between "tellurium oxychloride", prepared by dissolving tellurium in nitric acid, and the isomeric hydroxy(methyl)benzenes[296].

f) Condensation of organyl tellurium trichlorides with organic compounds: Organyl tellurium trichlorides are intermediates in the condensations of tellurium tetrachloride with organic compounds, which lead to diorganyl tellurium dichlorides. Drew[103] and Morgan and Burstall[297] have shown, that 4-ethoxyphenyl and 4-phenoxyphenyl tellurium trichloride when heated with the respective benzene derivative in the absence of a solvent to 160-180° gave the dichlorides (eq. 105). This method is not important for the preparation of symmetric tellurium dichlorides, which are available in a one step reaction from tellurium tetrachloride and the organic components, but presents an easy route to unsymmetric compounds.

(105) $R\text{-}TeCl_3 + X\text{-}C_6H_4\text{-}H \longrightarrow R\text{-}TeCl_2\text{-}C_6H_4\text{-}X$

Thus, 4-methoxyphenyl 4-ethoxyphenyl tellurium dichloride was prepared by heating ethoxyphenyl tellurium trichloride and 4-methoxybenzene at 160° for six hours[297]. Petragnani[340] made extensive use of this reaction and condensed 4-methoxy-, 4-ethoxy- and 4-phenoxyphenyl tellurium trichloride with acetone, acetophenone and 4-dimethylaminobenzene by mixing the neat reactants. In the case of resorcinol, methanol was used as a solvent. Some of the tellurium tribromides reacted under similar conditions and gave the unsymmetric diorganyl tellurium dibromides. The yields were lower than those obtained in the corresponding reactions with the chlorides. The triiodides were unreactive. Methylene bis(tellurium trichloride) refluxed with acetone in chloroform gave in quantitative yield the dichloride *(31)*[286].

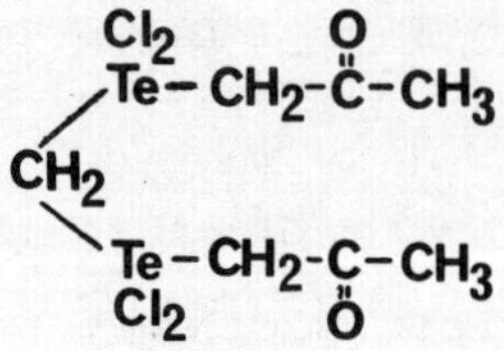

$$\mathrm{CH_2[TeCl_2{-}CH_2{-}C(=O){-}CH_3]_2}$$

(31)

g) Addition of tellurium tetrachloride to carbon-carbon double bonds: Tellurium tetrachloride adds to carbon-carbon double bonds forming the dialkyl tellurium dichlorides when the reagents are mixed in at least stoichiometric ratios. Ethylene[104], propene[12a,323b], 1-butene[323,323b], 2,2-diphenyl-4-pentenoic acid[304] and cyclohexene[123,140,323,323b] gave the expected bis(haloalkyl) tellurium dichlorides. Tellurium tetrachloride produced in a reaction with butadiene in carbon tetrachloride *via* a 1,4-addition bis(4-chlorobut-2-enyl) tellurium dichloride[12a].

Two isomeric dichlorides were detected in the product of the 1-butene reactions. A list of olefins, which did not react at all with tellurium tetrachloride or gave only indefinite products can be found in section IV-B-2. Tellurium tetrabromide did not undergo addition reactions[121] with olefins, but reacted with carbethoxymethylene triphenylphosphorane forming the phosphonium salt *(32)*[345].

(106) $$2(C_6H_5)_3P{=}CH{-}C(=O){-}OC_2H_5 + TeBr_4 \longrightarrow \left[\begin{array}{c}(C_6H_5)_3P{-}CH{-}C(=O){-}OC_2H_5\\ |\\ TeBr_2\\ |\\ (C_6H_5)_3P{-}CH{-}C(=O){-}OC_2H_5\end{array}\right]^{++} 2Br^-$$

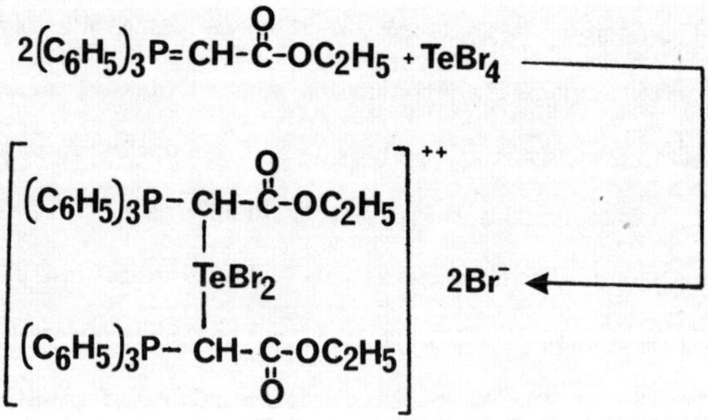

(32)

h) Addition of organyl tellurium trichlorides to carbon-carbon double bonds: The addition of 2-chlorocyclohexyl, phenyl, 4-methoxy-4-ethoxy- and 4-phenoxyphenyl, 1-naphthyl and 2-naphthyl tellurium trichloride to cyclohexene, 2,2-diphenyl-4-pentenoic acid[300,304] and 2,2-(2',2"-biphenylylene)-4-pentenoic acid[304] produced unsymmetric diorganyl tellurium dichlorides (eq. 107).

(107) $$RTeCl_3 + >C=C< \longrightarrow R\text{-}TeCl_2\text{-}C\text{-}CCl$$

The organyl tellurium trichlorides are less reactive than tellurium tetrachloride, combining with cyclohexene only at its reflux temperature and not at all with acetylenic compounds[300]. The pentenoic acids formed γ-lactones *(33)* after the tellurium-carbon bond had been formed[304] (eq. 108).

(108) $$R'TeCl_3 + CH_2{=}CH\text{-}CH_2\text{-}CR_2\text{-}COOH \longrightarrow R'TeCl_2\text{-}CH_2\text{-}CHCl\text{-}CH_2\text{-}CR_2COOH$$

$$\xrightarrow{-HCl} R'TeCl_2\text{-}CH_2\text{-}CH\text{—}CH_2\text{ (ring: }CH\text{-}O\text{-}C(=O)\text{-}CR_2\text{-}CH_2)$$

(33) R,R': see Table VII-6

i) The reaction of diorganyl tellurides with elemental halogens: Diorganyl tellurides are almost always quantitatively converted to the crystalline dihalides by passing a stream of chlorine through a solution of the telluride in diethyl ether, benzene, petroleum ether or chloroform, or by mixing equimolar amounts of the telluride and bromine or iodine in any of the same solvents. The dihalides precipitate immediately in pure form. Benzyl 2-naphthyl and benzyl 4-methoxyphenyl tellurides undergo cleavage of the benzyl group upon addition of bromine or iodine forming aryl tellurium trihalides and benzyl halide[426]. Literature references are given in Tables VII-5 and VII-6.

j) Sulfuryl halides, thionyl chloride, metal halides and tris(4-methylphenyl) bismuth difluoride as halogenating agents for tellurides: Rheinboldt and Vicentini[363] found that sulfuryl and thionyl chloride were better suited for the synthesis of dichlorides from tellurides than elemental chlorine. Sulfuryl bromide was employed for bromination[340]. The dihalides were obtained in quantitative yield as well formed crystals.

Iron(III) chloride, copper(II) chloride and mercury(II) chloride heated with bis(4-methylphenyl) telluride in glacial acetic acid in a sealed tube, were reduced to the lower valent metal chlorides transforming the telluride into the dichloride[306,412]. Silver chloride and iron(III) sulfate did not produce tetravalent tellurium compounds. The metal salts must be anhydrous, otherwise the dichlorides will be hydrolyzed to the hydroxide chlorides or their anhydrides as shown in scheme (109).

(109)
$$2\,FeCl_3 + R_2Te \longrightarrow R_2TeCl_2 + 2\,FeCl_2$$
$$R_2TeCl_2 + H_2O \longrightarrow R_2Te(OH)Cl + HCl$$
$$2\,R_2Te(OH)Cl \longrightarrow R_2\overset{Cl}{\overset{|}{Te}}-O-\overset{Cl}{\overset{|}{Te}}R_2 + H_2O$$

Tris(4-methylphenyl) bismuth difluoride was employed to prepare diorganyl tellurium difluorides from bis(4-methoxyphenyl)[317], bis(4-methylphenyl)[318] and di-1-naphthyl telluride[316]. The corresponding polonium derivatives were obtained similarly.

k) Reaction of diorganyl tellurides with *vic*-dibromides, α-halocarboxylic acids and benzoyl chloride: De Moura Campos and coworkers[301] investigated the action of *vic*-dibromides on diorganyl tellurides. They isolated diorganyl tellurium dibromides in 66-86 per cent yield and the olefins, formed by debromination of the organic bromides, in 58-100 per cent yield. The reaction conditions and pertinent data are given in Table VII-7. When diphenyl telluride was gently refluxed for 30 minutes in *vic*-dibromoethane the telluronium salt *(34)* was isolated (eq. 110).

(110) $$2(C_6H_5)_2Te + BrCH_2CH_2Br \longrightarrow [(C_6H_5)_2Te\ CH_2CH_2\ Te(C_6H_5)_2]^{++}\ 2Br^-$$

(34)

$$(34) \xrightarrow{BrCH_2CH_2Br} 2(C_6H_5)_2TeBr_2 + 2CH_2{=}CH_2$$

$$(34) \xrightarrow{\text{glac. acetic acid}} (C_6H_5)_2TeBr_2 + (C_6H_5)_2Te + CH_2{=}CH_2$$

Vigorous refluxing of the telluronium salt in dibromoethane or recrystallization from glacial acetic acid gave diphenyl tellurium dibromide[301]. It is likely, that all of the debromination reactions proceed *via* telluronium salts.

Butyl bromide and diphenyl telluride heated in a sealed tube at 180° for two days did not react, while methyl iodide gave only the telluronium salt[301]. Lederer[221] isolated diphenyl tellurium dihalides as by-products from reactions of diphenyl telluride with iodoacetic acid ethyl ester and α-bromopropionic acid.

Benzoyl chloride did not directly transfer the chlorine atom to the tellurium atom of bis(4-methylphenyl) telluride, since bibenzyl was never isolated. To account for the formation of bis(4-methylphenyl) tellurium dichloride and dibenzoate,which were obtained from the reaction mixture, the reactions (111)-(113) were proposed[306].

(111) $$2\,C_6H_5COCl + 2\,H_2O \longrightarrow 2HCl + 2\,C_6H_5COOH$$

(112) $$2HCl + R_2Te + 1/2\,O_2 \longrightarrow R_2TeCl_2 + H_2O$$

Table VII-7

Diorganyl Tellurium Dihalides from Diorganyl Tellurides and Organic Halides

R-Te-R'		Crganic Halide	Solvent	Reaction Temper-ature	Reaction Time/ hours	RR'TeX$_2$ Yield %	Te free Compound (Yield %)	Reference
R	R'							
phenyl	phenyl	$BrCH_2CH_2Br$	$BrCH_2CH_2Br$	reflux	2	86	--- ---	301
		$(C_6H_5CHBr)_2$	none	150	-	85	*trans*-$C_6H_5CH{=}CHC_6H_5$ (94)	301
		1,2-dibromo-cyclohexane	1,2-dibromo-cyclohexane	reflux	-	84	--- ---	301
		dibromo-cholesterol	xylene	reflux	-	77	cholesterol (58)	301
			none	95	-	71	cholesterol (93)	301
		$C_6H_5(CHBr)_2COOH$	none	100	-	74	$C_6H_5CH{=}CHCOOH$ (100)	301
		$CH_2{=}CH{-}CH_2Br$	$CH_2{=}CH{-}CH_2Br$	180	3	-	$(CH_2{=}CH{-}CH_2)_2$	301
		$ICH_2CO_2C_2H_5$	$ICH_2CO_2C_2H_5$	25	-	-	---	221
		$BrCH(CH_3)CO_2H$	$BrCH(CH_3)CO_2H$	60	-	-	---	221
4-methylphenyl	4-methylphenyl	C_6H_5COCl	none	150	6	86	---	306
		$BrCH_2COOH$	ether	25	120	-	---	221
4-methoxyphenyl	1-naphthyl	$BrCH_2CH_2Br$	$BrCH_2CH_2Br$	reflux	-	70	---	301
	2-naphthyl	$BrCH_2CH_2Br$	$BrCH_2CH_2Br$	reflux	-	66	---	301

(113) $$2\,C_6H_5COOH + R_2Te + 1/2O_2 \longrightarrow R_2Te(OOCC_6H_5)_2\ \ H_2O$$

$R = 4\text{-}CH_3C_6H_4$

The reaction did not proceed when carried out in a nitrogen atmosphere.

l) The oxidation of diorganyl tellurides by nitric acid and benzoyl peroxide: Nitric acid, ranging in concentration from dilute to 6N, oxidized diorganyl tellurides upon gentle heating to diorganyl tellurium dinitrates.

(114) $$3R_2Te + 8HNO_3 \longrightarrow 3R_2Te(NO_3)_2 + 2NO + 4H_2O$$

It is possible that the primary oxidation product is the diorganyl telluroxide which then combines with nitric acid to give the dinitrate. The compounds prepared in this way are listed in Table VII-5. Benzoyl peroxide transformed pentyl carbethoxymethyl[18] and butyl carbomenthoxymethyl telluride[17] into the dibenzoates.

m) Halogen exchange and replacement in diorganyl tellurium dihalides: Silver halides, potassium halides, hydrohalic acids, nitric acid, perchloric acid, silver carboxylates, silver cyanide and aqueous sodium hydroxide have been employed in halogen exchange and replacement reactions. Figure VII-4 summarizes these reactions. Literature references to specific compounds can be found in Table VII-5. The hydrolysis of dichlorides and dibromides by water and basic solutions yields diorganyl tellurium dihydroxides, listed in Table VII-5, diorganyl telluroxides (see section VII-C), and diorganyl tellurium halide hydroxides or their anhydrides listed in Table VII-8. Diorganyl tellurium diiodides are not affected by water[214].

Diorganyl tellurides were converted to the dichlorides[214,245a,306] and dibromides[214] by refluxing the solutions containing concentrated hydrohalic acid in presence of atmospheric oxygen. The telluride is first oxidized to the telluroxide, R_2TeO, or the tellurium

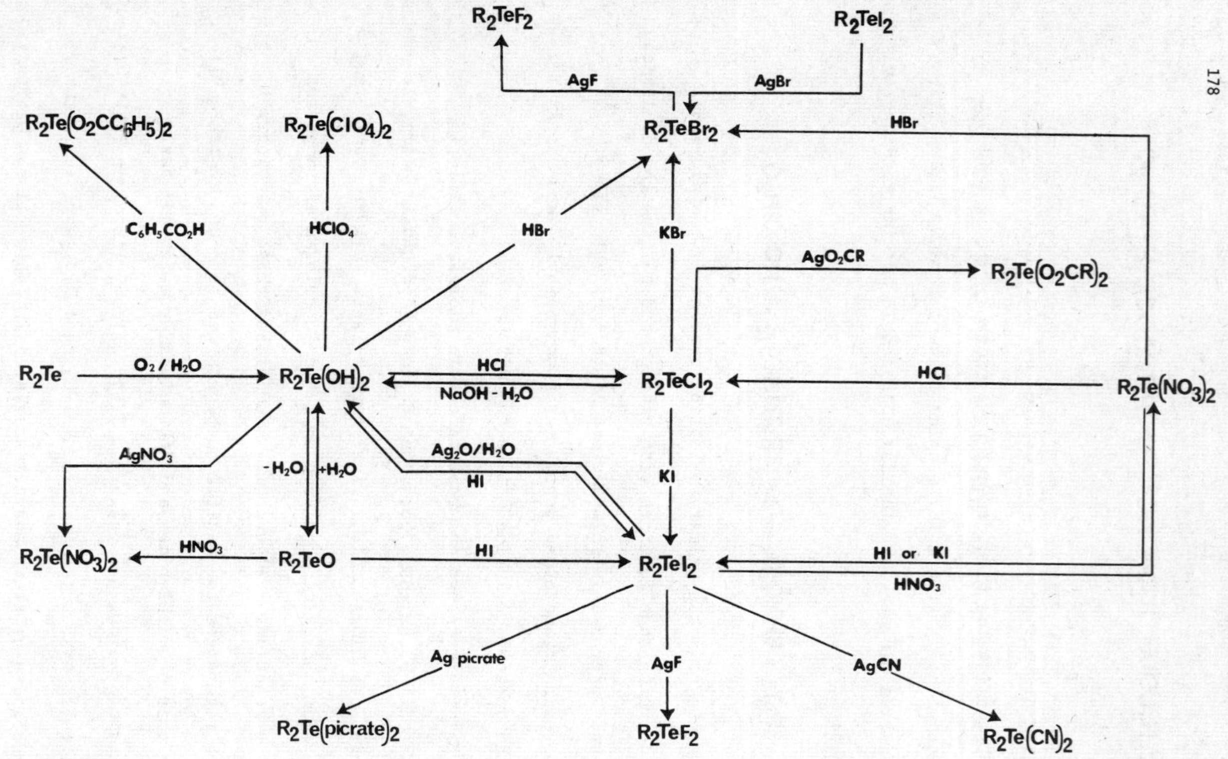

Fig. VII-4: Halogen Exchange and Replacement in Diorganyl Tellurium Dihalides

Table VII-8

Diorganyl Tellurium Compounds RR'TeXY

RR'TeXY				R_2TeX_2 + Reagent			
R	R'	X	Y	X or Compound	Reagent	mp. °C	Reference
methyl	R	Cl	OH	Cl	NH_3	---	170,445
				OH	HCl	---	142,248
		Br	OH	Br	NH_3	---	170,248
		I	OH	I	H_2O	---	248
		I	I_3	I	I_2	---	248,425
			CN	I	$R_2Te(CN)_2$	---	249
			ClO_4	I	$R_2Te(ClO_4)_2$	---	249
		CO_3H	OH	$R_2Te(OH)Cl$	Ag_2CO_3.	---	170
				R_2TeO	CO_2	---	170
		PO_3H_2	OH	R_2TeO	H_3PO_4	---	170
		$C_2O_{4/2}$	OH	$R_2Te(OH)Cl$	$Ag_2C_2O_4$	---	170
		HCO_2	OH	$R_2Te(OH)Cl$	$(HCO_2)_2Pb$	---	170
		CH_3CO_2	OH	$R_2Te(OH)Cl$	CH_3COOAg	---	170

Table VII-8 (cont'd)

methyl (cont'd)	R	Cl	$R_2Te(Cl)(O-)$	Cl	NaOH	---	321
		I	$R_2Te(I)(O-)$	I	NaOH	120	142,249,425
		SO_4H	OH	$R_2Te(OH)Cl$	Ag_2SO_4	---	170
				R_2TeO	H_2SO_4	---	170,445
		I	$(CH_3)_2Te(O-)(OTe(CH_3)_2)$	$R_2Te(OH)_2$	R_2TeI_2	152 (dec)	142,425
ethyl	R	Cl	OH	Cl	NH_3, KOH	---	170,444
				Cl	$R_2Te(OH)_2$	---	144
		Br	OH	Br	NH_3	---	170,444
				Br	$R_2Te(OH)_2$	---	144
		I	OH	I	NH_3	---	170,444
				I	$R_2Te(OH)_2$	---	144
		I	I_3	I	I_2	98	144
		CN	OH	$R_2Te(OH)Cl$	AgCN	---	170
		CO_3H	OH	$R_2Te(OH)Cl$	Ag_2CO_3	---	170
		SO_4H	OH	R_2Te	H_2SO_4/PbO_2	---	258

Table VII-8 (cont'd)

ethyl (cont'd)	R	SO_4H	OH	$R_2Te(OH)Cl$	Ag_2SO_4	---	170,444
		$C_2O_4/2$	OH	$R_2Te(OH)Cl$	$Ag_2C_2O_4$	---	170,444
		HCO_2	OH	$R_2Te(OH)Cl$	$(HCO_2)_2Pb$	---	170
		CH_3CO_2	OH	$R_2Te(OH)Cl$	CH_3COOAg	---	170
		I	$R_2Te(I)O-$	I	NH_3	107	144
butyl	R	NO_3	OH	R_2Te	HNO_3 dil.	84	17
pentyl	R	SO_4H	OH	OH	H_2SO_4	---	446
phenyl	R	Cl	OH	Cl	H_2O	233-4	214
		Br	OH	RTe(OH)Cl	NaBr	264-5	214
		Br	OH	Br	H_2O	264-5	214
		I	OH	RTe(OH)X	KI	214-5	214
		Cl	$R_2Te(Cl)O-$	RTe(OH)Cl	heat,150°	233-4	214
		Br	$R_2Te(Br)O-$	---	---	224	415
		I	$R_2Te(I)O-$	$R_2Te(OH)I$	heat,180°	216-7	214
		NO_3	$R_2Te(NO_3)O-$	NO_3	H_2O	218	227

Table VII-8 (cont'd)

phenyl	R	$-OOC-CH_2CH_2'-COO*$		Cl	$AgOOC(CH_2)_2COOAg$	255-7	245a
		$-OOC-(CH_2)_8'-COO-*$		Cl	$AgOOC(CH_2)_8COAg$	164-6	245a
4-methoxyphenyl	R	Cl	OH	Cl	H_2O	---	226
		Cl	$R_2Te(Cl)O-$	Cl	H_2O	250	415
4-ethoxyphenyl	R	Cl	$R_2Te(Cl)O-$	Cl	H_2O	---	297
4-methylphenyl	R	Cl	$R_2Te(Cl)O-$	R_2Te	$FeCl_3/H_2O$	260-5	306,412
			$R_2Te(Cl)O-$	$R_2Te(OH)Cl$	heat,150°	261-3	188,214
			NO_3	R_2Te	H_2O/HNO_3	237-8	227
			OH	Cl	H_2O	261-3	214
		Br	OH	$R_2Te(OH)Cl$	KBr	---	214
			$R_2Te(Br)O-$	$R_2Te(OH)Br$	heat,160°	---	214
		I	OH	$R_2Te(OH)X$	KI	203-4	214
		$C_5H_{11}CO_2$	OH	Cl	RCO_2Ag	136-7	306
		$C_9H_{19}CO_2$	OH	Cl	RCO_2Ag	96-9	306
3-methylphenyl	R	Cl	OH	Cl	H_2O	87	225
2-methylphenyl	R	Cl	$R_2Te(Cl)O-$	Cl	H_2O	220-2	214

Table VII-8 (cont'd)

2-methylphenyl	R	Br	$R_2Te(Br)(O-)$	Br	H_2O	224-5	214
		NO_3	OH	NO_3	H_2O	---	227
2,4-dimethylphenyl	R	Cl	$R_2Te(Cl)(O-)$	Cl	H_2O	239-40	223
2,5-dimethylphenyl	R	Cl	OH	Cl	H_2O	---	223
		I	$R_2Te(I)(O-)$	$R_2Te(OH)Cl$	KI	70	223
2,4,6-trimethylphenyl	R	Cl	OH	Cl	H_2O	237	224
		I	$R_2Te(I)(O-)$	Cl	KI/H_2O	100	224
$-CH_2CO_2$-menthyl	C_4H_9-	NO_3	OH	R_2Te	HNO_3	42	17
4-methylphenyl	phenyl	Cl	$RR'Te(O-)(Cl)$	Cl	H_2O	243-4	229
		Br	$RR'Te(O-)(Br)$	$(RR'TeCl)_2O$	KBr	---	229
		I	$RR'Te(O-)(I)$	$(RR'TeCl)_2O$	KI	---	229

* polymer

dihydroxide, $R_2Te(OH)_2$, which ten combines with the acids forming the dihalides.

Diphenyl tellurium dichlorides and the silver salts of the dicarboxylic acids $HOOC(CH_2)_nCOOH$ (n = 2,8) produced polymeric substances[245a] (see chapter XIII).

n) The oxidation of tellurides to diorganyl tellurium hydroxides: The oxidation of diorganyl tellurides produces diorganyl telluroxides which, in the presence of water, are easily hydrated to the dihydroxides.

$$2R_2Te + O_2 \longrightarrow 2R_2TeO \xrightarrow{H_2O} 2R_2Te(OH)_2 \quad (115)$$

The dihydroxides were isolated only a few cases[17,424]. They are easily dehydrated. Analytical results were in most cases not reported making the exact composition of these substances uncertain.

o) Disproportionation of an organyl tellurium trichloride: De Moura Campos and Petragnani[305] observed, that 2-chloro-2-phenylvinyl tellurium trichloride when heated in glacial acetic acid or ethanol disproportionated forming tellurium tetrachloride and the diorganyl tellurium dichloride.

$$2C_6H_5CCl{=}CH\ TeCl_3 \longrightarrow TeCl_4 + (C_6H_5CCl{=}CH)_2TeCl_2 \quad (116)$$

Such a behavior has not been observed with other organyl tellurium trichlorides.

p) The reaction of tellurium tetrachloride or organyl tellurium trichlorides with organyl mercury chlorides: Tellurium tetrachloride and organyl mercury chlorides are capable of producing diorganyl tellurium dichlorides when brought together in a 1:2 molar ratio. However, only 5-dichlorotellurobis(2,2-diphenyl-4-pentanolactone) was prepared in this way[304].

The reaction of organyl tellurium trichlorides with organyl mercury chlorides is the only general method known for the synthesis of unsymmetric diorganyl tellurium dichlorides[304,363,426]. The trichlorides are prepared as described in section VI-C-1. The condensation is best carried out in refluxing dioxane. The mercuric chloride formed is precipitated as the dioxane adduct. The dichlorides are separated upon addition of dilute hydrochloric acid as yellow oils, which solidify on trituration with the mother liquor. Purification is accomplished by reduction to the telluride with sodium sulfide removing all the mercury salts as the insoluble mercury sulfide. In the case, that the dichloride was not isolated, but immediately reduced to the unsymmetric telluride, the compound is not listed in Table VII-5, but included in Table VII-2 in section VII-A-2d.

q) The reaction of elemental tellurium or tellurium tetrachloride with arenediazonium chlorides and the reaction of tellurium with diphenyliodonium chloride: The reaction of elemental tellurium with diazonium chlorides was employed by Waters[437] and Tanijama and coworkers[415] to prepare symmetric diorganyl tellurium dichlorides in yields of approximately 15 per cent. Tellurium powder reacted with benzenediazonium chloride in the presence of chalk in cold acetone, whereas sulfur and selenium combined only when the solutions were heated[437]. According to Waters[437] the diazonium salts form radicals which then react with tellurium according to equations (117) and (118).

(117) $$C_6H_5N_2^+Cl^- \longrightarrow C_6H_5\cdot + N_2 + Cl\cdot$$

(118) $$Te + 2C_6H_5 + 2Cl\cdot \longrightarrow (C_6H_5)_2TeCl_2$$

Arenediazonium hexachlorotellurates(IV), $[RC_6H_4N_2]_2^+ TeCl_6^{--}$ (R = H, 4-CH_3, 4-CH_3O, 4-Cl), prepared from tellurium tetrachloride and the diazonium chlorides in concentrated hydrochloric acid, formed diaryl tellurium dichlorides in yields ranging from 21 to 55 per cent, when treated with copper powder in an acetone suspension at -15°[374b].

Sandin and coworkers[375] suggested, that the reaction of diphenyl iodonium chloride with elemental tellurium, which yields diphenyl tellurium dichloride, proceeds at least in part *via* a radical mechanism.

r) The reaction of bis(4-methoxyphenyl) ditelluride with methyl iodide: Bis(4-methoxyphenyl) ditelluride is cleaved by methyl iodide according to equation (119)[288,358].

(119) $$R_2Te_2 + 3CH_3I \longrightarrow R\text{-}\underset{I_2}{Te}\text{-}CH_3 + [RTe(CH_3)_2]^+ I^-$$

After refluxing the ditelluride in excess methyl iodide the more soluble diiodide is isolated from the mother liquor. This reaction does not seem to be of general applicability, since bis(4-ethoxyphenyl) ditelluride decomposed when refluxed in methyl iodide[288].

s) Miscellaneous reactions: The only example of a reaction in which one of the organic groups in diorganyl tellurium dichlorides was modified, was reported by Petragnani[340]. He esterified the aromatic hydroxyl groups in the dichlorides *(35)* with acetic acid anhydride in the presence of sulfuric acid and obtained aryl bis(2,4-acetoxyphenyl) tellurium dichlorides in good yields.

$$RO\text{-}C_6H_4\text{-}\underset{Cl_2}{Te}\text{-}C_6H_3(OH)_2 \qquad R = CH_3,\ C_2H_5,\ C_6H_5$$

(35)

Finally, diorganyl tellurium dichlorides could theoretically be the products of reactions that take place between tellurium tetrachloride or organyl tellurium trichlorides and Grignard reagents.

(120) $$TeCl_4 + 2RMgBr \longrightarrow R_2TeCl_2 + 2MgBrCl$$

Petragnani and de Moura Campos[344] state, however, that the reactions described by equations (121) and (122) do not take place.

(121) $TeCl_4 + RMgBr \longrightarrow RTeCl_3 + MgClBr$

(122) $RTeCl_3 + RMgBr \longrightarrow R_2TeCl_2 + MgClBr$

Lederer[213,214,239], however, isolated diorganyl tellurium dihalides as by-products of reactions between tellurium tetrachloride and Grignard reagents. Further studies are needed to clarify these questions. For other products isolated from these reaction mixtures please consult sections VII-A-1j and IV-B-3a.

2) Diorganyl tellurium compounds, R_2TeXY

A large number of compounds R_2TeXY which are listed in Table VII-8, have been synthesized. Diorganyl tellurium dihalides undergo hydrolysis when treated with water, ammonia or sodium hydroxide solutions. The nature of the isolated products depends upon the hydrolyzing reagent, the dihalide used and the conditions employed. Scheme (123) presents these reactions.

(123)

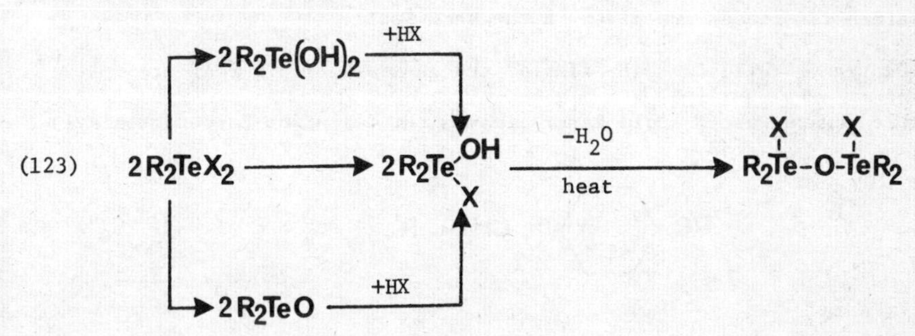

Hydrolysis with water to the hydroxide halides occurs only with dichlorides and dibromides. The diiodides are inert towards water[214,423]. The hydroxide halides ionize in solution releasing the halide ion[144,321]. They lose water rather easily converting to the anhydride *(36)*.

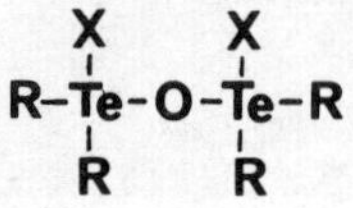

$$R-\underset{R}{\overset{X}{Te}}-O-\underset{R}{\overset{X}{Te}}-R$$

(36)

$$[R_2TeI]^+I_3^-$$

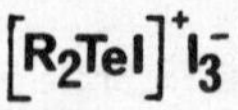

(37)

The halide ion in the hydroxide halide and in the anhydride can be exchanged for other anions using alkali metal salts or silver salts. Mixing the dihydroxide with a dihalide produces the hydroxide halide. Diorganyl tellurium diiodides combine with one mole of iodine to form compound *(37)* containing the triiodide anion. The reaction of diorganyl telluroxides with acids is reported to give diorganyl tellurium hydroxide compounds which have the acid anion incorporated into the molecule. The exact composition of these substances, which have been prepared by Woehler and coworkers, and by Heeren and Mallet, is unknown. Some of the formulas given in Table VII-8 are therefore tentative. The individual compounds, the method of their preparation and the pertinent literature references are given in the same Table.

Diorganyl tellurium dihalides of the general formula *(38)* are unstable toward water. Cold water cleaves the tellurium-carbon(oxoalkyl)

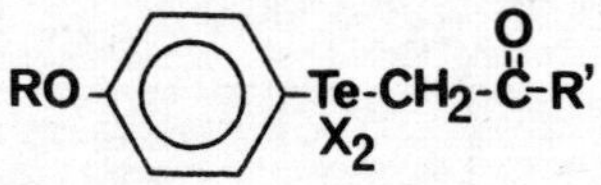

$$RO-C_6H_4-\underset{X_2}{Te}-CH_2-\overset{O}{\overset{\|}{C}}-R'$$

$R = CH_3, C_2H_5, C_6H_5$

$R' = CH_3, C_6H_5$

(38)

bond in acetonyl derivatives, while boiling water is necessary for the phenacyl derivatives[340].

3) Reactions of diorganyl tellurium dihalides

The reactions of diorganyl tellurium dihalides are summarized in Fig. VII-3.

The reduction of diorganyl tellurium dihalides by sodium sulfide, sodium hydrogen sulfite, sodium sulfite, potassium disulfite or elemental zinc to diorganyl tellurides is discussed in section VII-A-1c through VII-A-1e. Aryl acetonyl, aryl phenacyl and aryl 2,4-dihydroxyphenyl tellurium dichlorides when treated with sodium sulfide, sodium hydrogen sulfite, zinc in chloroform or hydrazine sulfate gave diaryl ditellurides (section VI-E-1h). Bis(2-chloropropyl) tellurium dichloride generated propene and tellurium upon treatment with reducing agents[323b].

Grignard reagents and diaryl tellurium dichlorides produce triorganyl telluronium halides (section VIII-1c), and symmetric (section VII-A-1f) or unsymmetric tellurides (section VII-A-2j) depending upon the reaction conditions and the nature and amount of Grignard reagent present. Diorganyl tellurium dihalides are reduced to the tellurides by diorganyl polonides[314,318]. Phenyl lithium and diphenyl tellurium dichloride yielded tetraphenyl tellurium (chapter IX).

Nefedov investigated exchange reactions in the following systems:

$$R_2Te^* + R_2TeX_2 \underset{}{\overset{CHCl_3}{\rightleftarrows}} R_2Te^*X_2 + R_2Te$$

$Te^* = {}^{127m}Te$; $X = Cl$[319a,d]; $X = F, Br, I$[319d]

$$R_2TeI_2^* + I_2 \overset{CHCl_3}{\rightleftarrows} R_2TeI_2 + I_2^*$$

$I^* = {}^{131}I$[319c]

$$R_2TeX_2 + 2X^{*-} \overset{CHCl_3/H_2O}{\rightleftarrows} R_2TeX_2^* + 2X^-$$

$R = C_6H_5$; $X^* = {}^{36}Cl, {}^{82}Br, {}^{131}I$[319c]

The components present in the reaction mixtures were separated by thin-layer chromatography[319b]. Values of 8 to 10 kcal/mole were obtained for the activation energies.

The halogen exchange reactions, which have been employed to prepare R_2TeY_2 from R_2TeX_2, are presented in section VII-B-1n. The hydrolytic behavior of the dihalides is discussed in section VII-B-2. It should be noted, that bis(2-chloropropyl) and bis(4-chlorobut-2-enyl) tellurium dichlorides lose in basic solution the organic groups as olefin[12a].

Diorganyl tellurium dibromides convert diorganyl ditellurides into tellurides and elemental tellurium (section VII-A-1r). Dimethyl tellurium diiodide formed a series of unstable ammonia addition compounds of the type $R_2TeI_2 \cdot nNH_3$ with n = 1 to 6[425]. The milky white semifluid mass formed reaches a limiting composition corresponding to $R_2TeI_2 \cdot 6NH_3$. The process is reversible. The iodide can be recovered unchanged when dry air is passed over the adduct (eq. 124).

(124) $$R_2TeI_2 + nNH_3 \rightleftharpoons R_2TeI_2 \cdot nNH_3$$

The stability of the compounds decreases with increasing n.

Reactions, which affect the organic part in the molecule, are discussed in section VII-B-1s.

Tellurium-carbon bond cleavage occurs in certain diorganyl tellurium dihalides. The introduction to chapter VII deals with the conversion of dimethyl and diethyl tellurium dihalides to triorganyl telluronium salts and organyl tellurium trihalides. Dibutyl tellurium diiodide treated with hydriodic acid in acetone gave butyl tellurium triiodide[17]. Equations (125) and (126) were proposed to account for the isolated product[17].

(125) $$R_2TeI_2 \xrightarrow{HI} RTeI + RI$$

(126) $$RTeI + 2HI + 1/2O_2 \longrightarrow RTeI_3 + H_2O$$

Bis(4-ethoxyphenyl) tellurium dihalides were decomposed by concentrated nitric acid to halogenated nitroethoxybenzenes[297]. Aqueous sodium nitrite in dilute hydrochloric acid cleaved the carbon-tellurium bonds in bis(4-dimethylaminophenyl) tellurium dichloride forming 4-nitrosodimethylaminobenzene and tellurium dioxide[295]. Bis(2-thienyl) tellurium dibromide[204], bis(2,4-dimethylphenyl) tellurium dichloride[288] and bis(4-phenoxyphenyl) tellurium dichloride[103] were decomposed by a potassium hydroxide solution and hot potassium carbonate solution.

Aryl methyl tellurium dichlorides and aryl butyl tellurium dichlorides were converted to diaryl ditellurides upon refluxing in pyridine[348b] (see section VI-E-1i). The bis(chloroalkyl) tellurium dichlorides, R_2TeCl_2 (R = $ClCH_2CH_2$[12a], $CH_3CHClCH_2$[12a,323b], $ClCH_2CH{=}CHCH_2$[12a], 2-chlorocyclohexyl[323b]), decomposed thermally to olefins, various chloroalkanes, chloroolefins, hydrogen chloride and inorganic tellurium compounds.

An attempt to reduce diphenyl tellurium dibromide with lithium aluminum hydride to the dihydride failed. Only the telluride was obtained[211].

C) Diorganyl Telluroxides

All the known diorganyl telluroxides, R_2TeO, are white solids. The water soluble compounds give a basic solution, probably forming the dihydroxide[214]. Nylen[321] determined the base strength of dimethyl telluroxide. Data collected in titrations of dimethyl tellurium dichloride and dimethyl tellurium hydroxide nitrate with aqueous sodium hydroxide solutions gave a pK value of +6 for the dissociation of $[(CH_3)_2TeOH]^+$ according to equation (127).

$$(127) \qquad [R_2TeOH]^+ \xrightarrow{H_2O} R_2TeO + H_3O^+$$

Klofutar and coworkers[193] showed that the basicity of the oxides increases in the series sulfoxide, selenoxide, telluroxide. Jensen[189] found a dipole moment of 3.93D for bis(4-methylphenyl) telluroxide in dioxane solution at 40°. This is indicative of a polar tellurium-oxygen bond. Rheinboldt and Giesbrecht[361] investigated the binary systems formed by diphenyl and bis(4-methylphenyl) telluroxide and the corresponding sulfoxides, selenoxides, sulfones and selenones.

Diorganyl telluroxides are prepared from diorganyl tellurium compounds in the following ways:

a) Hydrolysis of diorganyl tellurium dihalides: Diorganyl tellurium dibromides triturated with ammonia solutions or with five per cent aqueous sodium hydroxide solutions gave white telluroxides, which were purified by recrystallization from an organic solvent[223]. Bis(4-ethoxyphenyl) tellurium dichloride and dibromide were converted to the oxide by 2N sodium hydroxide[297]. Silver oxide in aqueous suspension affected the same conversion by precipitation of the halide ions[143,258].

b) Dehydration of diorganyl tellurium dihydroxides: Diorganyl tellurium dihydroxides are reported to lose water on standing forming the telluroxides[214,425]. The reaction is reversible. Telluroxides can be kept pure only under anhydrous conditions.

c) Oxidation of dialkyl tellurides to telluroxides: Diorganyl tellurides are oxidized by atmospheric oxygen to telluroxides[214,258,298,358]. $KMnO_4$ was also used as an oxidizing agent[55]. The aliphatic tellurides seem to be more susceptible towards oxidation. Hellwinkel and Fahrbach[173] obtained 2,2'-biphenylylene telluroxide from the telluride and chloramine-T (eq. 128).

(128) Te + H_3C–C_6H_4–SO_2NClNa ⟶ Te=NSO_2–C_6H_4–CH_3 *(39)* —H_2O⟶ TeO

The tellurimine *(39)* hydrolyzed easily to the telluroxide.

d) Conversion of diorganyl tellurium hydroxide halides into telluroxides: Diorganyl tellurium hydroxide halides, $R_2Te(OH)X$, and their anhydrides are hydrolyzed by sodium hydroxide[298]. The halide ions can be removed by silver oxide[170]. Diphenyl tellurium hydroxide iodide was converted to the telluroxide and the tellurium diiodide upon heating in methanol[214].

The compounds, which were prepared by these methods are listed in Table VII-9 together with yields, melting points and pertinent references. Aliphatic telluroxides are thermally unstable. Upon heating an intermolecular alkyl group migration occurs. Trialkyl telluronium salts are thus formed (see section VII-B). In the oxidation of tellurides telluroxides are expected and sometimes observed as products, but further

Table VII-9

Diorganyl Telluroxides, RR'TeO

R	R'	Method	Yield %	mp. °C	Reference
methyl	R	a	–	–	424
		b	–	–	425
		d	–	–	170
ethyl	R	a, c	–	–	170, 258, 444
c-tetramethylene		a, c, d	–	241 (dec)	298
c-pentamethylene		a	–	–	143
phenyl	R	a	90	191	361
		a	–	185 (dec)	199
		b, c, d	–	–	214
4-methylphenyl	R	a	84	167	214, 361
		a	–	173-4	189
3-methylphenyl	R	a	–	163-4	225
2-methylphenyl	R	a	–	205-6 (dec)	214
2,4-dimethylphenyl	R	a	–	216-7	223
2,5-dimethylphenyl	R	a	100	225-6	223

Table VII-9 (cont'd)

2,4,6-trimethylphenyl	R	a	100	204-5	224
4-methoxyphenyl	R	a	-	190-1	226
3-methoxyphenyl	R	a	-	87	235
2-methoxyphenyl	R	a	-	205-6	236
4-ethoxyphenyl	R	a	-	181 (dec)	234, 297
2-ethoxyphenyl	R	a	-	205-6	232
4-hydroxyphenyl	R	c	-	93	358
1-naphthyl	R	a	-	224-5 (dec)	233
4-methylphenyl	phenyl	a	-	154-5	229-234
2-methylphenyl	phenyl	a	-	216-7	238
2,2'-biphenylylene		c	34	214-5	173
4-methyl-2,2'-biphenylylene		c	-	-	55

oxidation to tellurinic acids can take place (see section VI-D-A). Balfe and coworkers[17] oxidized tellurides with air or hydrogen peroxide and isolated addition products of telluroxides with tellurinic acids of the general formula nRR'TeO·mRTeOOH (R = R' = C_4H_9: n = 1, m = 3; n = 2, m = 1; R = C_4H_9, R' = $CH_2CO_2C_2H_5$: n = 1, m = 2, all three compounds decomposing at 180°; R = C_4H_9, R' = CH_2CO_2-menthyl: n = m = 1, decomposing at 220°). The formation of tellurinic acids in these oxidation reactions was attributed to the instability of the telluroxides in their enol form *(40)*. The enols were further oxidized to the tellurinic and carboxylic acids. Butyric acid was in fact isolated as an oxidation product of

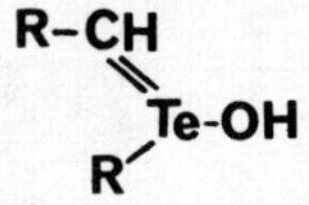

R-CH=Te(R)-OH

(40)

dibutyl telluride. The association of telluroxides and tellurinic acids *via* a hydrogen bond should prevent enolization and further oxidation of the telluroxide[17]. There is, however, no firm experimental evidence for the existence of the oxide in the enol form.

Heeren[170] and Mallet[258] found, that dialkyl telluroxides are converted to dihalides when treated with hydrohalic acids. Nitric acid gave the dinitrates[277]. Dimethyl telluroxide was reduced to the telluride by sulfurous acid[170].

D) Diorganyl Tellurones

Diorganyl tellurones, R_2TeO_2, the tellurium analogs of the well known sulfones, are ill-defined substances. It is doubtful whether a substance corresponding to the formula R_2TeO_2 has ever been obtained in pure form.

Vernon[424] claimed to have isolated dimethyl tellurone by oxidation of dimethyl telluride or its dihydroxide with hydrogen peroxide as an insoluble, white, amorphous powder. This substance possessed the characteristics of a peroxide. It was explosive, oxidized hydrohalic acids to the elemental halogens and decolorized potassium permanganate[424]. Lowry and Gilbert[144,249] prepared diethyl tellurone by air oxidation of the dihydroxide. Since treatment of the oxidation product with hydriodic acid gave the adduct $RTeI_3 \cdot R_3TeI$, the structure $[R_3Te]^+[RTeO_4]^-$ was proposed for the tellurone[249]. Balfe and coworkers[17] oxidized a number of dialkyl tellurides (see section VII-C) and obtained telluroxides, tellurinic acids and adducts of these two compounds. Bis(carbomenthoxymethyl) telluride, however, treated with 30 per cent hydrogen peroxide produced a white solid, which liberated iodine from potassium iodide and bromine from hydrobromic acid. Potassium permanganate was reduced. These authors thought that their compound was the hydrogen peroxide adduct of the telluroxide, $R_2Te(OH)(OOH)$.

Telluracyclohexane 1,1-dioxide monohydrate was obtained as a white, amorphous powder by oxidation of telluracyclohexane with hydrogen peroxide in methanolic solution[294,P-3]. It had the same properties as the other substances already described. It exploded on rapid heating.

When phenoxtellurine, dissolved in acetone or glacial acetic acid, was treated with an excess of 30 per cent hydrogen peroxide, an amorphous precipitate with properties characteristic of peroxides was isolated. The results of elemental analysis for the product, which had been heated in vacuum for one hour at 110°, suggested structure *(41)*[55,104].

Telluroisochroman *(42)* oxidized with hydrogen peroxide in boiling ethanol produced a white amorphous compound, which was insoluble in all common solvents[180]. The carbon analysis suggested that the tellurone had been formed.

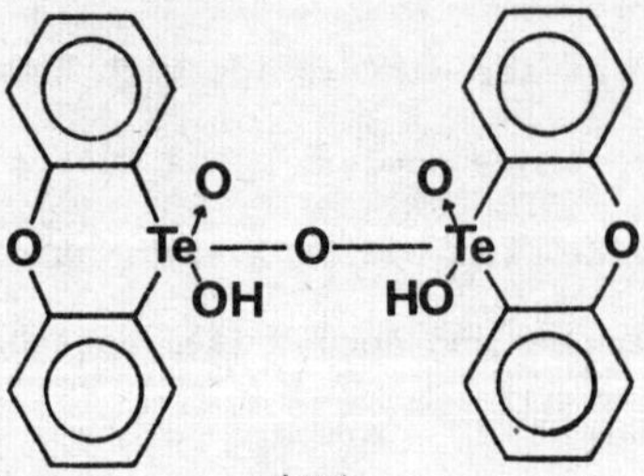

(41)

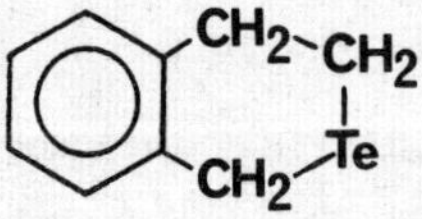

(42)

VIII. TRIORGANYL TELLURONIUM COMPOUNDS, $[R_3Te]^+X^-$

Many triorganyl telluronium salts, $[R_3Te]^+X^-$, have been described in the literature. Trialkyl, triaryl, dialkyl aryl and alkyl diaryl derivatives, including compounds with three different organic groups are known. Telluronium salts containing a tellurium-hydrogen bond are not stable at room temperature. Dimethyl telluride was protonated in liquid hydrogen chloride forming the cation $[(CH_3)_2TeH]^+$, which decomposed at -80° in vacuum[333]. A similar protonation occurred when telluracyclohexane was dissolved in liquid sulfur dioxide containing fluorosulfonic acid[210].

The following anions have been combined with the telluronium cations: F^-, Cl^-, Br^-, I^-, OH^-, NO_3^-, $B_3H_8^-$, BBr_4^-, $(C_6H_5)_4B^-$, SiF_6^{--}, CrO_4^{--}, $Cr_2O_7^{--}$, $ZnCl_3^-$, $HgCl_3^-$, $HgBr_3^-$, HgI_3^-, $PtCl_6^{--}$ and picrate$^-$. Table VIII-1 lists the triorganyl telluronium compounds, $[R_3Te]^+X^-$. Table VIII-2 contains the telluronium salts of the type $[R_2R'Te]^+X^-$. The unsymmetric telluronium compounds, $[RR'R''Te]^+X^-$ have been collected in Table VIII-3. The structure of the telluronium salts is discussed in section XIV-C.

1) The synthesis of triorganyl telluronium salts

Among the ten methods discussed below, only three have found wide application in the synthesis of triorganyl telluronium salts. The reaction of tellurium tetrachloride with Grignard reagents, or organic compounds of zinc and lithium, is the only direct route from an inorganic tellurium compound to a triorganyl telluronium salt. The combination of diorganyl tellurides with organic halides and the reaction of diorganyl tellurium dichlorides with Grignard reagents represent the other two convenient ways of synthesizing telluronium compounds. The preparation of the starting materials is covered in

Table VIII-1

Triorganyl Telluronium Compounds, $[R_3Te]^+ X^-$

R	X	Method	Yield %	mp. °C	Reference
methyl	F	c	-	128 (dec)	120
	Cl	c	-	-	54
	Br	c	-	>250 (dec)	77
		g	-	250-80 (dec)	107
	I	a	82	251-7	172
		b	55	240-1 (dec)	52,53,54,77,99, 391,424
		g	-	240 (dec)	107
	I_3	f	-	76.5	391
	OH	c	-	-	52,120
	BBr_4	f	-	215(dec)	77
	$SiF_6^{--}/2$	c	-	320-36 (dec)	120
	$PtCl_6^{--}/2$	c	-	-	52,53,54
ethyl	Cl	a	-	174	261
		c	-	-	21

Table VIII-1 (cont'd)

ethyl (cont'd)	Br	c	–	162	261
		g	–	215 (dec)	249
	I	b	–	92	21,52,261
		g	–	>180	144
		g	–	115	249
	OH	c	–	–	21,52,261
	$PtCl_6^{--}/2$	c	–	–	21,52
butyl	I	a	22	74	172
		e	55	74	172
phenyl	F	c	–	203 (dec)	120
	Cl	a	16	242-3	429
		c	–	244-5	212,213
		c[†]	–	242-3	314
		d	*	242-4	442
		e	–	–	429
	Br	a	–	260	364
		c	–	259-60	212,213,442

Table VIII-1 (cont'd)

phenyl (cont'd)	I	a	11	247-8	120,212,213,314†
		e	70	247-8	237
	OH	c	-	-	120,221
		d	96	-	442
	B_3H_8	c	95	57 (dec)	10
	$(C_6H_5)_4B$	c	-	218-9	442
		d	30.6	218-9	442
	$HgCl_3$	f	-	138-9	237,314†
	$HgBr_3$	f	-	143-4	237
	HgI_3	f	-	178	237
	picrate	c	-	160	221
4-methylphenyl	Cl	a†	25	-	318
		c	-	260-1	213
		e	-	-	429
	Br	c	-	265-6 (dec)	213
	I	a	32	232-3	213
		c†	-	-	318

Table VIII-1 (cont'd)

4-methylphenyl (cont'd)	OH	c	-	110	221
	picrate	c	-	194-5	221
3-methylphenyl	Cl	c	-	-	228
	Br	c	-	-	228
	I	a,c	26	160-1	228
	$HgCl_3$	f	-	159-60	228
	picrate	c	-	152-3	228
2-methylphenyl	Cl	c	-	175-6	221
	Br	c	-	197-8	221
	I	a	-	195-6	213
	OH	c	-	-	221
	picrate	c	-	182	221
2,4-dimethylphenyl	I	a	12	208-9	228
	picrate	c	-	138-9	228
2,5-dimethylphenyl	I	a	20	186-7	228
2,4,6-trimethylphenyl	Br	c	-	164	228
	I	a	16	169-70	228

Table VIII-1 (cont'd)

4-hydroxy-2-methylphenyl	Cl	h	–	184-5	296
4-NaO-2-methylphenyl	OH	h	–	137-8	296
2-hydroxy-5-methylphenyl	Cl	h	–	244-5 (dec)**	296
4-methoxyphenyl	Cl	c†	–	–	317
	Br	c†	–	–	317
	I	a	15	160	228,317†
	picrate	c	–	160	228
2-methoxyphenyl	Cl	c	–	–	239
	Br	c	–	202-3	239
	I	a	43	191	239
2-methoxyphenyl	$HgCl_3$	f	–	244-5	239
	$HgBr_3$	f	–	218	239
	HgI_3	f	–	234-5	239
	picrate	c	–	169-70	239
4-ethoxyphenyl	Br	c	–	218	231
	I	a	46	208-9	231
	picrate	c	–	178-9	231

Table VIII-1 (cont'd)

2-ethoxyphenyl	Br	c	-	203-3	231
	I	a	44	226	231
	picrate	c	-	164-5	231
pentafluorophenyl	Cl	a	-	188-90	197
1-naphthyl	Cl	c†	-	-	316
	Br	c†	-	-	316
	I	a†	-	-	316
2-thienyl	Br	e	-	244.5	204
	I	c	-	-	204

† mixed with tracer amounts of the corresponding Po compound

* see section VIII-1d

** mp. of dihydrate

sections VII-A-1, VII-A-2 and VII-B-1. Figure VIII-1 summarizes the synthetic routes to and the reactions of triorganyl telluronium compounds. The telluronium salts derived from heterocyclic tellurium compounds are treated in chapter XII. Triphenyl and tris(4-methylphenyl) ^{125m}Te-telluronium chlorides were formed by β-decay of various aryl ^{125}Sb - antimony compounds[312,315,439a].

a) The reaction of tellurium tetrachloride with organic compounds of magnesium, lithium or zinc: Lederer[212,213] was the first to show, that tellurium tetrachloride reacted in ether solution with an excess of a Grignard reagent over the amount required by equation (129) to give telluronium salts. The reaction mixture had to be immediately

$$(129) \qquad TeCl_4 + 3RMgX \longrightarrow R_3TeCl + 3MgClX$$

hydrolyzed at 0° to prevent the formation of dialkyl tellurides (see section VII-A-1j). The triorganyl telluronium salts were isolated as the rather insoluble iodides upon addition of potassium iodide to the aqueous solutions of the telluronium salts. The yields of triaryl telluronium iodides ranged from 15 to about 50 per cent. Only compounds of the type R_3TeX can be synthesized in this manner. Literature references to the later work of Lederer in this area and to triorganyl telluronium halides prepared by other authors using Lederer's method are included in Table VIII-1.

Hellwinkel and Fahrbach[172] obtained tributyl and trimethyl telluronium iodide by reacting tellurium tetrachloride and the appropriate alkyl lithium compound (1:4 molar ratio) and hydrolyzing the reaction mixture with an aqueous potassium iodide solution. Marquardt and Michaelis[261] isolated triethyl telluronium chloride as a product of the reaction between diethyl zinc and tellurium tetrachloride. It is almost certain, that the telluronium salts in all these reactions were

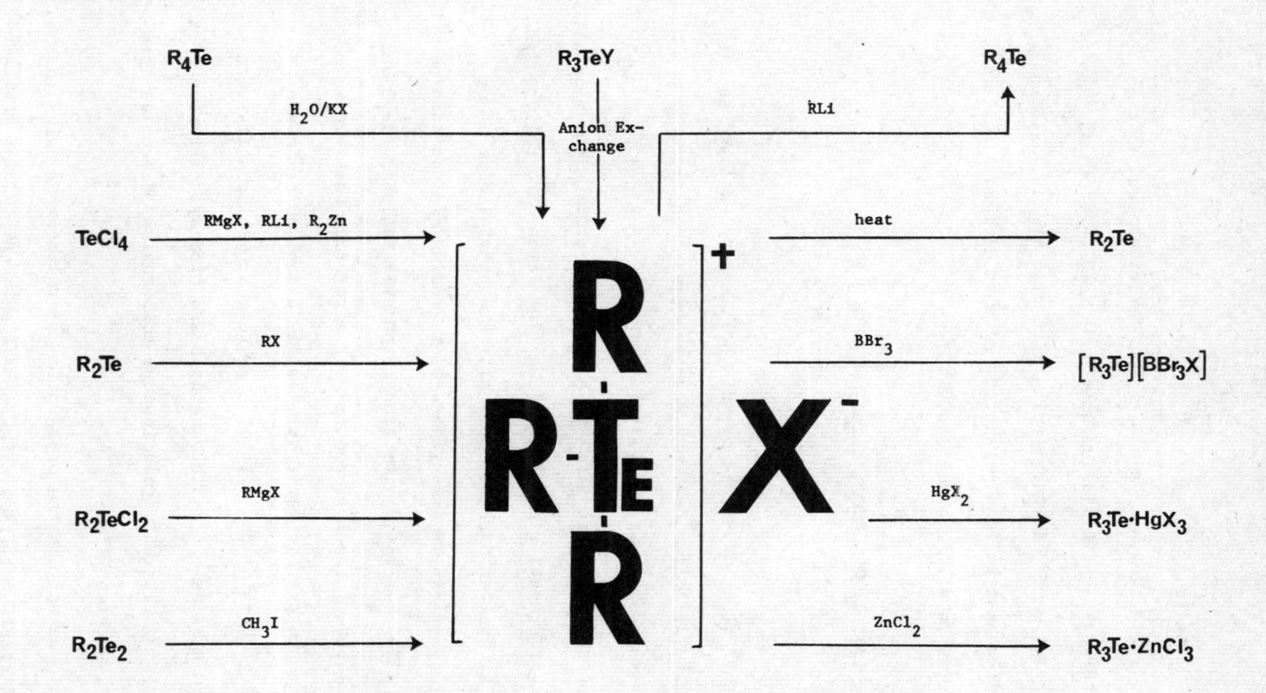

Fig. VIII-1: Syntheses and Reactions of Triorganyl Telluronium Compounds.

formed by hydrolytic cleavage of one carbon-tellurium bond in the intermediate tetraorganyl tellurium compounds (see chapter IX).

b) The reaction of diorganyl tellurides with organic halides: Triorganyl telluronium halides were produced when diorganyl tellurides were treated with an alkyl halide (eq. 130).

$$(130) \qquad R''\text{-Te-}R' + RX \longrightarrow \left[R\text{-}\overset{\overset{\displaystyle R'}{|}}{\underset{\underset{\displaystyle R''}{|}}{Te}} \right]^{+} X^{-}$$

Methanol, diethyl ether or excess organic halide were employed as solvents. Aliphatic tellurides were more reactive than aromatic compounds. Lower molecular weight dialkyl tellurides combined vigorously with methyl iodide, while aromatic and higher molecular weight aliphatic tellurides required heating and longer reaction times. No telluronium salts were obtained from diphenyl telluride and ethyl iodide[215] or butyl bromide[301]. Bis(2,4-dimethylphenyl) telluride did not combine with methyl iodide[223]. Bis(2,4,6-trimethylphenyl) telluride gave only a very small amount of the telluronium salt after four weeks at room temperature[224]. Methyl iodide, ethyl iodide, benzyl chloride and bromide, phenacyl bromide and aliphatic α-bromocarboxylic acids were used as organic halides. 1,2-Dibromoethane and diphenyl telluride formed upon gentle heating bis(methylene diphenyl telluronium) dibromide (see structure *(34)*, section VII-B-1k)[301].

It is not necessary in the preparation of telluronium salts to isolate the diorganyl telluride. Phenyl dimethyl telluronium iodide[348b] and 2-(diethoxy)methylphenyl dimethyl telluronium iodide[348c] were synthesized by treating the reaction mixture obtained from tellurium and the appropriate aryl lithium compound in diethyl ether with methyl iodide.

c) Anion exchange in telluronium salts: Once a telluronium salt has been formed, the anion can be exchanged easily. These metathetical reactions take advantage of the solubility differences between the various telluronium salts, of the insolubility of the silver halides, and the formation of water in reactions between telluronium hydroxides and acids. The solubility of the telluronium halides decreases in the order chloride > bromide > iodide. The chlorides can therefore be converted into the bromides and iodides, and the bromides into the iodides upon treatment of the aqueous solutions of the telluronium halides with the appropriate potassium halides. The picrates are less soluble than the halides and precipitate upon mixing of the aqueous solutions of the halides and picric acid or sodium picrate. The picrates can be used for the characterization of telluronium compounds.

The solubility differences between the silver halides make it possible to prepare telluronium bromides from iodides, and chlorides from bromides and iodides employing silver bromide and chloride, respectively. An aqueous suspension of silver oxide mixed with solutions of telluronium halides precipitates silver halides and forms telluronium hydroxides, which can be isolated upon evaporation of the filtered solutions. The hydroxides combine with acidic substances in a neutralization reaction according to equation (131).

$$(131) \qquad \left[R_3Te\right]^+ OH^- + HX \longrightarrow \left[R_3Te\right]^+ X^- + H_2O$$

$$X = F, Cl, Br, I, PtCl_6/2, SiF_6/2, \text{picrate}, (C_6H_5)_4B$$

In many instances the hydroxides were not isolated, but were reacted immediately with an acid. The alkaline solutions of telluronium hydroxides precipitate metal hydroxides and liberate ammonia from ammonia solutions. Carbon dioxide is not absorbed from the air . Pertinent literature references for these reactions are given in Tables VIII-1, VIII-2 and VIII-3. Figure VIII-2 presents a summary of the anion exchange reactions.

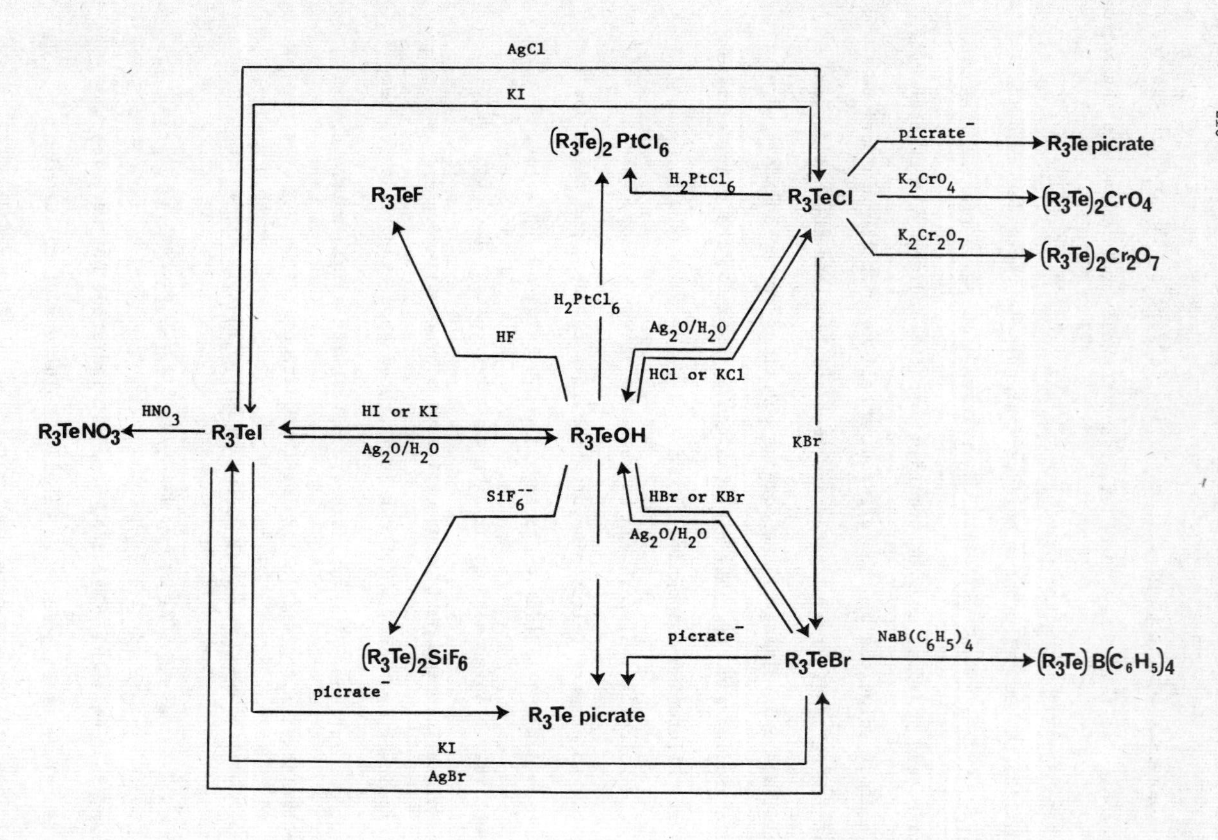

Fig. VIII-2: Anion Exchange Reactions of Triorganyl Telluronium Compounds.

d) Telluronium compounds from tetraorganyl tellurium compounds: One carbon tellurium bond in tetraorganyl compounds is easily cleaved by water forming a telluronium hydroxide. Additon of potassium or sodium iodide precipitates the telluronium iodide[172]. The reactions discussed in sections VIII-1a and VIII-1e probably produce the telluronium salts *via* hydrolysis of the intermediate tetraorganyl tellurium compounds.

Wittig and Fritz[442] obtained triphenyl telluronium chloride, upon dissolution of tetraphenyl tellurium in dichloromethane or chloroform at room temperature. With benzaldehyde in ether the telluronium chloride was isolated after hydrolysis of the reaction mixture with hydrochloric acid. Tetraphenyl tellurium donated one phenyl carbanion to triphenyl boron forming triphenyl telluronium tetraphenyl borate[442] (see chapter IX for more details).

e) The reaction of diorganyl tellurium dichlorides and triorganyl telluronium bromides with organometallic reagents: Diaryl tellurium dichlorides, which are accessible through methods discussed in section VII-B-1, react with a two to three fold molar excess of an aromatic Grignard reagent in toluene. The triaryl telluronium compounds are conveniently prepared by this method in yields as high as 77 per cent. Since aromatic halides do not combine with diorganyl tellurides, the reaction of dichlorides with Grignard reagents is the only easy route to mixed triaryl telluronium compounds, $[R_2R'Te]^+X^-$.

The dichloride solutions must be poured rapidly into the Grignard solutions. The stirred mixture must be hydrolyzed immediately. The telluronium salts are best isolated as the iodides after addition of potassium iodide[229,237].

Butyl lithium reacts with triphenyl telluronium bromide forming the tributyl derivative[172].

f) The reaction of telluronium halides with boron

tribromide, zinc dichloride, mercuric halides: Telluronium halides release their anion to coordinatively unsaturated metal halides. Boron tribromide[77] and zinc dichloride[215] have been employed in a few cases as the anion acceptors. Lederer (for references see Table VIII-1 and VIII-2) prepared a large number of addition compounds between telluronium halides and mercuric halides by mixing either aqueous or ethanolic solutions of the components. Some of the adducts were obtained only as amorphous solids, as gummy products or as oils. The analytical data suggest $[R_3Te]^+[HgX_3]^-$ as the stoichiometric composition of the adducts. Since mercury(II) is generally surrounded by four or six ligands, the possibility of a direct tellurium-mercury bond cannot be ruled out. Gold(III) chloride[213,215,216], tin(II) chloride[213] and copper(II) chloride[215] gave precipitates with telluronium salts.

g) Telluronium halides from "Vernon's β-base": "Vernon's β-base" (see section VII-B), obtained by heating dialkyl tellurium dihydroxides, has the formula $[R_3Te]^+[RTe(OH)_4]^-$ in its highest hydrated state. Treatment of such a compound with hydrohalic acids produced triethyl and trimethyl telluronium halides[107,249]. This reaction does not have preparative importance.

h) The reaction of "tellurium oxychloride" with cresols: Morgan and Burgess[296] condensed a tellurium compound, which they claim was tellurium oxychloride, $TeOCl_2$, with the isomeric cresols. With *p*-cresol the telluronium salt *(43)* and the corresponding telluronium chloride

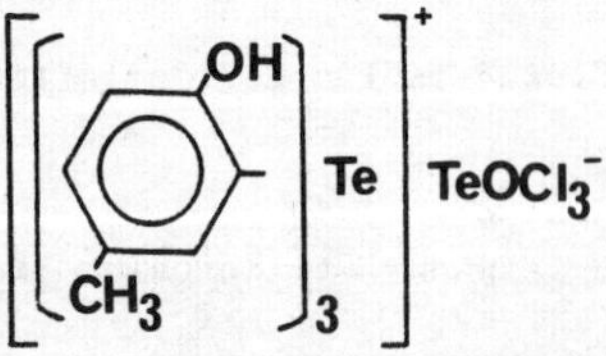

OH; CH3; Te; TeOCl3⁻

(43)

Table VIII-2

Triorganyl Telluronium Compounds, $[R_2R'Te]^+ X^-$

R	R'	X	Method	Yield %	mp. °C	Reference
methyl	benzyl	Cl	b	*	-	35,62a
		picrate	c	-	121	35,62a
	$C_6H_5COCH_2$	Br	b	-	90-1	35
	carbethoxymethyl	Br	b	-	137.5	35
	phenyl	I	b	-	-	348b
	4-methoxyphenyl	I	i	-	170-2	288,358
		picrate	c	-	126-7	288
	2-formylphenyl	I	j	90	125-30	348c
	2-(diethoxy)methylphenyl	I	b	40	110-15	348c
2-cyanoethyl	ethyl	I	b	-	58-9	191
butyl	methyl	I	b	83	159 (dec)	17
	carboxymethyl	Br	b	†	-	17
			b	-	94-5	172
	carbethoxymethyl	Br	b	87	62	17

Table VIII-2 (cont'd)

butyl (cont'd)	carbomenthoxymethyl	Br	b	79	90	17
	$C_6H_5COCH_2$	Br	b	–	87	17
pentyl	methyl	I	b	–	70	18
	carbethoxymethyl	Br	b	–	50	18
	carbomenthoxymethyl	Br	b	–	†	18
	$C_6H_5COCH_2$	Br	b	–	84	18
phenyl	methyl	Cl	c	–	129-30 (dec)	215
		Br	c	–	137-8	215
		I	b,c	100	123-4	215
		NO_3	c	–	168-9	215
		OH	c	–	–	215
		$CrO_4/2$	c	–	151	215
		$Cr_2O_7/2$	c	–	153 (dec)	215
		$ZnCl_3$	f	–	149-50	215
		$HgCl_3$	f	–	135-6	215
		$PtCl_6/2$	c	–	–	215
		picrate	c	–	93-4[††]	215

Table VIII-2 (cont'd)

phenyl (cont'd)	benzyl	Br	b	–	90-1	219
		OH	c	–	–	219
		picrate	c	–	–	219
	carboxymethyl	Br	b	100	not solid	216
		OH	c	–	117-8	216
			j	–	117-8	216
	carbomethoxymethyl	Cl	c	–	115-6	216
		Br	b	91	105-6	216
		$CrO_4/2$	c	–	72-3	216
		$Cr_2O_7/2$	c	–	115	216
		$HgCl_3$	f	–	35-6	216
	carbomethoxymethyl	$PtCl_6/2$	c	–	60 (dec)	216
		picrate	c	–	144-5	216
	carbethoxymethyl	Br	b	100	63-4 (dec)	216
		I	b	–	110 (dec)	221
	1-carboxyethyl	Br	b	–	98	221

Table VIII-2 (cont'd)

phenyl (cont'd)	1-carbomethoxyethyl	Br	b	–	~130	217
	1-carbethoxyethyl	Br	b	–	~125	217
	1-carbopropoxyethyl	Br	b	9	~99	217
	2,2-carbomethoxypropyl	Br	b	–	~116	217
	2,2-carbethoxypropyl	Br	b	–	130	221
	1-carboxypropyl	Br	b	small	84-5	221
	1-carbethoxypropyl	Br	b	–	142-3	221
	4-methylphenyl	Cl	c	–	–	237
		Br	c	–	228-9	237
		I	e	63	219-20	237
		HgI_3	f	–	222-3	237
		picrate	c	–	132-3	237
	3-methylphenyl	Cl	c	–	**	237
		Br	c	–	202-3	237
		I	e	60	202	237
		$HgCl_3$	f	–	**	237
		$HgBr_3$	f	–	**	237

Table VIII-2 (cont'd)

phenyl (cont'd)	3-methylphenyl	HgI_3	f	–	134-5	237
		picrate	c	–	105-6	237
	2-methylphenyl	Cl	c	–	**	237
		Br	c	–	203	237
		I	e	36	175-6	237
		$HgCl_3$	f	–	210-1	237
		HgI_3	f	–	184	237
		picrate	c	–	127-8	237
	2,5-dimethylphenyl	Cl	c	–	210-1	237
		Br	c	–	220-1	237
		I	e	36	213-4	237
		$HgCl_3$	f	–	176-7	237
		$HgBr_3$	f	–	174-5	237
		HgI_3	f	–	110	237
		picrate	c	–	170-1	237
	2,4-dimethylphenyl	I	e	40	103	237
		HgI_3	f	–	201	237

Table VIII-2 (cont'd)

phenyl (cont'd)	3,4-dimethylphenyl	I	e	22	117-8	237
	2,4,6-trimethylphenyl	I	e	10	153-4	237
		HgI_3	f	-	93-4	237
	4-methoxyphenyl	Cl	c	-	**	237
		Br	c	-	**	237
		I	e	70	-	237
		$HgCl_3$	f	-	**	237
		$HgBr_3$	f	-	**	237
		HgI_3	f	-	89-90	237
		picrate	c	-	126-7	237
	3-methoxyphenyl	I	e	6	95	237
	2-methoxyphenyl	Cl	c	-	**	237
		Br	c	-	220-1	237
		I	e	68	230-1	237
		HgI_3	f	-	218-9	237
		picrate	c	-	165-6	237
	4-ethoxyphenyl	I	e	-	131	237

Table VIII-2 (cont'd)

phenyl (cont'd)	4-ethoxyphenyl	HgI_3	f	–	76-7	237
	2-ethoxyphenyl	Cl	c	–	**	237
		Br	c	–	178-9	237
		I	e	54	247-8	237
		$HgCl_3$	f	–	**	237
		HgI_3	f	–	183-4	237
		picrate	c	–	**	237
	1-naphthyl	Cl	c	–	**	237
		Br	c	–	**	237
		I	e	77	148	237
		HgI_3	f	–	126	237
4-methylphenyl	methyl	Cl	c	–	**	215
		Br	c	–	73-4**	215
		I	b	100	85-6 (dec)	215
		OH	c	–	–	215
		$CrO_4/2$	c	–	–	215
		$Cr_2O_7/2$	c	–	54-5	215

Table VIII-2 (cont'd)

4-methylphenyl (cont'd)	methyl	$HgCl_3$	f	–	149-50	215
		$PtCl_6/2$	c	–	104-5	215
		picrate	c	–	157-8	215
	carboxymethyl	Br	b	–	**	221
	carbomethoxymethyl	Br	b	56	92-3	221
	carbethoxymethyl	Br	b	20	102-3	221
	phenyl	Br	c	–	230-1	229
		I	e	75	209-10	229
		$HgCl_3$	f	–	–	318
		picrate	c	–	133	229
3-methylphenyl	methyl	Cl	c	–	*	225
		I	b	100	121-2	225
		$PtCl_6/2$	c	–	–	225
		picrate	c	–	–	225
2-methylphenyl	methyl	Cl	c	–	–	215,221
		Br	c	–	134-5	215
		I	b	100	125-6	215-221

Table VIII-2 (cont'd)

2-methylphenyl (cont'd)	methyl	NO_3	c	–	155-7	215
		OH	c	–	-	215
		$CrO_4/2$	c	–	161-2	215
		$Cr_2O_7/2$	c	–	171-2 (dec)	215
		$Zn(OH)Cl_2$	f	–	186-7	215
		$HgCl_3$	f	–	134-5 (dec)	215
		$PtCl_6/2$	c	–	186	215
		picrate	c	–	143-4	215
2,5-dimethylphenyl	methyl	I	b	–	137	223
		picrate	c	–	170	223
2,4,6-trimethylphenyl	methyl	I	b	–	168-9	224
4-methoxyphenyl	methyl	I	b	–	108-9 (dec)	226
		picrate	c	–	153-4	226
3-methoxyphenyl	methyl	I	b	–	**	235
2-methoxyphenyl	methyl	I	b	–	124-5	236
4-ethoxyphenyl	methyl	I	b	–	69	234
2-ethoxyphenyl	methyl	I	b	–	138-40	232

Table VIII-2 (cont'd)

1-naphthyl	methyl	I	b	10	146	233
2-thienyl	methyl	I	b	-	106-15	205

* The chloride was not isolated, but converted to the picrate.

** Compounds were impure.

† Compounds are not crystalline.

†† mp. of monohydrate

was isolated. Compound *(43)* was converted into the telluronium chloride, $[R_3Te]^+Cl^-$, when treated with alcohols. Bis(4-hydroxy-2-methylphenyl) tellurium dichloride, obtained from *m*-cresol and tellurium oxychloride, was transformed into the telluronium chloride upon refluxing in ethanol or upon treatment with a sodium carbonate solution. When this telluronium chloride was dissolved in warm 2N sodium carbonate or 4N sodium hydroxide, compound *(44)* was formed.

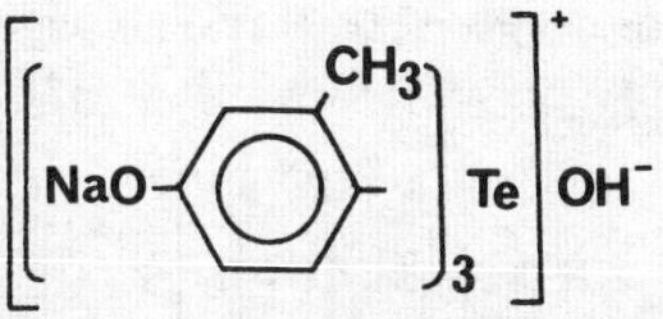

$$\left[\left(NaO-C_6H_3(CH_3)-\right)_3 Te\right]^+ OH^-$$

(44)

i) The reaction of bis(4-methoxyphenyl) ditelluride with methyl iodide: Morgan and Drew[288] and Reichel and Kirschbaum[358] reacted bis(4-methoxyphenyl) ditelluride with methyl iodide and obtained as one of the products dimethyl 4-methoxyphenyl telluronium iodide [see equation (119) in section VII-B-1r]. This reaction is very likely not of general applicability, since bis(4-ethoxyphenyl) ditelluride was decomposed by methyl iodide[288].

j) Modification of the organic moiety in telluronium compounds: Lederer[216] reported, that carbomethoxymethyl diphenyl telluronium bromide when suspended in water and treated with silver oxide, was converted into the telluronium hydroxide with concomitant hydrolysis of the ester group. Carboxymethyl diphenyl telluronium hydroxide was isolated as a crystalline solid. The corresponding ethyl ester behaved similarly. Dimethyl 2-(diethoxy)methylphenyl telluronium iodide was treated with 4N hydrochloric acid. Dimethyl 2-formylphenyl telluronium iodide was isolated in 90 per cent yield[348c].

2) Racemic and optically active triorganyl telluronium salts, $[RR'R''Te]^+ X^-$

The tellurium atom in a tetrahedral environment will become asymmetric, when it is bonded to three different organic groups. The lone electron pair occupies the fourth position. Success or failure in isolating the optically active ℓ- and d-salt will depend on the rate of racemization. Racemic telluronium salts are prepared through addition of an alkyl halide to an unsymmetric telluride, R-Te-R' (eq. 132). Table VIII-3 lists the known compounds.

$$\text{R–Te–R'} + \text{R''X} \longrightarrow \left[\begin{array}{c}\text{R'}\\ \text{R–}\overset{|}{\underset{|}{\text{Te}}}\\ \text{R''}\end{array}\right]^+ X^- \tag{132}$$

Lederer first synthesized such compounds. His attempts to separate methyl 4-methylphenyl phenyl[229] and methyl 2-methylphenyl phenyl[238] telluronium iodides into their optical antipodes failed. Lowry and Gilbert[250] reacted methyl 4-methylphenyl phenyl telluronium iodide with silver d-α-bromocamphorsulfonate (d-BCS) and obtained ℓ-$[R_3Te]^+$ d-$[BCS]^-$ with a molecular rotation $[M]^{18}_{5461}$ of +262° in ethyl acetate. The optically active salt showed rapid mutarotation in a variety of solvents soon reaching the value of $[M]^{18}_{5461}$ = +347°, the molecular rotation of d-BCS. The d-$[R_3Te]^+$ d-$[BCS]^-$ salt could not be isolated in these experiments. An impure salt of the d-telluronium cation was, however, obtained from the racemic iodide and d-10-camphor-sulfonate. The highest value for the widely varying molecular rotation was $[M]^{18}_{5461}$ = +162°. Mutarotation took place only slowly.

From the more reliable data on ℓ-$[R_3Te]^+$ d-$[BCS]^-$ a minimum value of -85° for the molecular rotation of the ℓ-telluronium cation was calculated. The active iodides were isolated from the camphorsulfonate salts. They also showed rapid mutarotation. The optically active

Table VIII-3

Triorganyl Telluronium Compounds, $[RR'R''Te]^{+} X^{-}$

R	R'	R"	X	Method	mp. °C	Rotation (Solvent)	Reference
methyl	carboxymethyl	2-formylphenyl	Br	b	88-92	-	348e
methyl	phenyl	4-methylphenyl	I	b	73-4	-	229,250
ℓ-methyl	phenyl	4-methylphenyl	I	c	67	-	250
d-methyl	phenyl	4-methylphenyl	I	c	70-2	-	250
ℓ-methyl	phenyl	4-methylphenyl	d-BCS†	c	-	$[M]^{18}_{5461}$ + 262° ($CH_3CO_2C_2H_5$)	250
d-methyl	phenyl	4-methylphenyl	d-CS††	c	-	$[M]^{18}_{5461}$ + 162° ($CH_3CO_2C_2H_5$)	250
methyl	phenyl	2-methylphenyl	I	b	119-20	-	238
methyl	ethyl	4-methoxyphenyl	I	b	168	-	358
ℓ-methyl	ethyl	4-methoxyphenyl	d-BCS†	c	-	$[M]_D$ + 198.58° ($CHCl_3$)	358
butyl	carbomenthoxymethyl	methyl	I	b	*	-	17

Table VIII-3 (cont'd)

butyl (cont'd)	carbomenthoxymethyl	$C_6H_5COCH_2$	Br	b	106	$[\alpha]^{20}_{5461}$ - 23.7° ($CHCl_3$)	17
	carbethoxymethyl	methyl	I	b	*	-	17
	carbethoxymethyl	$C_6H_5COCH_2$	Br	b	89	-	17
4-chlorophenacyl			Br	b	161-3		180
ℓ-4-chlorophenacyl			d-BCS[†]	c	147-58	$[M]^{16}_D$ - 99° (CH_3OH)	180
d-4-chlorophenacyl	CH_2 CH_2 CH_2		d-BCS[†]	c		$[M]^{16}_D$ + 597° (CH_3OH)	180
4-chlorophenacyl			picrate	c	198-9 (dec)	-	180
ℓ-4-chlorophenacyl			picrate	c	-	$[M]^{16}_D$ - 750° (acetone)	180
d-4-chlorophenacyl			picrate	c	-	$[M]^{16}_D$ + 575° (acetone)	180
d-phenyl	4-methylphenyl	hydroxy	d-CS	c	93	$[M]^{18}_D$ + 450.05° (acetone)	358

† BCS = bromocamphorosulfonate †† *d*-CS = *d*-10-camphorsulfonate * not crystalline

telluronium cations had a half-life of a few minutes[250].

Reichel and Kirschbaum[358] similarly obtained *ℓ*-methyl ethyl 4-methoxyphenyl telluronium *d*-bromocamphorsulfonate. The reaction of the sulfonate with potassium iodide gave only an inactive compound. The molecular rotation in chloroform for the telluronium cation was calculated to be $[M]_D = -71.42°$.

Holliman and Mann[180] resolved the telluronium cation *(45)* with *d*-BCS into the *d-d* and *d-ℓ* salts by taking advantage of the solubility differences in ethanol. The samples, however, were still optically impure. Mutarotation in methanol and ethanol was slow at room temperature but proceeded rapidly in the boiling solvents. The optically active but impure picrates mutarotated very slowly.

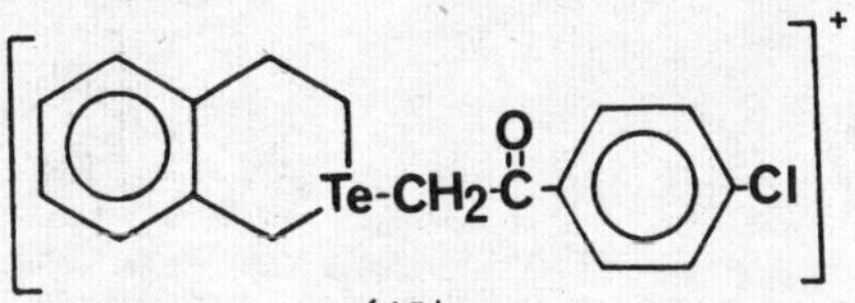

Te-CH₂-C(=O)-C₆H₄-Cl

(45)

Lowry and Gilbert[250] attempted the resolution of 4-methylphenyl phenyl telluroxide without obtaining definite results. Reichel and Kirschbaum[358] reacted the oxide in water solution with *d*-camphorsulfonate and isolated the *ℓ*-phenyl 4-methylphenyl telluronium salt. Treatment of the active compound with sodium carbonate solution gave an inactive telluroxide.

These results with telluroxides suggest, that diorganyl tellurium compounds of the type RR'TeXY are at least in solution telluronium compounds and should be formulated as $[RR'TeX]^+ Y^-$. Lowry and Huether[251] demonstrated, that optically active compounds can be obtained from unsymmetric diorganyl tellurium dihalides, $RR'TeX_2$. The reaction sequence is presented in scheme (133).

$$\mathrm{R'RTeBr_2} \xrightarrow{\text{Ag } d\text{-BCS}} \ell,d\text{-}\mathrm{R'RTe(BCS)_2}$$

ℓ,d-R'RTe(BCS)$_2$ $\xrightarrow{H_2O/\text{methanol}}$ d-[R'RTe-OH]$^+$ d-BCS$^-$ $\xrightarrow{OH^-}$ d-[R'RTe-OH]$^+$ OH$^-$

ℓ,d-R'RTe(BCS)$_2$ $\xrightarrow{H_2O \text{ acetone}}$ ℓ-[R'RTe-OH]$^+$ d-BCS$^-$ $\xrightarrow{OH^-}$ ℓ-[R'RTe-OH]$^+$ OH$^-$

R = C_6H_5, R' = 4-$CH_3C_6H_4$, BCS = bromocamphorsulfonate

While the absolute values of the molecular rotations of these salts are rather unreliable, the observed mutarotation towards smaller angles for the *d*-telluronium compounds and towards larger angles for the ℓ-compounds indicate the presence of optically active telluronium cations.

Ter Horst[182], investigating the same compounds, showed that the disulfonate, RR'Te(d-BCS)$_2$, mutarotated with a half-life of two minutes. The dibromides and diiodides prepared from the optically active disulfonate showed also mutarotation approaching 0° after approximately ten minutes. The mutarotations towards larger angles in the case of the disulfonate and from negative angles towards zero in the case of the diiodide and dibromide indicate that the ℓ-telluronium cation was present in these compounds and possessed a molecular rotation of $[M_D] = -158°$ (disulfonate) and -185° (diiodide). On the basis, that optical activity can be caused by the four groups R, R, R', R" arranged in an irregular tetrahedral manner around a central atom, ter Horst[182] investigated the compounds $(C_6H_5)_2Te(d\text{-}BCS)_2$, $(C_6H_5)_2Te(OH)(d\text{-}BCS)$ and $(C_6H_5)_2TeBr_2$ and observed a small rotation indicative of ℓ-telluronium cations. The tellurium atom, according to these results, must be surrounded by two phenyl groups, one anion and one lone electron pair forming an irregular tetrahedron. A reinvestigation of this problem should show whether the optical effects observed in the diphenyl series are real or were caused by factors not connected with telluronium cations.

Balfe and coworkers[17] prepared the telluronium compounds $[RR'R''Te]^+ Br^-$ containing the ℓ-menthoxy group, which imparted optical activity to the telluronium salt. In addition to the menthoxy compounds in Table VIII-3, dibutyl carbo-ℓ-menthoxymethyl telluronium bromide (mp. 90°, $[\alpha]^{17.5}_{5461}$ -23.6, $CHCl_3$) was also synthesized. For optically active, heterocyclic tellurium compounds consult section XII-C.

3) Reactions of triorganyl telluronium compounds

Almost all of the reactions given by triorganyl telluronium compounds affect the tellurium atom. The three carbon-tellurium bonds are kept intact in most cases. The important reactions are summarized in Figure VIII-1.

The exchange of anions in telluronium salts and their reactions with metal halides are discussed in section VIII-1-c and VIII-1-f.

Triorganyl telluronium halides react with organic zinc, magnesium and lithium compounds. Triethyl telluronium chloride and iodide dissolve in diethyl zinc at room temperature without reaction. When the mixture was heated to 100°, diethyl telluride and butane was formed[261]. Lederer[213,239], claimed, that he had obtained diphenyl telluride from triphenyl telluronium chloride and phenyl magnesium bromide.

Petragnani and de Moura Campos[344,347], however, state, that triorganyl tellurium halides are inert towards Grignard reagents. The following equations (134)-(138) describe the reactions of the telluronium salts with organic lithium compounds. A more extensive discussion of these reactions is given in chapter IX in connection with tetraorganyl tellurium compounds.

(134) $(CH_3)_3TeI \xrightarrow{CH_3Li} (CH_3)_4Te \xrightarrow{H_2O/NaI} (CH_3)_3TeI + CH_4$

(135) $(C_4H_9)_3TeI \xrightarrow{C_4H_9Li} (C_4H_9)_4Te \xrightarrow{heat} (C_4H_9)_2Te + C_8H_{18}$

(136) $(C_6H_5)_3TeCl \xrightarrow{C_6H_5Li} (C_6H_5)_4Te$

(137) $(CH_3)_3TeI$ + 2,2'-dilithiobiphenyl (Li Li) → dibenzotellurophene (Te) + C_2H_6 + $LiCH_3$ + LiI

(138) $(C_6H_5)_3TeX \xrightarrow{C_4H_9Li} (C_4H_9)_4Te \xrightarrow{H_2O/NaI} (C_4H_9)_3TeI$

[(equation) reference: (134) 172; (135) 172; (136) 442; (137) 172; (138) 132,17.

Triorganyl telluronium halides can be cleaved into organic halide and diorganyl telluride. The decomposition takes place upon heating the telluronium salt with or without a solvent. Heal[169] collected the decomposition data for trimethyl telluronium salts from the literature. He found, that the thermal stability increases in the series F, Cl, SO_4, Br, I, SiF_6. Detailed thermoanalytical investigations in this area have not been carried out. The following salts lost alkyl halide under the conditions indicated.

$[(C_6H_5)_2Te(CH_3)]^+ X^-$

boiling ethanol[215] (X = I, Br)

boiling dimethylaminobenzene[215] (X = I)

vacuum 60°[215] (X = I)

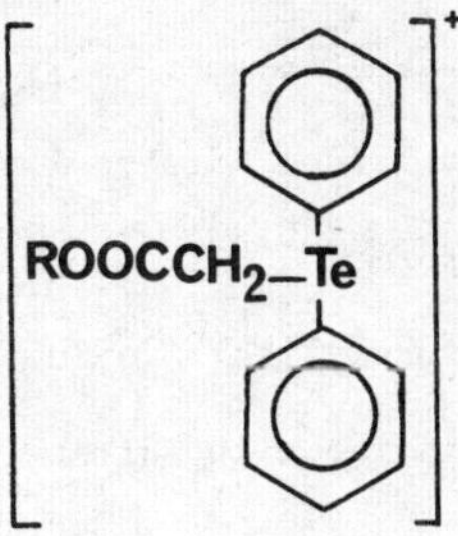

$\left[ROOCCH_2-Te(C_6H_5)_2\right]^+ Br^-$

boiling ethanol[216] ($R = CH_3$)

boiling water[216] ($R = C_2H_5$)

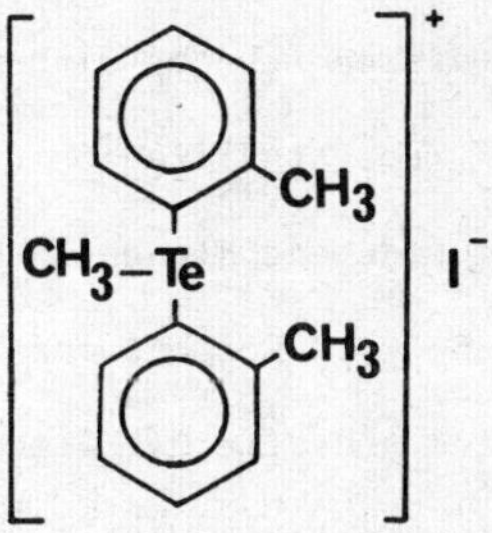

$\left[CH_3-Te(C_6H_4CH_3\text{-}o)_2\right]^+ I^-$

boiling ethanol[215]

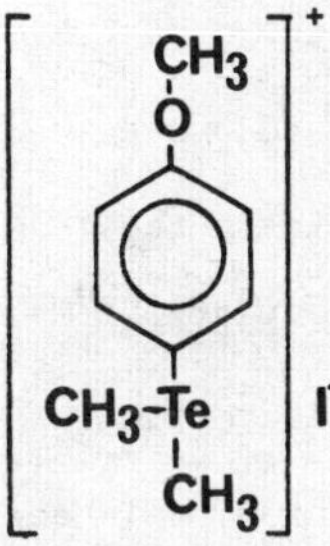

$\left[(CH_3)_2Te-C_6H_4OCH_3\text{-}p\right]^+ I^-$

warm water[288]

$\left[(CH_3)_2Te-C_6H_4R\text{-}o\right]^+ I^-$

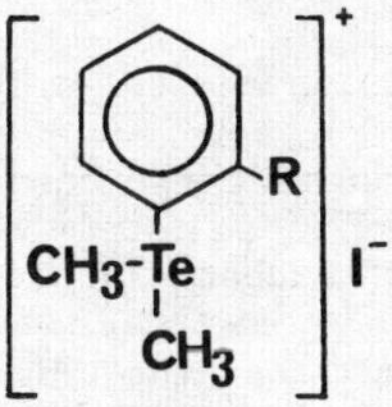

refluxing pyridine

($R = H$[348b], $-CHO$[348c])

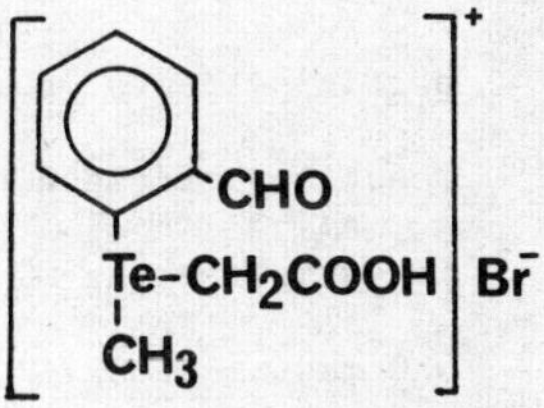

$\left[CH_3(HOOCCH_2)Te-C_6H_4CHO\text{-}o\right]^+ Br^-$

170°/0.2 mm Hg[348e]

The course of the thermal decomposition of dialkyl carbalkoxymethyl telluronium bromides[17] in vacuum is shown in equation (139).

(139) $\left[R_2TeCH_2COOR'\right]^+ Br^-$

$\xrightarrow[R' = C_2H_5]{R = C_4H_9} RBr + RTeCH_2COOR'$

$\xrightarrow[R' = C_2H_5]{R = C_5H_{11}} RBr + RTeCH_2COOR'$

$\xrightarrow[100° (20 \text{ mm})]{R = C_4H_9,\ R' = \ell\text{-menthyl}} RBr + RTeCH_2COOR' + Te(CH_2COOR')_2$

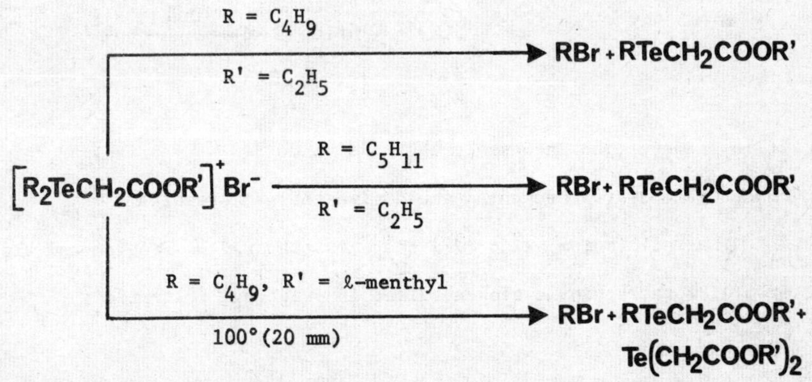

Bis(methylene diphenyl telluronium) dibromide produced in refluxing 1,2-dibromoethane diphenyl tellurium dibromide and ethylene, while upon recrystallization from glacial acetic acid ethylene, diphenyl telluride and diphenyl tellurium dibromide were formed[301] (see section VII-B-1k). Reichel and Kirschbaum[359] treated triphenyl telluronium iodide and dimethyl 4-methoxyphenyl telluronium iodide with molten sodium sulfide nonahydrate and observed the formation of tellurides and aromatic hydrocarbons.

The triorganyl sulfonium, selenonium, and telluronium cations are much more stable than the oxonium compounds. This extra stability has been attributed to hyperconjugation using the empty low energy d-orbitals of the sulfur, selenium and tellurium atom[270]. In order to prove the participation of sulfur, selenium and tellurium d-orbitals in the reactions of these trimethyl onium compounds, Doering and Hoffman[99] investigated the DO^- catalyzed exchange of hydrogen for deuterium in D_2O medium at 62°. They found, that the trimethyl sulfonium, selenonium and telluronium cations contained after three hours 98.0, 13.2 and 3.99 atom per cent deuterium, respectively. It was proposed that the exchange took place according to equation (140) requiring d-orbital participation.

$$(140)\quad \oplus\text{Te}-\text{CH}_3 + \text{OD}^- \xrightarrow{-\text{HOD}} \oplus\text{Te}-\text{CH}_2^{\ominus} \longleftrightarrow \text{Te}=\text{CH}_2\ (46) \xrightarrow{+\text{DOD}} \text{OD}^- + \oplus\text{Te}-\text{CH}_2\text{D}$$

It is, however, not necessary that the ylide *(46)* be formed. A concerted one-step mechanism would give the same results.

The telluronium ylide *(47)* has been prepared by Freeman and Lloyd[132a] from diazotetraphenylcyclopentadiene and diphenyl telluride.

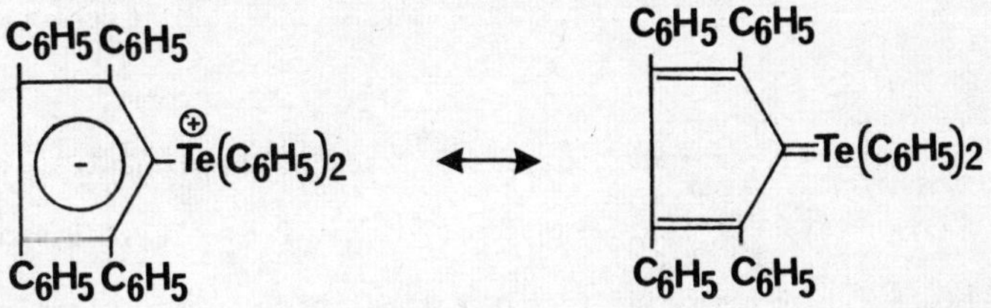

C_6H_5 C_6H_5 C_6H_5 C_6H_5

$\overset{\oplus}{Te}(C_6H_5)_2$ $\longleftrightarrow$ $=Te(C_6H_5)_2$

C_6H_5 C_6H_5 C_6H_5 C_6H_5

(47)

This ylide was decomposed by strong acids. Definite products were not obtained from its reactions with 4-nitrobenzaldehyde and nitrosobenzene.

A theoretical account of the catalytic activity of telluronium salts in liquid phase oxidation reactions has been published[323c]. The equivalent conductance of trimethyl telluronium hydroxide[45a] in aqueous solution at 25° was determined to be 205 Ohm^{-1} cm^2. The dipole moments for triphenyl telluronium fluoride, chloride and iodide in benzene at 25° are 8.77D and 4.69D[192a]. The thin layer chromatographic behavior of triphenyl telluronium chloride was investigated[319b]. Bowd[38e] reviewed the analytical applications of triphenyl telluronium salts.

IX. TETRAORGANYL TELLURIUM COMPOUNDS, R_4Te

Marquardt and Michaelis[258] made an unsuccessful attempt to prepare a tetraorganyl tellurium compound in 1888. They obtained only triethyl telluronium chloride from the reaction between tellurium tetrachloride and diethyl zinc. This telluronium salt when heated with diethyl zinc gave diethyl telluride and butane. Wittig and Fritz[442] succeeded in preparing tetraphenyl tellurium in 1952. They synthesized this compound from tellurium tetrachloride, from triphenyl telluronium chloride (33.6 per cent yield) and from diphenyl tellurium dichloride (51.8 per ce yield) and an excess of phenyl lithium as yellow crystals, which melted at 104-6° with decomposition. The reactions must be carried out in anhydrous ether under an inert atmosphere of nitrogen. The phenyl lithiu solution must not contain lithium bromide. The excess of phenyl lithium necessary in this reaction above the stoichiometrically required amount suggests $[(C_6H_5)_5Te]^- Li^+$ as an intermediate, pentacoordinated species. Hellwinkel and Fahrbach[171,173] reacted hexamethoxy tellurium, tetramethox tellurium and tellurium tetrachloride with 2,2'-biphenylylene dilithium and isolated bis(2,2'-biphenylylene) tellurium *(48)* in yields ranging from 40-55 per cent.

Te

(48)

This compound, which crystallized as yellow needles and melted at 214°, is much more stable than tetraphenyl tellurium. A hexavalent tellurium derivative was not formed when hexamethoxy tellurium was employed as starting material. The same authors reported in later publications[172,173] that the reactions described by equations (141)-(144) had also produced *(48)* in acceptable yields.

(141) [dibenzotellurophene–Te–CH_3]$^+$ I^- + 2,2'-dilithiobiphenyl $\xrightarrow{52\%}$ *(48)* + LiI + $LiCH_3$

(142) dibenzotellurophene–$TeCl_2$ + 2,2'-dilithiobiphenyl $\xrightarrow{36\%}$ *(48)* + 2LiCl

Bis(4,4'-dimethyl-2,2'-biphenylylene) tellurium was prepared according to equation (142) in 48 per cent yield. The yellow compound decomposed between 204 and 219°[173].

(143) $TeCl_4$ + 2,2'-dilithiobiphenyl $\xrightarrow{31-44\%}$ *(48)* + 4LiCl

(144) dibenzotellurophene + CH_3–C_6H_4–$SO_2N(Cl)Na$ $\xrightarrow{-NaCl}$ dibenzotellurophene Te=NSO_2–C_6H_4–CH_3

$\xrightarrow[44\%]{\text{2,2'-dilithiobiphenyl}}$ CH_3–C_6H_4–SO_2NLi_2 + *(48)*

Nefedov[315] and Wheeler[439a] obtained ^{125m}Te-tetraphenyl tellurium by β-decay of ^{125}Sb in $Sb(C_6H_5)_5$ and $(C_6H_5)_4SbCl$. Tetrakis(pentafluorophenyl) tellurium which decomposes at 210°, was the product of the reaction between tellurium tetrachloride and pentafluorophenyl lithium[83].

Tetraalkyl and dialkyl diaryl tellurium compounds are thermally rather unstable and have not yet been isolated. Experiments by Hellwinkel and Fahrbach[172] prove, however, that these substances exist in solution. The alkylation of tellurium tetrachloride (eq. 145) and of trialkyl telluronium iodides (eq. 146) with lithioalkanes gave these tetraalkyl tellurium compounds.

(145) $$TeCl_4 + 4RLi \longrightarrow R_4Te + 4LiCl$$

$R = C_4H_9, CH_3$

(146) $$R_3TeI + RLi \longrightarrow R_4Te + LiI$$

$R = C_4H_9, CH_3$

In the reaction between trimethyl telluronium iodide and 2,2'-dilithiobiphenyl dimethyl 2,2'-biphenylylene tellurium was produced. The formation of this compound requires an exchange of a methyl group, which is removed as methyl lithium, for an aromatic group (eq. 147).

(147) $(CH_3)_3TeI$ + 2,2'-dilithiobiphenyl ⟶ (49) + $LiI + CH_3Li$

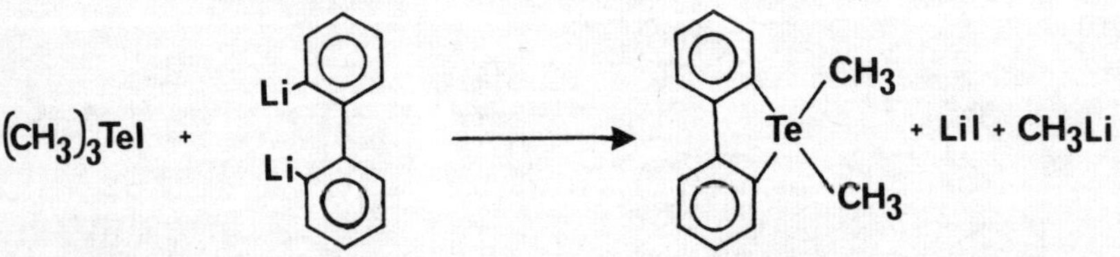

(49)

When triphenyl telluronium bromide or 2,2'-biphenylylene 2-biphenylyl telluronium chloride was acted upon by four moles of butyl lithium, all the aromatic groups were cleaved from the tellurium atom as aryl lithium compounds. Tetrabutyl tellurium was the reaction product (eq. 148,149).

(148) $(C_6H_5)_3TeBr + 4C_4H_9Li \longrightarrow (C_4H_9)_4Te + LiBr + 3C_6H_5Li$

(149) $[(C_{12}H_8)Te(C_{12}H_9)]^+ Cl^- + 4C_4H_9Li \longrightarrow C_{12}H_9Li + C_{12}H_8Li_2 + (C_4H_9)_4Te + LiCl$

Bis(2,2'-biphenylylene) tellurium *(48)* can also be employed to prepare solutions of tetrabutyl tellurium according to equation (150).

(150) $(C_{12}H_8)_2Te + 4C_4H_9Li \xrightarrow{k_1} (C_{12}H_8)Te(C_4H_9)(C_{12}H_8Li)$

$\xrightarrow{k_2}$ $-C_{12}H_8Li_2$ → $(C_{12}H_8)Te(C_4H_9)_2$ (also directly from $(C_{12}H_8)_2Te$: $+2C_4H_9Li$, $-C_{12}H_8Li_2$)

$(C_{12}H_8)Te(C_4H_9)_2 \xrightarrow{k_3} (LiC_{12}H_8)Te(C_4H_9)_3 \xrightarrow{k_4} C_{12}H_8Li_2 + (C_4H_9)_4Te$

$(C_{12}H_8)Te(C_4H_9)_2 \xrightarrow{heat} (C_{12}H_8)Te + C_8H_{18}$

The rate constants for the four reaction steps have been estimated to increase in the order $k_1 < k_3 < k_4 \approx k_2$. The exchange reactions can either proceed according to an S_N2 mechanism or *via* penta-coordinated tellurium intermediates. The aryl lithium compounds formed in the exchange reactions were reacted with benzophenone and the expected hydroxy compounds were isolated[172].

Methyl, phenyl and 4-dimethylaminophenyl lithium did not react with *(48)*. Franzen and Mertz[132] observed, that triphenyl telluronium iodide exchanged organic groups with butyl lithium in a fast reaction, while tetraphenyl tellurium reacted only slowly. These authors proposed, that the faster reaction with triphenyl telluronium iodide is due to a direct exchange on the telluronium cation without participation of a tetracoordinated tellurium species[132]. Although tetraalkyl tellurium compounds have not been isolated, the following reactions prove their existence in solution: Hydrolysis of solutions of tetraalkyl tellurium compounds gave trialkyl telluronium hydroxides. Upon distillation of the reaction mixture containing tetrabutyl tellurium decomposition to dibutyl telluride, butane and octane took place[172]. The mixed diaryl dialkyl tellurium compounds *(49)* (eq. 147) and *(50)* (eq. 150) lost the aliphatic groups under these conditions and produced 2,2'-biphenylylene telluride. The solution containing tetrabutyl tellurium was reacted with bis(2,2'-biphenylylene) arsonium and phosphonium iodide. The transfer of a butanide group from tellurium to the group V element with formation of *(51)* was taken as further evidence of the existence of tetrabutyl tellurium (eq. 151).

The thermal stability of the tetraorganyl tellurium compounds increases in the series alkyl, phenyl, 2,2'-biphenylylene. The tetra-alkyl derivatives decompose at room temperature while tetraphenyl tellurium is stable to approximately 115°. Tetrakis(pentafluorophenyl) tellurium[83]

(151) 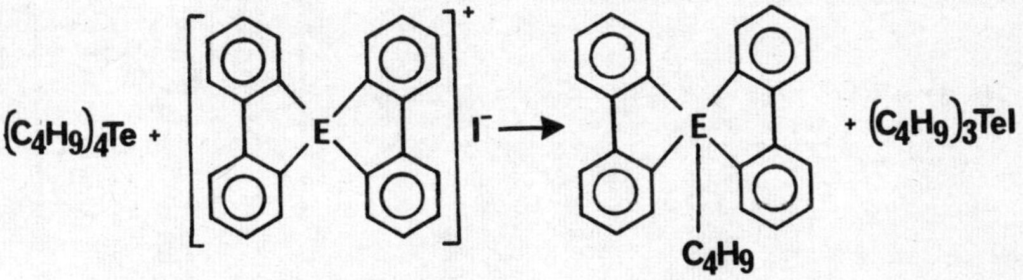 $(C_4H_9)_4Te$ + $[\ldots E \ldots]^+ I^-$ ⟶ ... + $(C_4H_9)_3TeI$

E = P, As (51)

produced the telluride and bis(pentafluorophenyl) when heated in a sealed tube at 200-220°. Bis(2,2'-biphenylylene) tellurium remains largely unchanged when heated to 210°. At 260° in vacuum decomposition takes place according to equation (152)[173].

(152) 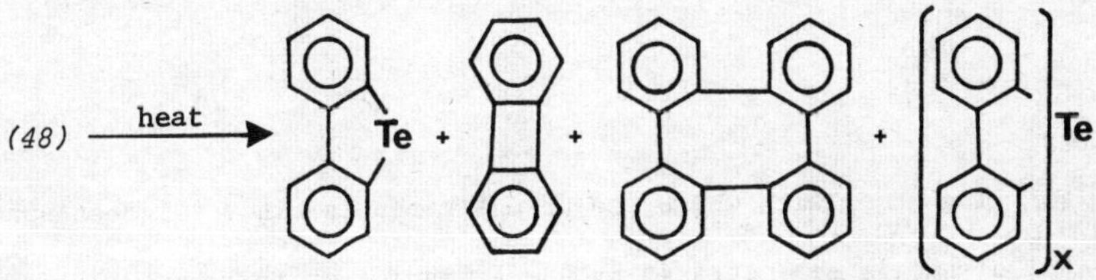 *(48)* $\xrightarrow{heat}$ ... Te + ... + ... + [...]$_x$Te

+ higher polyphenylene compounds

The corresponding 4-methyl substituted derivative behaves similarly[173].

While tetraphenyl tellurium is easily hydrolyzed by water, compound *(48)*[173] is relatively stable under these conditions. Organic solutions of *(48)* when kept in dry air remain unchanged for several days. In tetrahydrofuran containing oxygen-free water the telluronium hydroxide is formed. Hydrogen chloride in ethanol, or bromine and iodine in carbon tetrachloride convert *(48)* into the corresponding telluronium compounds *(52)* (eq. 153).

Data for these telluronium salts can be found in section XII-A. Compound *(52)* (X = H, Y = I) is also formed, when *(48)* is treated

(153) 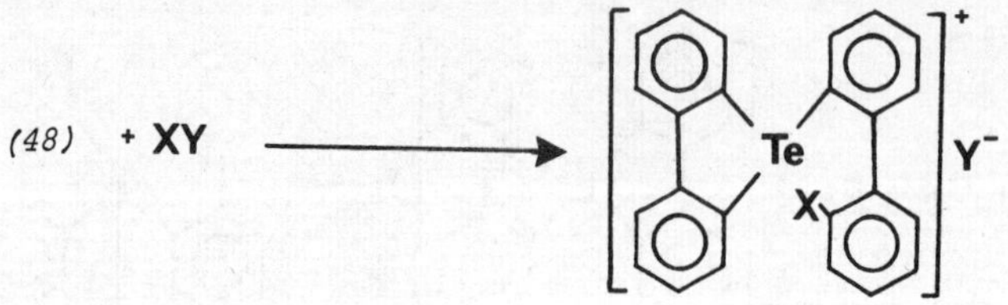 *(48)* + XY → $[(C_6H_4C_6H_4)Te(C_6H_4C_6H_4X)]^+ Y^-$ *(52)*

X, Y: H, OH; H, Cl; Br, Br; I, I

with excess methyl iodide at room temperature. A pentavalent tellurium intermediate has been proposed[173] for this reaction as shown in equation (154).

(154) 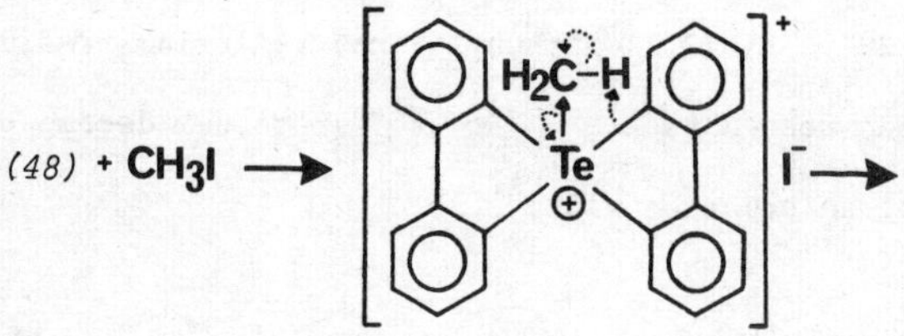 *(48)* + CH_3I → $[(C_6H_4C_6H_4)Te^{\oplus}(H_2C{-}H)(C_6H_4C_6H_4)]^+ I^-$ → *(52)*

$X = CH_3$, $Y = I$

The ease, with which one phenyl group is cleaved in tetraphenyl tellurium, is demonstrated by the formation of triphenyl telluronium tetraphenyl borate from triphenyl boron and tetraphenyl tellurium in ethereal solution. Tetraphenyl tellurium approaches the reactivity of an organometallic compound in its reactions with methylene chloride, chloroform (eq. 155) and benzaldehyde (eq. 156).

(155) $(C_6H_5)_4Te \xrightarrow{CH_2Cl_2} (C_6H_5)_3TeCl + C_6H_5CH_2Cl$

$(C_6H_5)_4Te \xrightarrow{CHCl_3} (C_6H_5)_3TeCl + C_6H_5CHCl_2$

(156) $R_3Te(R) + {}^{\ominus}\overline{O}{-}\overset{\oplus}{C}H{-}R \longrightarrow R_3Te{-}\overline{O}{-}CHR_2 \xrightarrow{HCl} R_3TeCl + R_2CHOH$

$R = C_6H_5$

Reaction (156) carried out in ether under nitrogen atmosphere gave a 47.3 per cent yield of diphenylmethanol[442].

Structural data are not available for tetraorganyl tellurium compounds. A trigonal bipyramidal arrangement of the four ligands and the lone electron pair can be postulated (see Fig. VII-2). The bidentate ligands 2,2'-biphenylylene must each occupy one equatorial and one axial position when the electron pair takes up the third equatorial position. Hellwinkel and Fahrbach[173] investigated the H^1 nmr spectrum of bis(4,4'-dimethyl-2,2'-biphenylylene) tellurium. The compound showed only one peak for the methyl hydrogen atoms in the range from -55° to room temperature. This result is inconsistent with the trigonal bipyramidal structure. The authors propose as the most likely explanation for the equivalence of the methyl groups a rapid interconversion of the trigonal bipyramid *via* a tetragonal bipyramid into the other trigonal bipyramid, in which the methyl groups formerly in axial position occupy now equatorial sites. A tetrahedral arrangement with a stereochemically inactive lone electron pair was also mentioned as a possible cause for the equivalence of the methyl groups.

X. TELLURIUM ANALOGS OF ALDEHYDES AND KETONES

The only known telluroaldehyde, telluroformaldehyde *(53)* was obtained by action of carbene, $:CH_2$, on tellurium mirrors. The carbene was generated by thermal decomposition of diazomethane[335,365,441] or methane[23] and by photodissociation of ketene[335].

(157) $$Te + \dot{C}H_2 \longrightarrow H_2C{=}Te$$

(53)

Telluroformaldehyde, first formed as a monomeric gas[335], polymerized upon condensation. Several modifications are reported to exist. A deep-red solid, insoluble in water, only slightly soluble in ethanol, ether and benzene was stable in air and sublimed at 100° under high vacuum with only slight decomposition[365]. The compound decomposed to elemental tellurium at 150°. The second modification is colorless at -70°. It turns grey-green at room temperature and decomposes when exposed to air. It is soluble in acetone, chloroform and ether[335]. Williams and Dunbar[441] proved with mass-spectroscopic and ir-techniques that the product of the reaction between tellurium mirrors and carbene generated by decomposition of diazomethane at 500° was tritelluroformaldehyd No evidence for polymers was found. Tellurformaldehyde reacted with bromine and iodine with formation of dihalomethanes[335].

1,3-Dimethylbenzimidazoline heated with two moles of tellurium to 100-200° for 10-60 minutes gave an 8.8 per cent yield of 1,3-dimethyl-2-telluroxobenzimidazoline[161] *(54)*. The compound melted at 199-200°.

Dialkyl telluroketones have been prepared by Lyons and Scudder[225] by passing hydrogen telluride for 24 hours at room temperature through a solution made up of equal volumes of the ketone and concentrated hydrochloric acid. The liquid telluroketones were purified by vacuum

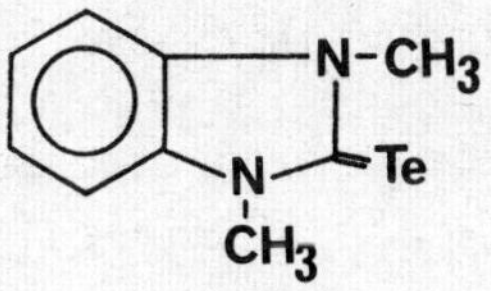

(54)

distillation. The following compounds, RC(Te)R', were prepared (R,R', bp./torr): CH_3, CH_3, 55-58°/10-13; CH_3, C_2H_5, 63-66°/9-10; C_2H_5, C_2H_5, 69-72°/8-11; C_3H_7, C_3H_7, decomposition. Benzil and benzophenone did not react under these conditions.

XI. ORGANIC TELLURIUM COMPOUNDS CONTAINING A TELLURIUM-METAL OR A TELLURIUM-METALLOID BOND

The tellurium atom is capable of forming bonds with metallic, metalloidal and non-metallic elements. All the compounds, which contain a tellurium-metal or metalloid bond and at least one tellurium-carbon bond, are treated in this chapter. In addition, substances, which have organic groups in their molecules, even if they do not possess a tellurium-carbon bond, are also included. Bis(triethylsilyl) telluride and triorganyl phosphine tellurides are examples for the latter class of tellurium compounds. The following elements form bonds with tellurium: lithium, sodium, magnesium, boron, aluminum, gallium, carbon, silicon, germanium, tin, lead, nitrogen, phosphorus, oxygen, sulfur, selenium, fluorine, chlorine, bromine, iodine, manganese, iron, molybdenum, palladium, cadmium, platinum, gold, mercury and uranium.

The organic compounds of tellurium with oxygen, sulfur and the group VII elements have already been treated in previous chapters. The discussion in this section will be organized according to the heteroatoms bonded to the tellurium atom. The compounds of the group I and II metals with tellurium are of the type R-Te-M (M = Na, Li, MgBr). Silicon, germanium, tin and lead form molecules having the general formula $R_3M\text{-}Te\text{-}R'$ or $R_3M\text{-}Te\text{-}M'R_3$ (M, M' = group IV element; R = organic group or H; R' = Li, H, or C_2H_5). The compounds with the other metallic elements are complexes, in which diorganyl tellurides, diorganyl ditellurides or organyltelluro groups serve as ligands.

A) Organic Compounds of Tellurium with Metals of Group I, II and III

The compounds R–C≡C–Te–Na were prepared by addition of elemental tellurium to the sodium acetylide in liquid ammonia (eq. 158).

$$\text{(158)} \qquad R\text{-}C{\equiv}C\text{-}Na + Te \longrightarrow R\text{-}C{\equiv}C\text{-}Te\text{-}Na$$

R = CH_3[38a,41,43]; R = $CH_2{=}CH$[348]; R = $CH_2{=}C(CH_3)$[355]; R = C_6H_5[38]

The alkynyl sodium tellurides were never isolated, but were reacted immediately with alkyl halides to give alkynyl alkyl tellurides (see section VII-A-2a).

Lithium compounds of the type $(C_6H_5)_3M$–Li (M = Ge[387], Sn[384,385], Pb[388]) and tellurium powder in tetrahydrofuran formed grayish-black solutions containing the compounds $(C_6H_5)_3M$–Te–Li. These solutions were very unstable. They deposited tellurium upon contact with air. When reacted with $(C_6H_5)_3M$–Cl the tellurides $[(C_6H_5)_3M]_2Te$ were produced.

Aryl lithium compounds reacted with ditellurides[342] and with elemental tellurium[343,348b,348c] producing aryl lithium tellurides. Butyl lithium and tellurium gave butyl lithium telluride[348d]. Lithium has been reported to cleave diethyl ditelluride to form ethyl lithium telluride[117a]. The reduction of diaryl ditellurides, $(RC_6H_4)_2Te_2$ (R = H, 4–CH_3[348b,d], 2–CH_3[348c,d], 4–Br[348b]) and of bis(2-bromo-4-methoxyphenyl) ditelluride with sodium borohydride in sodium hydroxide solution generated aryl sodium tellurides. None of these compounds was ever isolated. These air sensitive substances were either oxidized to ditellurides or treated with an alkylating agent.

Organic tellurium compounds with a tellurium-magnesium bond were the products of the insertion of a tellurium atom into the carbon-magnesium bond of a Grignard reagent. The reactions were carried out in ethereal solution. Phenyl bromomagnesium telluride was employed by several investigators[39,147,181,343,351,352] as an intermediate in telluride syntheses. Petragnani[342] obtained a pasty substance upon addition of petroleum ether to a solution containing phenyl bromomagnesium telluride, which had been prepared from a diorganyl ditelluride and phenyl magnesium bromide (eq. 159).

$$(159) \qquad R_2Te_2 + C_6H_5MgBr \longrightarrow RTeC_6H_5 + RTeMgBr$$

The highly unstable precipitate was easily hydrolized and oxidized. Dubien[111] reviewed the work on organic magnesium compounds performed during the period 1920-1926 and included those containing a magnesium-tellurium bond.

Dimethyl telluride acts as an electron pair donor toward boron trihalides forming the adducts $(CH_3)_2Te{\cdot}BX_3$ (X = Cl, Br, I), which melt with decomposition at 98-101°[379a], 138-142°[76,77,379a] and 143-145°[379a], respectively. These addition compounds were obtained by mixing benzene or carbon disulfide solutions of the components. When dimethyl ditelluride was treated with boron tribromide, elemental tellurium and the adduct $(CH_3)_2Te{\cdot}BBr_3$ were observed as products[76]. The compounds $[(CH_3)_2TeBr_3]_2{\cdot}BBr_3$, a colorless solid, which melted at 57-8°, was obtained from a benzene solution of the reactants[77]. Boron trichloride and dimethyl telluride in liquid hydrogen chloride gave a white precipitate of $[(CH_3)_2TeH]^+[BCl_4]^-$, which lost hydrogen chloride at -80° in vacuum forming $(CH_3)_2Te{\cdot}BCl_3$[333]. This adduct melted at 212-3°

Trimethyl aluminum and trimethyl gallium[80] formed 1:1 adducts with dimethyl telluride (eq. 160).

$$(160) \qquad (CH_3)_2Te + (CH_3)_3M \longrightarrow (CH_3)_2Te{\cdot}M(CH_3)_3$$

M = Al, Ga

The gallium adduct melted at -32° and a boiling point of 122° was obtained by extrapolating vapor pressure data. The enthalpy of dissociation calculated from the temperature dependence of the equilibrium constant is reported as ~-8 kcal/mole. The aluminum adduct, with an extrapolated boiling point of 146°, was almost completely dissociated in the temperature range from 100 to 150°. The donor strengths of the dimethyl chalcogenides decreased in the order O>Se>S=Te and O>S>Se>Te with the acceptors trimethyl gallium and aluminum, respectively.

B) Organic Compounds of Tellurium Containing a Tellurium-Group IV Element Bond

Compounds containing a bond between tellurium and germanium, tin or lead are synthesized from the lithium tellurides *(55)* and the halides *(56)* in tetrahydrofuran[384,385,387-389] according to equation (161) *(method a)*.

$$R_3M\text{-}Te\text{-}Li + Cl\text{-}M'R_3 \longrightarrow R_3M\text{-}Te\text{-}M'R_3 + LiCl \qquad (161)$$

(55) *(56)* *(57)*

$R = C_6H_5$; M, M' = Ge, Sn, Pb

The symmetric (M = M') and unsymmetric (M = M') tellurides *(57)* are yellow solids, which are rather stable towards heat and hydrolytic agents. The symmetric lead compound can be exposed for months to moist air without decomposition. The tin and germanium tellurides hydrolyze readily under these conditions. The stability of these compounds towards moisture and oxygen increases in the sequence $(R_3Ge)_2Te < (R_3Pb)_2Te$ and $(R_3Ge)_2Te < R_3GeTeSnR_3 < R_3GeTePbR_3$[388]. Unsymmetric aliphatic compounds *(57)* (M = M', R = alkyl) are reported to transform easily into the symmetric derivatives[390]. Table XI-1 list the known compounds.

Table XI-1

Organic Compounds Containing a Tellurium-Group IV Element Bond, $R_3M\text{-}Te\text{-}R'$

R	M	R'	Method	mp. °C	bp. °C(torr)	Yield %	Reference
phenyl	Ge	$Ge(C_6H_5)_3$	a	120(dec)	--	11	387
		$Sn(C_6H_5)_3$	a	142-6	--	48	387
		$Pb(C_6H_5)_3$	a	115-7	--	42	387
		Li	-	*	--	--	387
phenyl	Sn	$Sn(C_6H_5)_3$	a	148-50	--	59	384,385
		$Pb(C_6H_5)_3$	a	135(dec)	--	27	385
		Li	-	*	--	--	384,385
phenyl	Pb	$Pb(C_6H_5)_3$	a	128-9	--	60	388
		Li	-	*	--	--	388
H	Si	SiH_3	b	--	--	100	86
			b	--	49(50)	40-50	49
methyl	Si	$Si(CH_3)_3$	b	--	--	40	49
			c	--	40-2(0.25)	15	181
ethyl	Si	H	h	--	--	72.9	37

Table XI-1 (cont'd)

ethyl (cont'd)	Si	C_2H_5	d	--	52-3 (1)	11.5	431,432
		$Si(C_2H_5)_3$	d	--	97-9 (1)	55.4	432
			e	--	73-7 (45)	70.6	37,434
		$Sn(C_2H_5)_3$	f	--	109-12 (1)	90.8	431,432
H	Ge	GeH_3	h	-75	--	100	86
ethyl	Ge	C_2H_5	d	--	61(1)	39	431,432
			g	--	42(1)	89	117a
	Ge	$Ge(C_2H_5)_3$	d	--	112-5(1)	57.9	432
			e	--	112-5(1)	60-75	430,432
			h	--	--	78.8	434
cyclohexyl	Ge	$Ge(c\text{-}C_6H_{11})_3$	d	128-9	--	77.7	432
ethyl	Ge	$Sn(C_2H_5)_3$	f	--	126-8(1)	62	431,432
	Sn	$Sn(C_2H_5)_3$	d	--	119-21(1)	91	431,432
			d	--	--	72.6	434
			e	--	131-4(1.5)	64.4	435
			e	--	144-5(1.5)	-	430
			i	--	125-9(1)	71.6	37

* in solution only, not isolated.

Lithium telluride, prepared in liquid ammonia, reacted with silyl bromide[86] in liquid ammonia, and with silyl iodide and trimethylsilyl chloride in tetraline[49] to give the corresponding disilyl tellurides *(method b)*. Hooton and Allred[181] synthesized bis(trimethylsilyl) telluride in low yield by addition of trimethylsilyl chloride to diethyl ether containing phenyl bromomagnesium telluride *(method c)*. The mechanism for this reaction is unknown. Should the phenyl bromomagnesium telluride be the source of tellurium, then a tellurium-carbon(phenyl) bond must be cleaved under these mild conditions, perhaps in a disproportionation reaction of the unsymmetric telluride C_6H_5-Te-Si$(CH_3)_3$ into the symmetric tellurides.

The hydrides, R_3MH (R = C_2H_5, M = Si, Ge, Sn; R = *c*-C_6H_{11}, M = Ge), when heated with diethyl telluride cleaved one or both ethyl groups of the telluride as ethane *(method d)*. The reactivity of the hydrides increased in the order Si<Ge<<Sn. Thus, triethylsilane and diethyl telluride kept at 200° for seven hours produced the symmetric *(60)* and unsymmetric telluride *(59)*, R_3Si-Te-R' (R = C_2H_5 or SiR_3). The corresponding germanium hydride gave comparable yields of the two tellurides, when heated for seven hours at only 140°. With triethylstannane bis(triethylstannyl telluride was the sole product obtained, when the reagents were kept at 20° for only one hour (eq. 162).

$$2R_3MH + (C_2H_5)_2Te \longrightarrow \begin{cases} 2R_3M\text{-}Te\text{-}C_2H_5 + C_2H_6 & (59) \\ R_3M\text{-}Te\text{-}MR_3 + 2C_2H_6 & (60) \end{cases} \qquad (162)$$

(58)

The symmetric tellurides *(60)* were probably formed from *(59)* and an additional molecule of the hydride *(58)*. A disproportionation of *(59)* into *(60)* and diethyl telluride is also a likely reaction path[432].

Powdered tellurium heated with the hydrides $(C_2H_5)_3MH$ in sealed, evacuated tubes (M = Si, 22 hrs. at 280°; M = Ge, 18 hrs. at 210°;

M = Sn, 4 hrs. at 130°) gave the symmetric tellurides *(60)* and hydrogen *(method e)*. The unsymmetric tellurids, $(C_2H_5)_3M\text{-}Te\text{-}Sn(C_2H_5)_3$ (M = Si, Ge) were obtained from triethylstannane and the appropriate unsymmetric telluride *(59)* according to equation (163)*(method f)*.

$$(163)\quad (C_2H_5)_3M\text{-}Te\text{-}C_2H_5 + (C_2H_5)_3SnH \xrightarrow[20°]{24\ hrs.} (C_2H_5)_3M\text{-}Te\text{-}Sn(C_2H_5)_3 + C_2H_6$$

M = Si, Ge

Triethylgermyl lithium cleaved the tellurium-tellurium bond in diethyl ditelluride[117a]. Ethyl triethylgermyl telluride and ethyl lithium telluride were the products *(method g)*.

Ethyl lithium telluride, which was prepared from diethyl ditelluride and lithium in diethyl ether solution, reacted with chlorosilanes to give ethyl silyl tellurides[117a] (eq. 163a).

$$(163a)\quad nC_2H_5TeLi + Cl_nSiHR_{3-n} \longrightarrow (C_2H_5Te)_nSiHR_{3-n} + nLiCl$$

R, n, % yield, bp.(mm Hg): C_2H_5, 1 , 68%, 59-60°(5); C_2H_5, 2, 70%, 106°(3); –, 3, 32%, 125-130°(1); CH_3, 1, 48%, 56°(20)

Ethyl triethylgermyl telluride, which boiled at 42° at 1 mm Hg, was obtained in 89% yield from ethyl lithium telluride and triethylgermyl bromide[117a] in hexane solution.

Bis(triethylsilyl) telluride, treated with an equimolar amount of trifluoroacetic acid at 20° for one hour, underwent silicon-tellurium bond cleavage. Triethylsilyl hydrogen telluride was isolated as the major product[37] *(method h)* (eq. 164).

$$(164)\quad [(C_2H_5)_3Si]_2Te \xrightarrow{CF_3COOH} \underset{73\%}{(C_2H_5)_3Si\text{-}TeH} + \underset{15\%}{H_2/H_2Te} + \underset{10\%}{Te}$$

The tellurides, $(R_3M)_2Te$, exchanged the R_3M groups when heated with the hydrides $R_3M'H$ *(method i)*. The reactivity of the hydrides

(165) $$(R_3M)_2Te + 2R_3M'H \longrightarrow (R_3M')_2Te + 2R_3MH$$

M = Si, M' = Ge, 36 hrs. at 230°; M = Ge, M' = Sn, 1 hr. at 70°;

$R = C_2H_5$

with respect to these exchange reactions increased in the order Si<Ge<Sn[37,434]. Disilyl telluride, $(H_3Si)_2Te$, was converted quantitatively into the corresponding germanium compound when kept with an excess of germyl bromide[86] for 18 hours at 0°. Literature references are given in Table XI-1.

The tellurides, $(R_3M)_2Te$, in benzene or hexane solution, were transformed into the halides R_3MX by chlorine, bromine and iodine according to equation (166)[435]. The reaction of bis(trimethylsilyl)

(166) $$(R_3M)_2Te + X_2 \longrightarrow 2R_3MX + Te$$

M = Si, Ge, Sn, X = Cl, Br, I

telluride with silver iodide proceeded rapidly at room temperature. Trimethyliodosilane and silver telluride were isolated[181]. Vyazankin and coworkers[433] showed that tellurium is eliminated from the tellurides *(60)* upon treatment with acyl peroxides in benzene at room temperature (eq. 167).

(167) $$(R_3M)_2Te + R'\overset{O}{\overset{\|}{C}}\text{-OO-}\overset{O}{\overset{\|}{C}}\text{-}R'' \longrightarrow Te + R_3MO\overset{O}{\overset{\|}{C}}\text{-}R' + R_3MO\overset{O}{\overset{\|}{C}}\text{-}R''$$

$R = C_2H_5$; M, R', R": Si, C_6H_5, C_6H_5; Si, CH_3, C_6H_5; Sn, C_6H_5, C_6H_5

Bis(triethylgermyl) telluride and dicyclohexyl percarbonate produced triethyl cyclohexoxygermane, tellurium and carbon dioxide[433].

The tellurium atom in the tellurides $(R_3M)_2Te$ can be replaced by sulfur or selenium. Since the reactivity of the group VI elements in these exchange reactions decreases with increasing atomic number, sulfur can replace selenium and tellurium, and selenium can replace tellurium[37,434] (eq. 168).

$$(R_3M)_2Te + Y \longrightarrow (R_3M)_2Y + Te \tag{168}$$

$R = C_2H_5$; M, Y: Si, S, 12 hrs. at 170°; Sn, S, 90 hrs. at 20°; Sn, Se, 30 hrs. at 150°

Disilyl telluride decomposed easily on rough glass surfaces. The compound became brown on irradiation with blue light or upon heating to temperatures higher than 70°. With oxygen disilyl ether was formed[49]. Digermyl telluride gave with hydrogen iodide at room temperature germyl iodide and hydrogen telluride[86].

Egorochkin[117b] established by infrared spectroscopy, that the tellurium atom in the tellurides $(C_2H_5)_3M\text{-}Te\text{-}M'(C_2H_5)_3$ and $(C_2H_5)_3M\text{-}Te\text{-}C_2H_5$ (M, M' = group IV element) is hydrogen-bonded to the deuterium atom in deuteriochloroform[117b]. The gas chromatographic behavior of some of these tellurides has been investigated[38c].

Figure XI-1 summarizes the synthetic routes to and the reactions of the compounds containing a tellurium-group IV element bond.

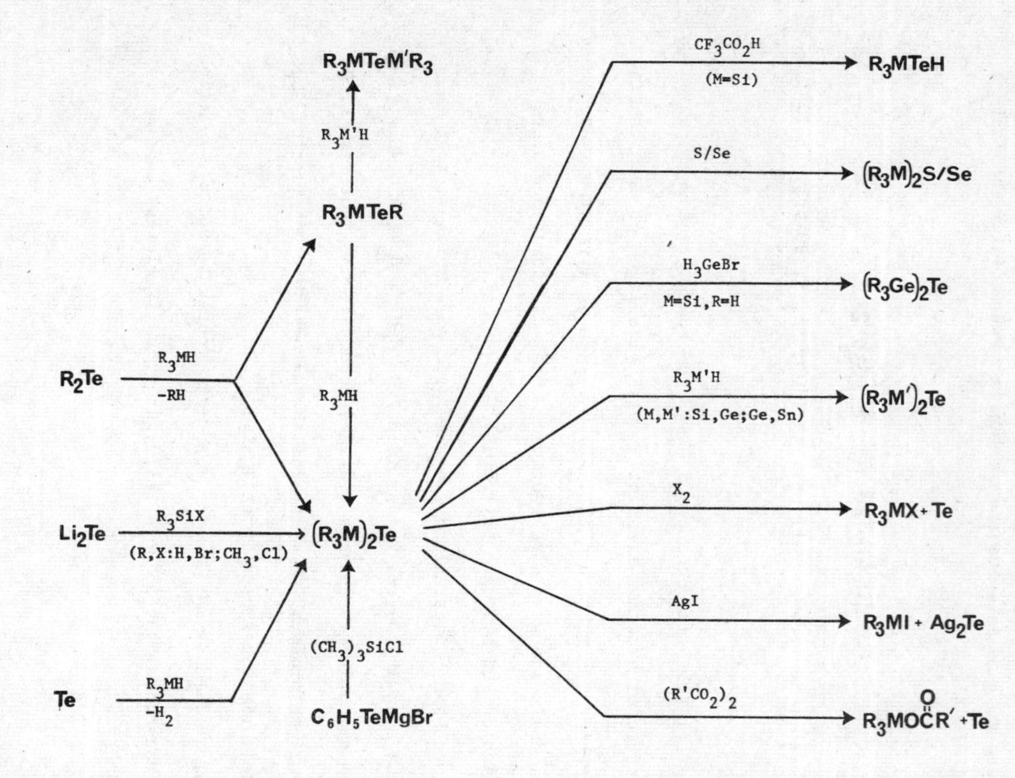

Fig. XI-1: Syntheses and Reactions of Compounds Containing Tellurium–Group IV Element Bonds

C) Organic Compounds Containing a Tellurium-Phosphorus or a Tellurium-Arsenic Bond

The first organic compound containing a tellurium-phosphorus bond was prepared by Foss[124] in 1950. He found, that tellurium dissolved in ethanolic solutions of potassium diethyl phosphite. Colorless, hygroscopic needles of potassium diethyl tellurophosphate were isolated, which darkened rapidly in moist air. Dilute acids or iodine liberated tellurium.

Elemental tellurium heated for four hours at 70-80° with diethyl allylphosphonite yielded diethyl (allyl)tellurophosphonate, an air, light and heat sensitive yellow liquid. It was purified by vacuum distillation. An attempt to isolate diethyl (ethyl)tellurophosphonate was unsuccessful[356]. Zingaro and coworkers[450,451] refluxed triorganylphosphines in anhydrous toluene with tellurium powder and obtained phosphine tellurides (see Table XI-2) as pale yellow, crystalline solids. Warming the solutions of these compounds in saturated aliphatic hydrocarbons resulted in the formation of shiny tellurium mirrors on the surface of the glass containers. The aminophosphine tellurides listed in Table XI-2 were prepared similarly[154]. A telluride could not be prepared from methyldichlorophosphine and tellurium[440]. The phosphine, $C[CH_2P(C_6H_5)_2]_4$, did not form a telluride, when heated at 180° with tellurium powder in the presence or absence of potassium cyanide in an evacuated bomb-tube[119].

Petragnani and de Moura Campos[347] reported that their attempts to prepare a compound with a P-Te-C bond had failed. In a recent short publication Russian investigators[396] claimed that such compounds can be synthesized from diphosphines and ditellurides as shown in equation (169). No further details with respect to the reaction conditions or individual compounds were given.

$$(169) \quad (CH_3)_2P\text{-}P(CH_3)_2 + R_2Te_2 \longrightarrow 2(CH_3)_2P\text{-}TeR$$

The dipole moment of $(C_8H_{17})_3PTe$ as the neat liquid at 50° was determined to be 5.95D. A value of 4.5D for the P-Te bond moment is given[38d].

Table XI-2

Organic Phosphorus Compounds Containing Tellurium

Compound	mp. °C bp. °C(torr)	Yield %	Reference
$(C_2H_5O)_2P(O)TeK$	--	--	124
$[(CH_3)_2N]_3PTe$	85	65.6	154,352a
$[(CH_3)_2N]_2(C_2H_5)PTe$	46-7	--	154
$[(CH_3)_2N]_2(C_6H_5)PTe$	67-8	--	154
$(C_2H_5O)_2(CH_2{=}CHCH_2)PTe$	65-6 (0.32)	37	356
$(CH_3)_2P\text{-}TeR$	--	--	395
$(C_2H_5)_3PTe$	76-8	--	451
$(C_3H_7)_3PTe$	41-2	--	451
$(C_4H_9)_3PTe$	35	--	450,451
$(c\text{-}C_6H_{11})PTe$	184-7	--	451
$(C_6H_5)(C_3H_7)_2PTe$	67-71	--	451
$(C_6H_5)(C_4H_9)_2PTe$	51-3	--	451

The only compound containing tellurium-arsenic bonds is the arsanthrene derivative *(61)*. It was prepared by treating the oxygen bridged derivative in ethanol with hydrogen telluride[7a].

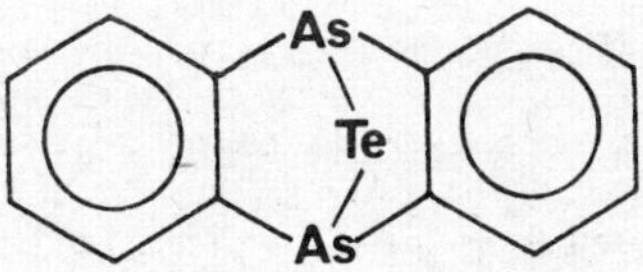

(61)

An X-ray structure analysis of *(61)* has been carried out[192b].

D) Organic Tellurium Compounds Containing a Tellurium-Selenium Bond

Tellurium bis(thioselenodiethylphosphinate) and tellurium bis(diselenodiethylphosphinate), $R_2P(Y)SeTeSe(Y)PR_2$ ($R = C_2H_5$, Y = S, Se), are the only compounds reported containing tellurium-selenium bonds and organic groups in the molecule, which have been isolated in pure form. They were prepared from the appropriate sodium phosphinate and tetravalent tellurium in hydrochloric acid solution[185]. They melted at 138-42 (dec.) and 141-3°, respectively. The tellurium atom in the thioseleno derivative is surrounded by two selenium and two sulfur atoms in an approximately planar arrangement. Only weak interactions exist between the tellurium atom and the two sulfur atoms belonging to two neighboring molecules[185a].

Pfisterer[348a] prepared dimethyl selenide telluride, CH_3-SeTe-CH_3, from Na_2SeTe and methyl iodide in liquid ammonia. Since this compound decomposed during vacuum distillation, it could not be separated from the by-product dimethyl diselenide.

E) Organic Tellurium Compounds as Ligands in Transition Metal Complexes

Diorganyl tellurides and diorganyl ditellurides form complexes with a variety of transition metal ions. These complexes have been prepared using metal carbonyls, metal carbonyl halides, salts of the type K_2MX_4 (M = Pd, Pt) or MX_n (M = Ag, Au, Cd, Hg) and the appropriate organic tellurium compound. Ligand exchange reactions in complexes already containing a tellurium ligand have also been employed. Diorganyl ditellurides experienced in most of these reactions tellurium-tellurium bond cleavage forming dinuclear complexes with bridging organyltelluro groups. The following transition metals have been investigated with respect to their tendency to form complexes with organic tellurium compounds.

Cr*	†Mn*	†Fe*	Co**	Ni**		
†Mo		†Ru*	Rh*	Pd*	Ag*	Cd*
W*	Re*			Pt*	Au*	††Hg*

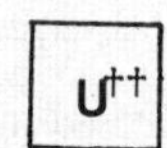

* Complexes with R_2Te

** Attempts to prepare complexes containing an organic tellurium compound were unsuccessful.

† Complexes containing RTe groups

†† Complexes with R_2Te_2

Table XI-3 lists the known complexes and their physical properties and gives pertinent literature references.

Octacarbonyldicobalt and diphenyl telluride gave only black $[Co_2Te(CO)_5]$. Tetracarbonylnickel precipitated metallic tellurium and nickel[176]. Hexacarbonylchromium did not react with diphenyl telluride. Hexacarbonylchromium did, however, react with bis(trimethylsilyl) telluride to give $[(CH_3)_3Si]_2Te{\cdot}Cr(CO)_5$. The corresponding tungsten complex was obtained similarly[190a].

Bis(π-cyclopentadienyltricarbonylmolybdenum), $[\pi\text{-cpMo(CO)}_3]_2$, reacted with diphenyl ditelluride to yield several complexes depending upon the reaction conditions[419]. Scheme (170) summarizes these reactions. The mononuclear complex was converted into the dinuclear species upon standing in solution at room temperature or by thermal decomposition. The completely decarbonylated, amorphous, insoluble complex *(62)* was not isolated.

Table XI-3

Organic Tellurium Compounds as Ligands in Transition Metal Complexes

Compound	Reactants	Conditions	mp. °C	Ref.
	Chromium			
$\{[(CH_3)_3Sn]_2TeCr(CO)_5\}$	$Cr(CO)_6 + R_2Te$	THF, irradiation (Hg lamp)	73(dec)	390a
	Molybdenum			
$[\pi\text{-}cpMo(CO)_3(TeC_6H_5)]$	$[\pi\text{-}cpMo(CO)_3]_2 + R_2Te_2$	benzene, 25°, ir lamp	80-2	419
$[\pi\text{-}cpMo(CO)_2(TeC_6H_5)]_2$	$[\pi\text{-}cpMo(CO)_3]_2 + R_2Te_2$	benzene, reflux, 14 hrs.	175-6	419
	$[\pi\text{-}cpMo(CO)_3(TeR)]$	thermal decomposition; on standing in solution	175-6	419
$[\pi\text{-}cpMo(TeC_6H_5)_2]_x$	$[\pi\text{-}cpMo(CO)_3]_2 + R_2Te_2$	xylene, reflux, 5hrs.	>190(dec)	419
	Tungsten			
$\{[(CH_3)_3Sn]_2TeW(CO)_5\}$	$W(CO)_6 + R_2Te$	THF, irradiation (Hg lamp)	71(dec)	390a
	Manganese			
$[Mn(CO)_4TeC_6H_5]_2$	$Mn_2(CO)_{10} + R_2Te$	xylene, 125°, 125 hrs.	--	176

Table XI-3 (cont'd)

$\{[(C_6H_5)_2Te]_2Mn(CO)_3\}Cl$	$Mn(CO)_5Cl + R_2Te$	ether,reflux,12 hrs.	--	176
$\{[(C_6H_5)_2Te]_2Mn(CO)_3\}Br$	$Mn(CO)_5Br + R_2Te$	ether,reflux,5 hrs.	--	176
$\{[(C_6H_5)_2Te]_2Mn(CO)_3\}I$	$Mn(CO)_5I + R_2Te$	ether,reflux,5 hrs.	--	176
$[(POT)_2Mn(CO)_3]Cl$*	$Mn(CO)_5Cl$ + POT	ethanol,50°	--	176
$[(C_6H_5)_2TeMn(NO)_3]$**	$[Mn(CO)_3TeR_2]Cl + NO$	benzene,20°,45 min.	--	176
$\{[(C_6H_5)_2Te]_2Mn(CO)_4\}^+ [Cr(SCN)_4(NH_3)_2]^-$	$[Mn(CO)_3(TeR_2)_2]Cl + CO$	benzene/$AlCl_3$,1 hr.	85(dec)	206
$[(C_6H_5)_2TeMn(CO)_3AAB]Br$*	$[Mn(CO)_3(TeR_2)_2]Br$ + AAB	benzene,reflux 6 hrs.	116(dec)	178
	<u>Rhenium</u>			
cis-$\{[(C_2H_5)_2Te]Re(CO)_4Cl\}$	$[Re(CO)_4Cl]_2 + 2R_2Te$	CCl_4	48	120a,c
fac-$\{[(C_2H_5)_2Te]_2Re(CO)_3Cl$	$[Re(CO)_4Cl]_2 + 4R_2Te$	CCl_4	42	120a,c
	<u>Iron</u>			
$[(C_6H_5)_2TeFe(CO)_4]$	$[Fe(CO)_4]_3 + R_2Te$	cyclohexane,reflux, 15 hrs.	--	176
$[(C_6H_5)_2TeFe(CO)_3Br_2]$	$Fe(CO)_4Br_2 + R_2Te$	CH_2Cl_2,-38°,2 hrs.	--	176
$[(C_6H_5)_2TeFe(CO)_3I_2]$	$Fe(CO)_4I_2 + R_2Te$	ether,10°,5 hrs.	--	176

Table XI-3 (Cont'd)

$[(C_6H_5)_2TeFe(CO)(NO)_2]$	$Fe(NO)_2(CO)_2 + R_2Te$	ether,20°,60 hrs.	--	176
$[Fe(CO)_3(TeC_6H_5)]_2$	$[Fe(CO)_4]_3 + R_2Te_2$	benzene,reflux	104-6(dec)	197,378b
$\{Fe(CO)_3[Te(4\text{-}CH_3OC_6H_4)]\}_2$	$[Fe(CO)_4]_3 + R_2Te_2$	benzene,reflux,5 hrs.	--	176
$[Fe(CO)_3(TeC_6F_5)]_2$	$[Fe(CO)_4]_3$ or $Fe(CO)_5 + R_2Te_2$	--	--	197,357
$[\pi\text{-}cpFe(CO)_2(TeC_6H_5)]^*$	$[\pi\text{-}cpFe(CO)_2]_2 + R_2Te_2$	benzene,reflux,3 hrs.	66	378a
$[\pi\text{-}cpFe(CO)(TeC_6H_5)]_2^{*\dagger}$	$[\pi\text{-}cpFe(CO)_2]_2 + R_2Te_2$	benzene,reflux	--	378a
$\{Fe(NO)_2[Te(4\text{-}CH_3OC_6H_4)]\}_2$	$[Fe(NO)_2(CO)_2] + R_2Te_2$	benzene,20°,few days	--	177
	<u>Ruthenium</u>			
$\{[(C_4H_9)_2Te]_3Ru(CO)I_2\}^{**}$	$[Ru(CO)_2I_2]_n + R_2Te$	benzene	--	178a
$\{[C_4H_9)_2Te]_2Ru(CO)_2I_2\}$	$[Ru(CO)_2I_2]_n + R_2Te$	benzene,120°,12 hrs.	68	178a
$\{[(C_6H_5)_2Te]_3Ru(CO)Cl_2\}^{**}$	$RuCl\cdot nH_2O + R_2Te$	ethanol + CO	177	178a
$\{[(C_6H_5)_2Te]_3Ru(CO)I_2\}^{**}$	$[Ru(CO)_2I_2]_n + R_2Te$	benzene	--	178a
$\{[(C_6H_5)_2Te]_2Ru(CO)_2Cl_2\}\cdot 1/2CH_2Cl_2$	$RuCl_3\cdot nH_2O + R_2Te$	ethanol + CO	168	178a
$\{[(C_6H_5)_2Te]_2Ru(CO)_2Br_2\}$	$[Ru(CO)_2Br_2]_n + R_2Te$	benzene,80°,6 hrs.	215	178a
$\{[(C_6H_5)_2Te]_2Ru(CO)_2I_2$	$[Ru(CO)_2I_2]_n + R_2Te$	benzene,130°,8 hrs.	238	178a
$[Ru(CO)_2(TeC_6H_5)_2]_n$	$Ru_3(CO)_{12} + R_2Te_2$	benzene,60°,4 hrs.	>200(dec)	378b
$[Ru(CO)_3(TeC_6H_5)]_2$	$Ru_3(CO)_{12} + R_2Te_2$	benzene,60°,4 hrs.	105-7(dec)	378b

Table XI-3 (Cont'd)

	Cobalt			
$[Co_2(CO)_5Te]_n$	$[Co(CO)_4]_2 + Te(C_6H_5)_2$	benzene or acetone, 20°, 40 hrs.	--	176
	Rhodium			
trans-$\{[(C_2H_5)_2Te]_2Rh(CO)Cl\}$	$[Rh(CO)_2Cl_2]_2 + R_2Te$	pentane	30	120b
$\{[(C_2H_5)_2Te]_2Rh(CO)Cl_3\}$	$[L_2Rh(CO)Cl] + Cl_2$	pentane	64-75(dec)	120b
	$[L_2Rh(CO)Cl] + HCl$	pentane	--	120b
$\{[(C_2H_5)_2Te]_2Rh(CO)ClBr_2\}$	$[L_2Rh(CO)Cl] + Br_2$	pentane	>30(dec)	120b
$\{[(C_2H_5)_2Te]_2Rh(CO)ClI_2\}$	$[L_2Rh(CO)Cl] + I_2$	CH_2Cl_2	62-5	120b
$\{[(C_2H_5)_2Te]_2Rh(CO)(CH_3)ClI\}$	$[L_2Rh(CO)Cl] + CH_3I$	room temperature	71-8(dec)	120b
$\{[(C_2H_5)_2Te]_2Rh(CO)(CH_3CO)ClI\}$	$[L_2Rh(CO)(CH_3)ClI] + CO$	CH_2Cl_2	oil	120b
$\{[(C_2H_5)_2Te]_2Rh(CO)(C_6H_5SO_2)Cl_2\}$	$[L_2Rh(CO)Cl] + C_6H_5SO_2Cl$	benzene	>71(dec)	120b
	Palladium			
trans-$(CH_3)_2TePdI_2$	$K_2PdI_4 + R_2Te$	water	--	8a,9
trans-$[(CH_3)_2TePdCl_2]_2$	$K_2PdCl_4 + R_2Te$	water	--	8a,9
trans-$[(CH_3)_2TePdBr_2]_2$	$K_2PdBr_4 + R_2Te$	water	--	8a,9
cis-$[(C_2H_5)_2Te]_2PdCl_2$	$(NH_4)_2PdCl_4 + R_2Te$	water	99	72

Table XI-3 (Cont'd)

trans-$[(C_2H_5)_2TePdCl_2]_2$	Na_2PdCl_4 + R_2Te	ethanol	110-25(dec)	72
	$(R_2Te)_2PdCl_4$ + R_2Te	ethanol	110-25(dec)	72
trans-$[(C_2H_5)_2Te]_2PdCl(C_6H_5)$	*trans*-L_2PdCl_2 + RMgCl	ether,-78°	32-3	392c
trans-$[(C_2H_5)_2Te]_2PdCl(4\text{-}ClC_6H_4)$	*trans*-L_2PdCl_2 + RMgCl	ether,-78°	99-106	392c
trans-$[(C_2H_5)_2Te]_2PdBr(C_6H_5)$	*trans*-L_2PdBr_2 + RMgBr	ether,-78°	88-92	392c
trans-$[(C_2H_5)_2Te]_2PdBr(2\text{-}CH_3C_6H_4)$	*trans*-L_2PdBr_2 + RMgBr	ether	97-102	392c
trans-$[(C_2H_5)_2Te]_2PdBr[2,4,6\text{-}(CH_3)_3C_6H_2]$	*trans*-L_2PdBr_2 + RMgBr	ether,-78°	97-100	392c
trans-$[(C_2H_5)_2Te]_2PdBr(4\text{-}FC_6H_4)$	*trans*-L_2PdBr_2 + RMgBr	ether,-78°	110-5	392c
trans-$[(C_2H_5)_2Te]_2PdBr(2\text{-}ClC_6H_4)$	*trans*-L_2PdBr_2 + RMgBr	ether,-78°	69-71	392c
trans-$[(C_2H_5)_2Te]_2PdBr(4\text{-}ClC_6H_4)$	*trans*-L_2PdBr_2 + RMgBr	ether,-78°	112-24	392c
trans-$[(C_2H_5)_2Te]_2PdI(C_6H_5)$	*trans*-L_2PdI_2 + RMgI	ether,-78°	87-90	392c
trans-$[(C_2H_5)_2Te]_2PdI(4\text{-}CH_3C_6H_4)$	*trans*-L_2PdI_2 + RMgI	ether,-78°	87	392c
trans-$[(C_2H_5)_2Te]_2Pd(CNS)(C_6H_5)$	*trans*-$L_2PdBr(C_6H_5)$ + KCNS	--	--	392c
trans-$[(C_3H_7)_2TePdCl_2]_2$	Na_2PdCl_4 + R_2Te	ethanol	132(dec)	72
	Platinum			
cis-$[(CH_3)_2Te]_2PtCl_2$	K_2PtCl_4 + R_2Te	water	--	8a,9
trans-$[(CH_3)_2Te]_2PtBr_2$	K_2PtBr_4 + R_2Te	water	--	8a,9

Table XI-3 (Cont'd)

trans-$[(CH_3)_2Te]_2PtI_2$	$K_2PtI_4 + R_2Te$	water	--	8a,9
cis-$[(C_2H_5)_2Te]_2PtCl_2$	$K_2PtCl_4 + R_2Te$	ethanol,20°	--	68,187
trans-$[(C_2H_5)_2Te]_2PtCl(C_6H_5)$	*trans*-$L_2PtBr(C_6H_5) + LiCl$	CH_3OH	60-4	392b
trans-$[(C_2H_5)_2Te]_2PtBr(C_6H_5)$	*cis*-$L_2PtCl_2 + C_6H_5MgBr$	boiling ether	83	392b
trans-$[(C_2H_5)_2Te]_2Pt(2\text{-}CH_3C_6H_4)_2$	*cis*-L_2PtCl_2 + RMgBr	ether/THF	86-92	392b
trans-$[(C_2H_5)_2Te]_2PtCl[2,4,6(CH_3)_3C_6H_2]$	*trans*-$L_2PtBr(R) + LiCl$	CH_3OH	111-6	392b
trans-$[(C_2H_5)_2Te]_2PtBr(2,4,6(CH_3)_3C_6H_2]$	*cis*-L_2PtCl_2 + RMgBr	ether	111-4	392b
cis-$[(C_3H_7)_2Te]_2PtCl_2$	$K_2PtCl_4 + R_2Te$	ethanol,20°	--	68
trans-$(C_2H_5)_2Te$,piperidine $PtCl_2$	*trans*-$[R_2TePtCl_2]_2$ + piperidine	acetone	61	69
trans-$[(C_2H_5)_2TePtCl_2]_2$	*cis*-$(R_2Te)_2PtCl_2 + Na_2PtCl_4$	ethanol,20°, few days	142(dec)	68
trans-$[(C_3H_7)_2TePtCl_2]_2$	$R_2Te + Na_2PtCl_4$	ethanol,20°,16 hrs.	120-31(dec)	68
cis-$[(C_6H_5CH_2)_2Te]_2PtCl_2$	K_2PtCl_4 or $[(C_3H_7)_3NH]_2PtCl_4 + R_2Te$	ethanol/water or absolute ethanol	>200	137,138
cis-$[(C_6H_5)_2Te]_2PtCl_2$	$K_2PtCl_4 + R_2Te$	ethanol/water	~200(dec)	187
	Silver			
$[(CH_3)_2Te]_2\cdot AgI$	$2R_2Te + AgI$	acetone/water/KI	73-4	80
[illegible]	[illegible]	[illegible]	[illegible]	[illegible]

Table XI-3 (Cont'd)

	Uranium			
$UCl_5 \cdot (C_6H_5)_2Te_2$	$UCl_5 \cdot Cl_2C{=}CClC(O)Cl + R_2Te_2$	benzene	151(dec)	392,392a

* POT = phenoxtellurine, AAB = 4-aminoazobenzene, cp = cyclopentadienyl

** not isolated

† mixture of two isomers

$$\text{(170)}\quad [\pi\text{-cpMo(CO)}_3]_2 \begin{cases} \xrightarrow[+\ (C_6H_5)_2Te_2]{\text{benzene, reflux, 14 hrs.}} [\pi\text{-cpMo(CO)}_2\text{TeR}]_2 \\ \xrightarrow[+\ (C_6H_5)_2Te_2]{\text{benzene, 25°, 2.5 hrs.}} [\pi\text{-cpMo(CO)}_3\text{TeR}] \\ \xrightarrow[+\ (C_6H_5)_2Te_2]{\text{toluene, reflux, 12 hrs.}} [\pi\text{-cpMo(TeR)}_2]_x \end{cases}$$

(62)

Bis(pentacarbonylmanganese) reacted slowly with diphenyl telluride at 125° in xylene (eq. 171) to give di-μ-phenyltellurobis-(tetracarbonylmanganese)*(63)*.

$$\text{(171)}\quad [Mn(CO)_5]_2 + 2(C_6H_5)_2Te \longrightarrow [Mn(CO)_4(TeC_6H_5)]_2 + (C_6H_5)_2 + 2CO$$

The manganese atoms in this diamagnetic compound, which contains two bridging phenyltelluro groups, are octahedrally coordinated.

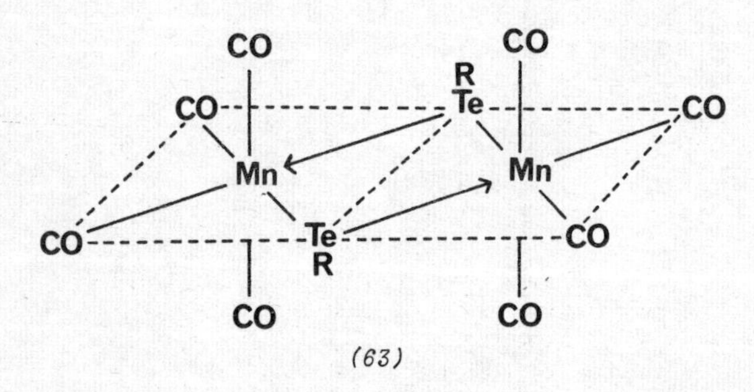

(63)

The dipole moment of 1.45 ± 0.09D does not allow to make a definite statement concerning the *cis*- or *trans*- arrangement of the two phenyl groups[176].

Two carbonyl groups are replaced by diphenyl telluride ligands, when pentacarbonylmanganese halides are refluxed in diethyl ether with diphenyl telluride (eq. 172) or phenoxtellurine.

(172) $$Mn(CO)_5X + 2(C_6H_5)_2Te \longrightarrow Mn(CO)_3(TeR_2)_2X + 2CO$$

X = Cl, Br, I

The reactivity of the carbonyl halides decreases in the sequence Cl>Br>I. The diamagnetic compounds are non-electrolytes, thermally stable and resistent to atmospheric agents. The strong *trans*-effect of the carbonyl groups makes it likely that two equatorial CO ligands in the pentacarbonylmanganese halide have been replaced. This leads to the formulation of the telluride containing complexes with two CO groups *trans* to the telluride ligands as shown in structure *(64)*.

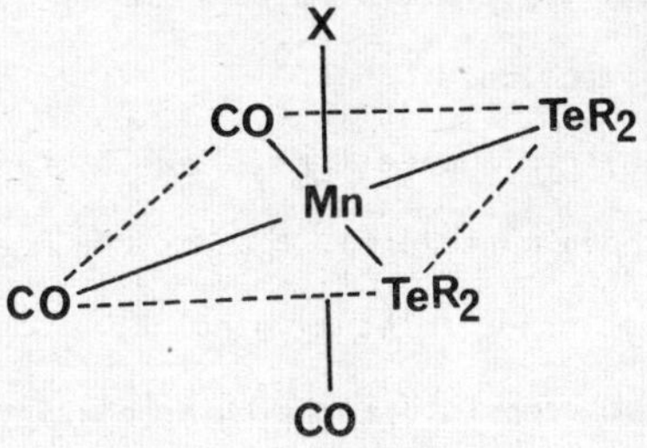

X
CO TeR$_2$
Mn
CO TeR$_2$
CO

(64)

The arrangement of the ligands in these complexes, however, could not be ascertained by infrared investigations[176].

It was not possible to replace only one or more than two carbonyl groups. The diaryl telluride complexes are stabilized by $(Mn)d_{\pi} \rightarrow (Te)d_{\pi}$ bonding made possible by the electron- withdrawing phenyl groups. The electron-donating groups in dialkyl tellurides increase the electron density on the tellurium atom and prevent $d_{\pi}-d_{\pi}$ bonding. The dibutyl telluride adduct - in line with these considerations - could not be isolated. Pentacarbonylmanganese chloride, when refluxed in ether in the presence of dibutyl telluride, quantitatively dimerized with concomitant loss of carbon monoxide (eq. 173).

$$(173)\quad 2\,Mn(CO)_5Cl + 2(C_4H_9)_2Te \rightleftarrows 2\,Mn(CO)_4Te(C_4H_9)_2Cl + 2CO$$
$$2\,Mn(CO)_4Te(C_4H_9)_2Cl \rightarrow 2(C_4H_9)_2Te + [Mn(CO)_4Cl]_2$$

The dibutyl telluride adduct was only of transitory existence[176].

The complex $Mn(CO)_3[Te(C_6H_5)_2]_2Cl$ took up one carbonyl ligand when carbon monoxide was bubbled through a benzene solution of this complex containing aluminum chloride. The cationic manganese complex was isolated as the salt $\{Mn(CO)_4[Te(C_6H_5)_2]_2\}^+[Cr(SCN)_4(NH_3)_2]^-$. In the presence of zinc chloride at 70° under 300 atm. carbon monoxide pressure a mixture of salts containing the two cations $\{Mn(CO)_4(TeR_2)_2\}^+$ and $[Mn(CO)_5TeR_2]^+$ was the reaction product[206].

Carbonyl and diphenyl telluride ligands can be replaced by nitrosyl groups and organic nitrogen containing ligands. When tricarbonylbis(diphenyl telluride)manganese(I) chloride in benzene solution was treated with nitrogen monoxide at room temperature all three carbonyl groups and one telluride ligand were exchanged according to equation (174).

$$(174)\quad 2\,Mn(CO)_3[Te(C_6H_5)_2]_2Cl + 6NO \longrightarrow 2\,Mn(NO)_3Te(C_6H_5)_2 + 6CO + (C_6H_5)_2Te + (C_6H_5)_2TeCl_2$$

The green nitrosyl complex could not be isolated in pure form, because it had the same solubility properties as diphenyl telluride[176]. Tricarbonylbis(diphenyl telluride)manganese(I) bromide exchanged both telluride ligands for amine ligands in reactions with pyrrolidine, piperidine, morpholine, piperazine, benzidine, hydrazobenzene, phenylhydrazine and benzaldehyde phenylhydrazone forming mononuclear or dinuclear manganese complexes. Only 4-aminoazobenzene gave a mononuclear, telluride containing complex[178] (eq. 175).

(175) $Mn(CO)_3[Te(C_6H_5)_2]_2Br + L \longrightarrow Mn(CO)_3[Te(C_6H_5)_2](L)Br + (C_6H_5)_2Te$

L = 4-aminoazobenzene

Infrared data[176,178], magnetic susceptibilities[176] and dipole moments[176] of these complexes were determined. Experiments with ^{14}CO in benzene at 30° showed, that the carbonyl groups in the complexes $Mn(CO)_3[Te(C_6H_5)_2]_2X$ exchanged with rate ratios of 15:7:1 for X = Cl, Br, I[175].

The halogen bridges in μ-dichlorobis[tetracarbonylrhenium(I)] are split by diethyl telluride to give *cis*- $[(C_2H_5)_2TeRe(CO)_4]$ and *fac*-$\{[(C_2H_5)_2Te]_2ClRe(CO)_3\}$ depending on the molar ratios of the reagents employed[120a].

Iron carbonyl complexes with organic tellurium compounds as ligands were prepared from $Fe_3(CO)_{12}$[176,357,378b], $Fe(CO)_5$[357], $Fe(CO)_4I_2$[176], $Fe(CO)_4Br_2$[176], $Fe(NO)_2(CO)_2$[177] and $[\pi\text{-cpFe(CO)}_2]_2$[378a]. Diphenyl telluride[176], diphenyl ditelluride[197,378a,b], bis(pentafluorophenyl) ditelluride[197,357] and bis(4-methoxyphenyl) ditelluride[176] were employed as the tellurium containing reactants. Diphenyl telluride replaced only one carbonyl group in reactions with the tetracarbonyliron(II) halides. The infrared data did not reveal the position of the telluride ligand in the octahedral molecule[176]. Tris(tetracarbonyliron) formed with diphenyl telluride the trigonal-bipyramidal complex $Fe(CO)_4[Te(C_6H_5)_2]$[176]. With ditellurides dinuclear, aryltelluro-bridged complexes $[Fe(CO)_3TeR]_2$ were isolated[176,197,357,378b]. Dinitrosyldicarbonyliron and diphenyl telluride in ether solution produced $Fe(NO)_2(CO)[Te(C_6H_5)_2]$[176], while bis(4-methoxyphenyl) ditelluride gave the dinuclear complex $[Fe(NO)_2(TeR)]_2$[177]. Bis[cyclopentadienyldicarbonyliron] and diphenyl ditelluride produced the mononuclear complex $[\pi\text{-cpFe(CO)(TeC}_6H_5)]$ upon refluxing in benzene. When the reaction mixture was irradiated with visible light and infrared radiation two isomers of the formula $[\pi\text{-cpFe(CO)(TeC}_6\text{H}_5)]_2$ were obtained, which could not be separated[378a].

Ruthenium carbonyl halides and tris(tetracarbonylruthenium) have been found to react with diorganyl tellurides and diphenyl ditelluride, respectively. When $[\text{Ru(CO)}_2\text{X}_2]_n$ and diphenyl telluride or dibutyl telluride were heated in benzene the complexes $[(\text{R}_2\text{Te})_2\text{Ru(CO)}_2\text{X}_2]$ (X = Br, I) were isolated, in which the telluride ligands are in *trans* position to each other. The halogens occupy positions *trans* to the carbonyl groups. The corresponding chloro complex was obtained from hydrated ruthenium trichloride, diphenyl telluride and carbon monoxide in ethanolic solution. The presence of trisubstituted derivatives $[(\text{R}_2\text{Te})_3\text{Ru(CO)X}_2]$ was detected by ir spectroscopy[178a]. Benzene solutions of tris(tetracarbonylruthenium) and diphenyl ditelluride were heated at 60°. The complexes $[\text{Ru(CO)}_3(\text{TeC}_6\text{H}_5)]_2$ and $[\text{Ru(CO)}_2(\text{TeC}_6\text{H}_5)_2]_n$ were formed. The polymeric compounds isolated from these systems had molecular masses corresponding to values of n = 6-7 or 12-14[378a].

The reactions of diethyl telluride with rhodium complexes, which lead to the formation of telluride containing derivatives[120b], are summarized in scheme (176).

Dimethyl[8a,9], diethyl[68,69,72,187,392b,c], dipropyl[68,72], dibenzyl[135,137,1 and diphenyl[187] tellurides form complexes with palladium(II) and platinum(II) halides (see Table XI-3). Chatt and coworkers[67-75] in their investigations of the *trans*-effect in inorganic complexes prepared mono- and dinuclear palladium and platinum complexes, $(\text{R}_2\text{Te})_2\text{MCl}_2$ and $[(\text{R}_2\text{Te})\text{MCl}_2]_2$ (M = Pd,Pt), and mixed platinum complexes of the type $(\text{R}_2\text{Te})(\text{am})\text{PtCl}_2$ (am = amine). A general outline of the reactions leading to these complexes is presented in scheme (177).

(176)

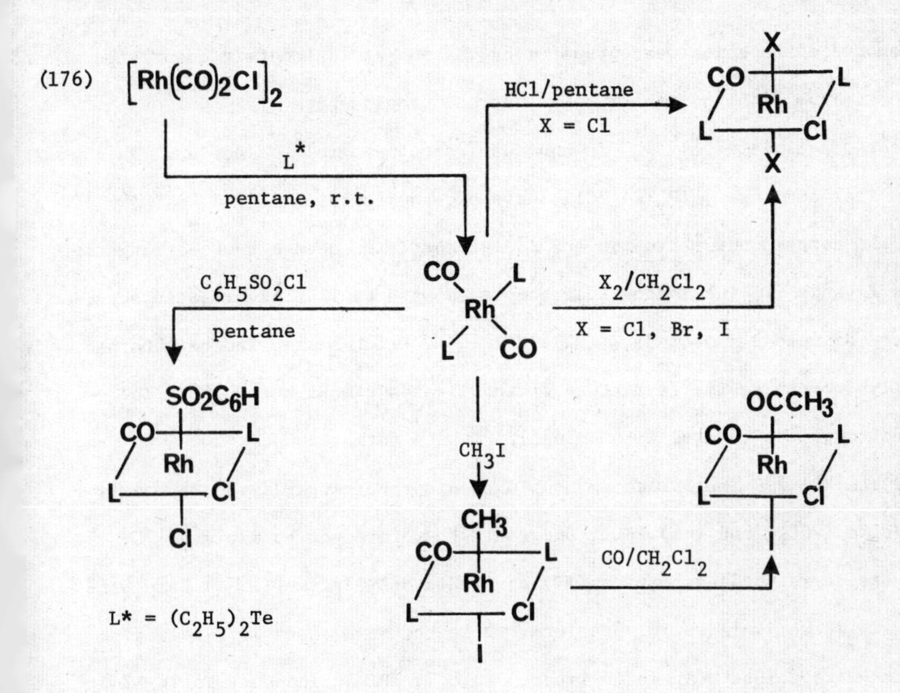

L* = $(C_2H_5)_2Te$

(177)

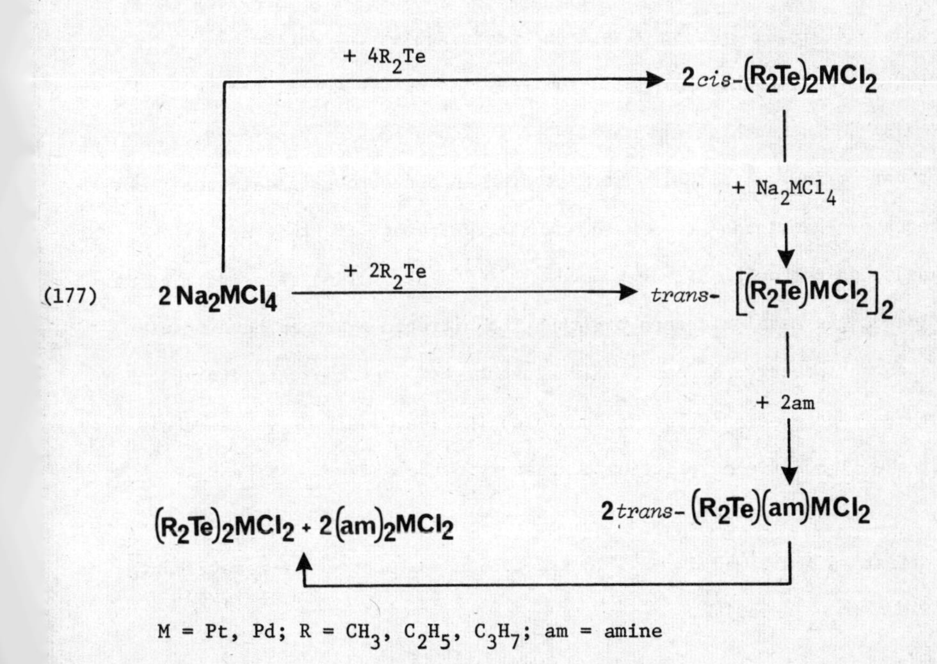

M = Pt, Pd; R = CH_3, C_2H_5, C_3H_7; am = amine

The mononuclear species were prepared by shaking stoichiometric quantities of the ligands with the alkali metal tetrachloroplatinate(II) [8a,68] or tetrachloropalladate(II)[8a,72] in aqueous[8a] or ethanolic[68,72] medium. The mononuclear complexes, $(R_2Te)_2MCl_2$, have *cis*-configuration[8a,68,72,137,138,187], while the corresponding bromide and iodide complexes have a *trans*-arrangement of the ligands[8a]. The mononuclear complexes were easily converted in solution into the dinuclear compounds, $[(R_2Te)MCl_2]_2$[68], which could also be obtained directly from an equimolar mixture of the ligands and Na_2MCl_4[8a,72]. The bridged complexes reacted with amines[69,73] with formation of *trans*-$[(R_2Te)(am)MCl_2]$. The palladium complexes disproportionated spontaneously into $(R_2Te)_2PdCl_2$ and $(am)_2PdCl_2$ and could therefore not be isolated[73]. Among the corresponding platinum(II) complexes only *trans*-$[(C_2H_5)_2Te$(piperdine)$PtCl_2]$ was isolated in pure form, while the more unstable *p*-toluidine complex was obtained only in an impure state[69]. Please consult Table XI-3 for individual compounds. The stability of the complexes $L_2Pt_2Cl_4$ for ligands containing a group VI element decreases in the series[71,72] $R_2O \lll R_2S \gg R_2Se < R_2Te$. For palladium the sequence is $R_2S > R_2Se > R_2Te$[71].

The N-H stretching frequencies of the complexes with primary or secondary amines, $L(am)MCl_2$, were studied in order to elucidate the effects of the ligands (L = R_2Te and a large number of other neutral ligands) on the metal-nitrogen bond[67,70,71,73,74]. The telluride ligands influence the metal-nitrogen and thus the nitrogen-hydrogen bond mainly through an inductive effect[71]. The investigation of the ultraviolet and visible spectra of a series of complexes [L(piperdine)$PtCl_2$] indicated, that the ligand field splitting decreases in the order $R_2S > R_2Se > R_2Te$. For the diethyl telluride compound the maximum of absorption caused by the first spin-allowed $d_{xy} \rightarrow d_{x^2-y^2}$ transition was located at a wavelength of 3400Å (ε_{max} = 350)[75].

Palladium complexes containing diethyl telluride ligands and an aryl group σ-bonded to the central metal atom have been synthesized from *trans*-{$[(C_2H_5)_2Te]_2PdX_2$} and the appropriate Grignard reagent[392c] in ether solution (eq. 178).

$$(178)\quad trans\text{-}\left(\left[(C_2H_5)_2Te\right]_2PdX_2\right) + RMgX \longrightarrow trans\text{-}\left(\left[(C_2H_5)_2Te\right]_2PdX(R)\right) + MgX_2$$

R, X: C_6H_5, Cl; 4-ClC_6H_4, Cl; H, Br; 2-$CH_3C_6H_4$, Br; 2,4,6-$(CH_3)_3C_6H_2$, Br; 4-FC_6H_4, Br; 2-ClC_6H_4, Br; 4-ClC_6H_4, Br; H, I; 4-$CH_3C_6H_4$, I

Treatment of the complex *(65)* (X = Br, R = C_6H_5) with KCNS gave the corresponding thiocyanate derivative[392c].

Analogous platinum complexes, *trans*-{$[(C_2H_5)_2Te]_2PtBr(R)$} were prepared from *cis*-{$[(C_2H_5)_2Te]\ PtCl_2$} and 2-methylphenylmagnesium bromide phenylmagnesium bromide or 2,4,6-trimethylphenylmagnesium bromide in boiling ether. The latter two complexes were converted to the chloro derivatives by treatment with lithium chloride in methanol[392b].

Coates[80] prepared the adducts $(R_2Te)\cdot 2AgI$ and $[(R_2Te)_2Ag]I$ (R = CH_3) by mixing acetone solutions of the telluride with solutions of silver iodide in nearly saturated aqueous potassium iodide. Dimethyl telluride was released on heating $[(R_2Te)_2Ag]I$ in vacuum. The formulation of this compound as shown was suggested by the fact, that silver iodide was precipitated upon addition of silver nitrate to the acetone solution of the complex. Gold(III) chloride, reacted with diphenyl telluride[147].

Cadmium iodide combined also with dimethyl telluride. The adduct was not analyzed[80].

Elemental mercury, shaken with a benzene solution of diphenyl ditelluride, formed bis(phenyltelluro) mercury[323d] in 83 per cent yield. Mercury(II) compounds coordinate with organic tellurium compounds. Diethyl tellurium diiodide gave with diphenyl mercury in chloroform solution the complex $R_2TeI_2\cdot Hg(C_6H_5)_2$, which melted at 94°.

Bis(4-methylphenyl) mercury was unreactive[144]. Diphenyl ditelluride[150] and bis(4-methoxyphenyl) ditelluride[293] formed mercuric chloride and iodide adducts, respectively.

A large number of mercuric halide complexes with diorganyl tellurides, $R_2Te \cdot HgX_2$ (X = Cl, Br, I) have been prepared by Lederer (see Table XI-4). These crystalline adducts were employed in the isolation and identification of diorganyl tellurides. The ethereal solutions of the tellurides were shaken with aqueous solutions of the mercuric halides, or solutions of the components in acetone or ethanol were combined. The adducts precipitated as crystalline solids. Some of these derivatives, however, were reported to be amorphous. The products were purified by recrystallization from ethanol, acetone, benzene or glacial acetic acid. The melting of the analytically pure products was generally preceeded by a 5-10° softening range. Some derivatives upon heating were slowly converted to viscous oils. Bis(2,4,6-trimethylphenyl) telluride did not form mercuric halide adducts[224]. Lederer[220] did not succeed in preparing adducts of tellurides with basic mercuric cyanide, thiocyanate and sulfate. Basic mercuric nitrate, however, produced the adducts $(2\text{-}CH_3C_6H_4)_2Te \cdot Hg(OH)NO_3$, which melted at 98-9°, and the oily $(C_6H_5)_2Te \cdot Hg(OH)NO_3$. The stereochemistry of these complexes is unknown. The diorganyl tellurides can be regenerated from the adducts by their treatment with sodium hydroxide solution[35].

Diphenyl ditelluride replaced the organic ligand in the complex $UCl_5 \cdot Cl_2C{=}CClCOCl$ in benzene solution. The compound $UCl_5 \cdot (C_6H_5)_2Te_2$ precipitated as a violet-black powder[392,392a].

Table XI-4

Diorganyl Telluride-Mercuric Halide Adducts, $RR'Te \cdot HgX_2$

R	R'	X = Cl		X = Br		X = I	
		mp. °C	Ref.	mp. °C	Ref.	mp. °C	Ref.
methyl	methyl	179(dec)	35,38	160-1(dec)	58	107-8(dec)	58,149,150
pentyl	pentyl	impure	18	88	18	viscous liquid	18
phenyl	methyl	132	150	124-5	150	89-90	150
phenyl	phenyl	160-1	147,218	148	220	146	220
	2-methylphenyl	--	--	--	--	133-4	238
	4-methylphenyl	91	229	54	229	74	229
2-methoxyphenyl	2-methoxyphenyl	143-4	236	84	236	80-1	236
3-methoxyphenyl	3-methoxyphenyl	89	235	114-5(dec)	235	122	235
4-methoxyphenyl	4-methoxyphenyl	90	220,226	77-8	220,226	63	226
2-ethoxyphenyl	2-ethoxyphenyl	174-5	232	160-1	232	90	232
4-ethoxyphenyl	4-ethoxyphenyl	150-1	234	155-6	234	123-4	234
2-methylphenyl	2-methylphenyl	212	218	199-200	220	142-3	220
3-methylphenyl	3-methylphenyl	116-7	225	53	225	indefinite	225

Table XI-4 (cont'd)

4-methylphenyl	4-methylphenyl	135-6*	218	85	220	65	220
2,4-dimethylphenyl	2,4-dimethylphenyl	106	223	99	223	107-8	223
2,5-dimethylphenyl	2,5-dimethylphenyl	179-80	223	169-70	223	166-7	223
1-naphthyl	1-naphthyl	187-8(dec)	233	178-9	233	152-3	233
2-thienyl	2-thienyl	dec.	205	--	--	--	--
cyclo-tetramethylene		146-7	298	--	--	--	--
2,2'-biphenylylene		248-50	173	--	--	--	--

* $R_2Te \cdot HgCl_2 \cdot 6C_2H_5OH$

XII. HETEROCYCLIC TELLURIUM COMPOUNDS

Heterocyclic tellurium compounds, in which the tellurium atom is part of a five- or six-membered ring system, have been synthesized. The single attempt to prepare telluracyclobutane was unsuccessful[121]. Morgan and coworkers investigated in great detail the telluracyclopentane and telluracyclohexane systems in the years between 1920 and 1930. Tellurophene, benzotellurophene, and dibenzotellurophene and their derivatives have received little attention. The easily synthesized phenoxtellurine has been investigated in more detail.

The transient molecules telluracyclopropane and methyltelluracyclopropane were detected by uv-spectroscopy and mass spectroscopy in reaction mixtures obtained by flash photolysis of dimethyl telluride with ethylene and propene, respectively[83a]. Epitellurobenzene was found as a fragment in the mass spectrum of benzotellurophene[51a].

A) Five-membered Heterocyclic Tellurium Compounds

Telluracyclopentane, tellurophene, benzotellurophene and dibenzotellurophene comprise the five-membered, heterocyclic tellurium compounds.

1) Telluracyclopentane and its derivatives

The telluracyclopentane ring system can be synthesized employing 1,4-diiodobutane and elemental tellurium[112,298] or 1,4-dibromobutane and sodium telluride[121] (eq. 179) or aluminum telluride[298].

(179) $$Br(CH_2)_4Br + Na_2Te \longrightarrow \text{(66)} + 2NaBr$$

(66): telluracyclopentane ring — H_2C_4—$_3CH_2$, H_2C^5—$^2CH_2$, Te^1

To the sodium telluride prepared from tellurium, sodium hydroxide and Rongalite[420] in aqueous medium was added a slight excess of the dibromide. The telluracyclopentane formed was reacted with bromine and isolated as the dibromide[121]. Aluminum telluride, Al_2Te_3, (0.07 moles) reacted exothermically with 1,4-dibromobutane (0.30 moles) at 125° in the absence of a solvent (eq. 180). Upon extraction of the black semisolid reaction mixture with acetone and ethanol *cyclo*tetramethylene 4-bromobutyl telluronium bromide *(73)* was obtained, while the aqueous extract deposited tetramethylene bis(*cyclo*tetramethylene telluronium bromide) *(75)*. Telluracyclopentane, which is possible primary product, might have formed the telluronium salts *(73)* and *(75)* with excess bromide.

(180)

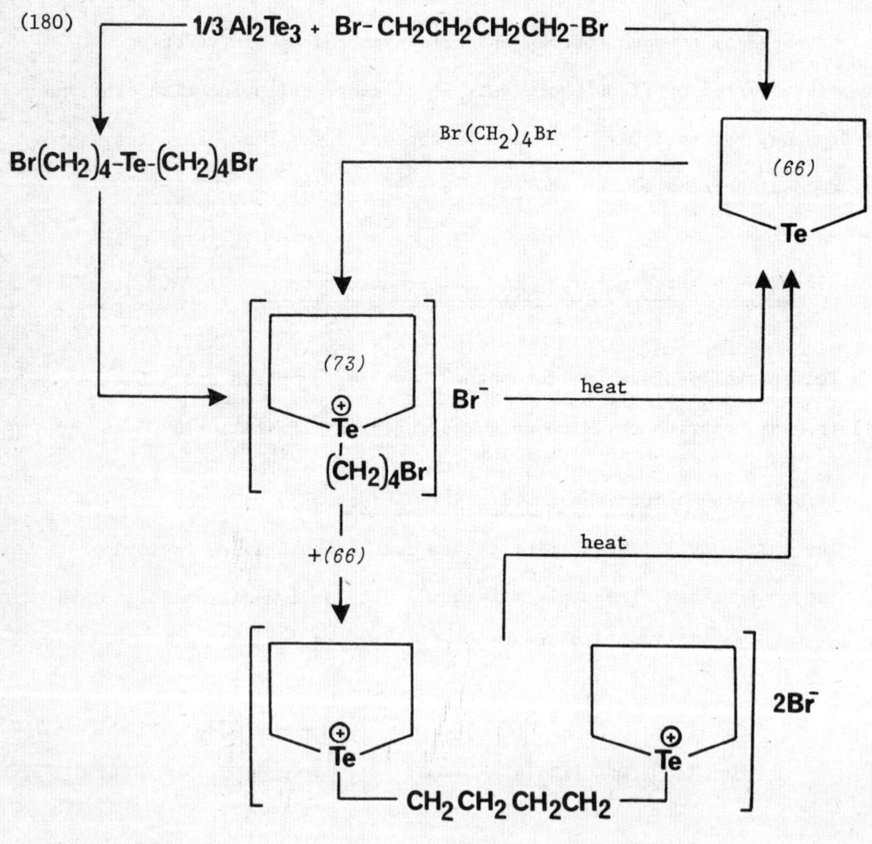

The telluronium bromides *(73)* and *(75)* were converted into the corresponding iodides *(74)* and *(76)* by treatment with potassium iodide. The telluronium salts *(73)* and *(75)* dissociated thermally above 153° to telluracyclopentane *(66)* and 1,4-dibromobutane[298]. Telluracyclopentane is best prepared by reduction of its diiodide *(67)* with sulfur dioxide[298]. The diiodide *(67)* can be obtained in quantitative yield from elemental tellurium and 1,4-diiodobutane[112,298]. Telluracyclopentane reacted in the same manner as the non-cyclic diorganyl tellurides. With elemental halogens dihalides were formed. The dichloride and dibromide dissolved in warm water forming acidic solutions, from which the original compounds crystallized unchanged on cooling. Table XII-1 summarizes the preparative methods and the reactions of telluracyclopentane compounds.

Buchta and Greiner[47,48] prepared ditellura-3,3'-bicyclopentyl *(78)* from 1,2,3,4-tetra(iodomethyl)butane and sodium telluride in 62 per cent yield. The compound, which melted at 145° sublimed at 160° at 0.01 torr.

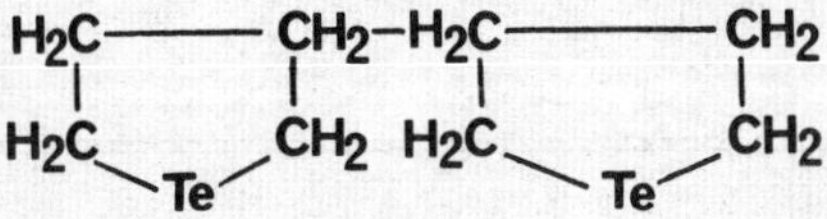

(78)

2) Tellurophene and its derivatives

Attempts to prepare tellurophene *(79)* from aluminum telluride and sodium succinate or acetylene have been unsuccessful[272].

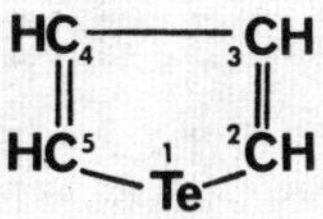

(79)

Diacetylene and substituted diacetylenes, RC≡C-C≡CR, however, reacted with sodium telluride in methanol[133b,257,P-13]. A likely reaction mechanism has been outlined in section IV-C (eq. 41). Fringuelli[133b]

Table XII-1

Telluracyclopentane and Its Derivatives

Compound	Method	Yield %	mp. °C bp. °C(torr)	Reference
(66) telluracyclopentane	$Na_2Te + Br(CH_2)_4Br$	70	--	121
	(67) + SO_2	--	105-6(122)	112,298
			166-7(761)	298
(67) telluracyclopentane 1,1-diiodide	$Te + I(CH_2)_4I$, *(66)* + I_2	100	149-50	112,298
(68) telluracyclopentane 1,1-dibromide	*(66)* + Br_2	70	128-31	121,298
(69) telluracyclopentane 1,1-dichloride	*(66)* + Cl_2	71	112-3	121,298
(70) telluracyclopentane 1-oxide	*(68,69)* + NaOH, *(66)*+ air	--	241(dec)	298
(71) telluracyclopentane·$HgCl_2$	*(66)* + $HgCl_2$	--	146-7	298
(72) *cyclo*tetramethylene methyl telluronium iodide	*(66)* + CH_3I	--	--	298
(73) *cyclo*tetramethylene 4-bromobutyl telluronium bromide	*(66)* + $Br(CH_2)_4Br$	--	152-3	298
(74) *cyclo*tetramethylene 4-bromobutyl telluronium iodide	*(73)* + KI	--	175-6(dec)	298
(75) tetramethylene bis(*cyclo*tetramethylene telluronium bromide)	*(66)* + $Br(CH_2)_4Br$	--	225*	298

Table XII-1

(76) tetramethylene bis(*cyclo*tetramethylene telluronium iodide)	*(75)* + KI	--	215(dec)	298
(77) anhydride of telluracyclopentane bromide hydroxide	*(68)* + NaOH or *(68)* + *(70)*	--	207(dec)	298

*

mp. of monohydrate

reported, that tellurophene was obtained in this manner with an average yield of 47 per cent, when oxygen and moisture was rigorously excluded and the butadiyne was used immediately after its preparation. For the successful synthesis of sodium telluride iron-free liquid ammonia and commercial metallic grey tellurium were necessary. Amorphous or partly oxidized tellurium did not react. Concentration of the methanolic reaction mixture at the end of the reaction at reduced pressure caused extensive decomposition of tellurophene. Better yields were obtained by addition of water to the methanolic solution followed by extraction with diethyl ether.

Tetrachlorotellurophene was the product of the reaction between finely powdered tellurium and hexachloro-1,3-butadiene. The reaction mixture was shaken for 40 hours at 250°. Whereas tellurium and organic halides usually gave the organic tellurium dihalides (section VII-B-1a), only tetrachlorotellurophene was isolated in this case. Tellurium tetrachloride, which was also formed in this reaction, was extracted with concentrated hydrochloric acid before the tellurophene was vacuum distilled[250]. Tetraphenyltellurophene[45] was synthesized according to equation (181).

$$\text{Li-C(R)=C(R)-C(R)=C(R)-Li} \xrightarrow{+TeCl_4} \text{tetraphenyltellurophene (}C_4R_4Te\text{)} \xleftarrow{+Li_2Te} \text{I-C(R)=C(R)-C(R)=C(R)-I} \quad (181)$$

$R = C_6H_5$

One would again expect a tellurophene dichloride to be formed in the reaction with tellurium tetrachloride. Data for individual compounds are listed in Table XII-2.

Table XII-2

Tellurophene and Its Derivatives

Compound	Method	mp. °C bp. °C(torr)	Yield %	Reference
(79) tellurophene	Na_2Te + HC≡C-C≡CH in MeOH	-36°; 91(100)	47	133b
		~-36°;148(714)	69	257
		150(atm)	65	P-13
	dibromide + $NaHSO_3$	--	--	257
2,5-dideuteriotellurophene	*(79)* + D_2SO_4/CH_3OD	--	--	257
tellurophene 1,1-dibromide	*(79)* + Br_2	125 (dec)	--	133b,257,P-13
(80) 2-lithiotellurophene	*(79)* + C_4H_9Li in ether/hexane	--	not isolated	133b
(81) 2-methyltellurophene	*(80)* + $(CH_3)_2SO_4$ in ether	108-10(100)	75	133b
	(82) + $N_2H_4 \cdot H_2O$ in glycol at 140°	--	40	133b
(82) 2-formyltellurophene	*(80)* + $HC(O)N(CH_3)C_6H_5$ in ether/hexane	90-92(2)	24	133b
(83) 2-acetyltellurophene	*(79)* + $(CH_3CO)_2O/SnCl_4$	134-6(15)	33	133b
2-(1'-hydroxyethyl)tellurophene	*(80)* + CH_3CHO in ether	133-5(15)	60	133b
	(83) + $LiAlH_4$ in ether	--	--	133b
(84) 2-carboxytellurophene	*(80)* + CO_2 in ether/hexane	110-1	37	133b

Table XII-2 (cont'd)

Compound	Preparation			
(85) 2-carbomethoxytellurophene	*(84)* + CH_2N_2 in ether	118-20(13)	96	133b
2-hydroxymethyltellurophene	*(85)* + $LiAlH_4$ in ether	--	25	133b
2-(2'-hydroxyisopropyl)tellurophene	Na_2Te + RC≡C-C≡CH	40	47	P-13
(86) 5-lithio-2-methyltellurophene	*(81)* + C_4H_9Li in ether/hexane	--	not isolated	133b
5-carboxy-2-methyltellurophene	*(81)* + CO_2 in ether/hexane	149-50	35	133b
(87) 2-carbomethoxy-5-acetyltellurophene	*(85)* + $(CH_3CO)_2O/SnCl_4$ in benzene	80-1	42	133b
	(88) + CH_2N_2 in ether	--	--	133b
(88) 2-carboxy-5-acetyltellurophene	*(87)* + 1N NaOH in ethanol	190-1	95	133b
2,5-bis(hydroxymethyl)tellurophene	Na_2Te + $(RC≡C)_2$	107	28	P-13
2,5-bis(2'-hydroxyisopropyl)tellurophene	Na_2Te + $(RC≡C)_2$	92	--	P-13
(89) 2,5-dibutyltellurophene	Na_2Te + $(RC≡C)_2$	80(0.15)	--	P-13
2,5-dibutyltellurophene 1,1-dibromide	*(89)* + Br_2	175-80 (dec)	--	P-13
(90) 2,5-diphenyltellurophene	Na_2Te + $(RC≡C)_2$	225	55	257,P-13
2,5-diphenyltellurophene 1,1-dibromide	*(90)* + Br_2	205 (dec)	--	P-13
(91) 2,5-bis(1'-pyrrolidinylmethyl)tellurophene	Na_2Te + $(RC≡C)_2$	150(0.2)	64	P-13
(91) dihydrochloride	*(91)* + HCl	330 (dec)	--	P-13
2,5-bis(acetoxymercuri)tellurophene	*(79)* + $Hg(O_2CCH_3)_2$	--	--	257

Table XII-2 (cont'd)

(92) tetrachlorotellurophene	Te + $Cl_2C{=}CCl{-}CCl{=}CCl_2$;	49.31(0.02)	14	256
	(93) + $NaHSO_3$	--	100	256
(93) tetrachlorotellurophene 1,1-dichloride	*(92)* + Cl_2	200	100	256
tetraphenyltellurophene	$TeCl_4$ + $[LiC(C_6H_5){=}C(C_6H_5)]_2$	239	56	45,P-68
	Li_2Te + $[IC(C_6H_5){=}C(C_6H_5)]_2$	239	82	45,P-67
tetraphenyltellurophene 1,1-dibromide	tetraphenyltellurophene + Br_2	243-5	92	45

That tellurophene exhibits aromatic character, is manifested by the similarity of its uv-spectrum with that of thiophene and by the observation, that the 2- and 5-positions are susceptible to electrophilic and nucleophilic substitution. Studies of the rates of formylation of tellurophene by phosgene and dimethylformamide, of acetylation by acetic anhydride catalyzed by tin(IV) chloride and of trifluoroacetylation showed, that tellurophene is more reactive than thiophene and selenophene but less reactive than furan[133a]. Tellurophene forms 2-lithiotellurophene when treated with butyl lithium in diethyl ether[133b]. Starting with such a lithium derivative 2-substituted derivatives of tellurophene have been prepared. Modification of the thus introduced groups lead to other new tellurophene compounds (scheme 182)[133b].

Strong mineral acids decompose tellurophene[133a,b]. Tellurophene, therefore cannot be nitrated[133b]. The alkylation of 2-lithiotellurophene with ethyl bromide or bromobenzene gave unsatisfactory results[133b]. 2,5-Dideuteriotellurophene and 2,5-bis(acetoxymercuri)tellurophene were obtained by reacting tellurophene with D_2SO_4/CH_3OD, and mercury(II) acetate in ethanol, respectively[257].

Tetraphenyltellurophene failed to undergo the Diels-Alder reaction with maleic anhydride even when heated at 220° for 21 hours. Tetraphenyltellurophene 1,1-dibromide, however, reacted with maleic anhydride with elimination of tellurium producing the polycyclic compounds given in scheme (183)[45].

As far as investigated (see Table XII-2), tellurophene and its derivatives added chlorine and bromine to give the 1,1-dihalides[45,256,257], which were quantitatively reduced by sodium hydrogen sulfite to the cyclic tellurides. Tetrachlorotellurophene 1,1-dichloride was rapidly decomposed by a basic aqueous solution as shown in equation (184), while tetrachlorotellurophene was uneffected[256].

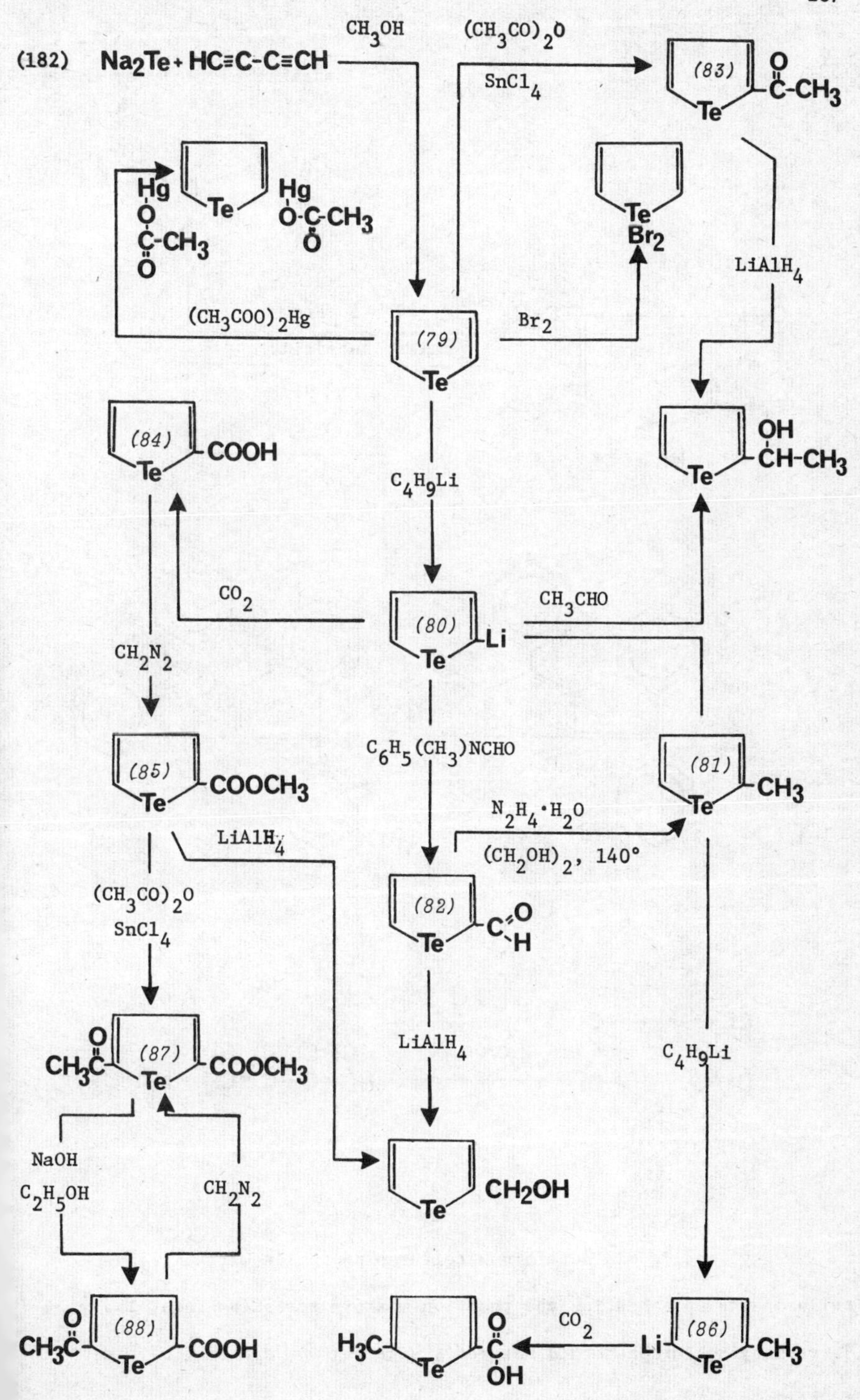
(182)
Na2Te + HC≡C-C≡CH
CH3OH
(CH3CO)2O
SnCl4
(83)
Te
C-CH3
O
Hg
Te
Hg
O
C-CH3
O
O-C-CH3
O
Te
Br2
LiAlH4
(CH3COO)2Hg
(79)
Te
Br2
(84)
COOH
Te
OH
CH-CH3
Te
C4H9Li
CO2
CH3CHO
(80)
Li
Te
CH2N2
C6H5(CH3)NCHO
(81)
CH3
Te
(85)
COOCH3
Te
N2H4·H2O
(CH2OH)2, 140°
LiAlH4
(CH3CO)2O
SnCl4
(82)
Te
C
O
H
LiAlH4
C4H9Li
(87)
CH3C
O
COOCH3
Te
NaOH
C2H5OH
CH2N2
CH2OH
Te
(88)
CH3C
O
COOH
Te
H3C
Te
C
O
OH
CO2
Li
(86)
Te
CH3

(183)

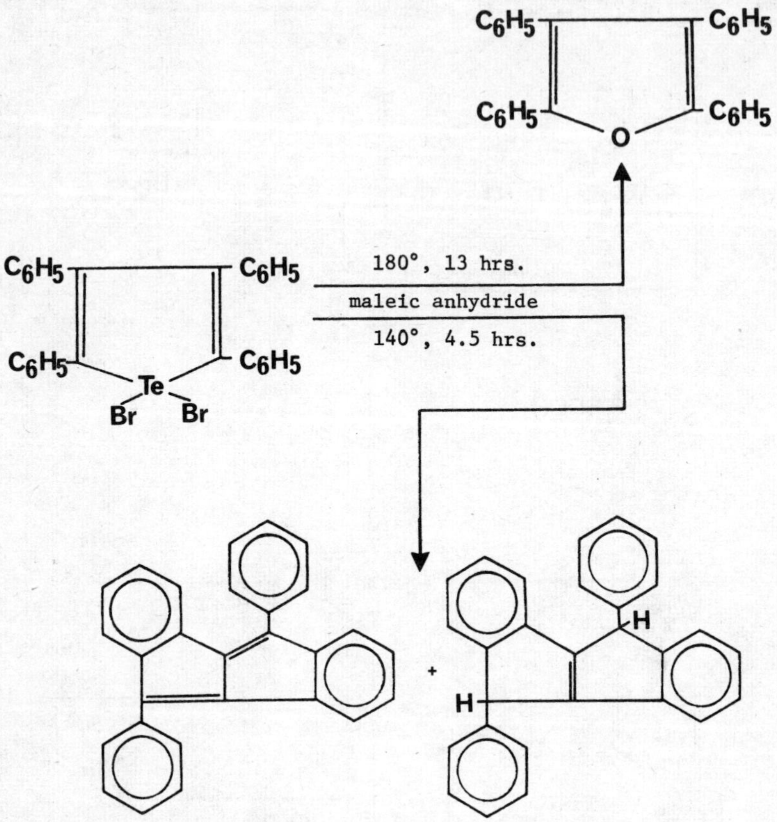

180°, 13 hrs.

maleic anhydride

140°, 4.5 hrs.

(184) H_2O/OH^- + H_2TeO_3

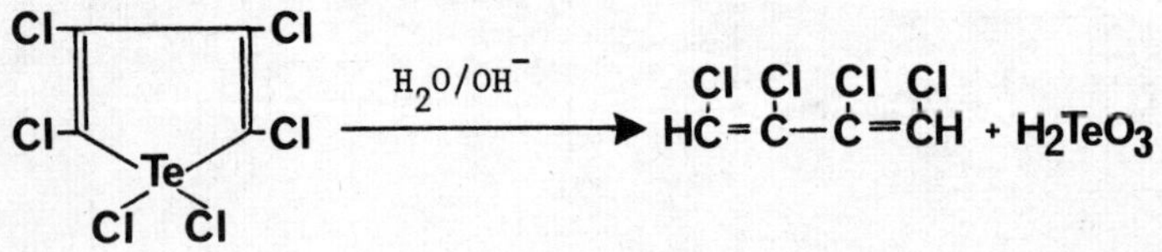

Methyl iodide did not form a telluronium iodide with tetraphenyltellurophene upon heating the reaction mixture for 19 hours at 150°[45]. Tetraphenyltellurophene did not react with $Fe(CO)_5$ and $Fe_2(CO)_9$, but when refluxed with $Fe_3(CO)_{12}$ in toluene/benzene it gave a small yield of the complex tetraphenyltellurophene·$Fe(CO)_3$. This experiment could not be repeated[45].

Tellurophene was found to be dimeric in benzene, while the 2,5-bis(hydroxymethyl) and 2,5-bis(2'-hydroxyisopropyl) derivatives dissolved as monomers[257].

For infrared and nmr data for tellurophene and its derivatives see sections XIV-A and XIV-B, respectively.

3) Benzotellurophene and its derivatives

Sadekov[374a] obtained 3-chloro-2-phenylbenzotellurophene 1,1-dichloride by heating 2-chloro-1,2-diphenylvinyl tellurium trichloride in 1,2,4-trichlorobenzene (eq. 185).

(185) [structure: C_6H_5–CCl=C(C_6H_5)–$TeCl_3$] $\xrightarrow{-HCl}$ [structure: 3-chloro-2-phenylbenzotellurophene with $TeCl_2$]

2-Carboxybenzotellurophene was formed, when 2-formylphenyl carboxymethyl telluride was refluxed with acetic acid anhydride in pyridine. Decarboxylation of this compound by refluxing it in quinoline in the presence of copper produced benzotellurophene *(94)* (R = R' = H)[348e].

[structure: benzotellurophene ring numbered 1 (Te) to 7, R at position 3, R' at position 2]

(94)

Benzotellurophene reacted with chlorine, bromine and iodine in chloroform to give the 1,1-dihalides. Unlike tetraphenyltellurophene, benzotellurophene formed a telluronium bromide when treated with methyl bromide.

Trinitrofluorenone and picric acid gave 1:1 adducts[348e]. Sodium sulfide nonahydrate reduced 3-chloro-2-phenylbenzotellurophene 1,1-dichloride to 3-chloro-2-phenylbenzotellurophene[374a]. Table XII-3 contains pertinent data for benzotellurophene and its derivatives.

Mazza and Melchionna[267] claimed to have prepared the benzotellurophene derivatives *(94)* (R = CH_3CO_2, OH, R' = H) from 2-carboxyphenyl carboxymethyl telluride. Since the synthesis of the starting telluride could not be repeated[122], Mazza's results remain questionable.

Table XII-3

Benzotellurophene and Its Derivatives

Compound	mp. °C	Yield %	Ref.
benzotellurophene	65-6	75	348e
3-chloro-2-phenylbenzotellurophene	--	95	374a
2-carboxybenzotellurophene	206-8	80	348e
benzotellurophene 1,1-dichloride	220 (dec)	100	348e
benzotellurophene 1,1-dibromide	240 (dec)	100	348e
benzotellurophene 1,1-diiodide	180 (dec)	100	348e
3-chloro-2-phenylbenzotellurophene 1,1-dichloride	--	51	374a
methobromide of benzotellurophene	195 (dec)		348
benzotellurophene·trinitrofluorenone	145-7	--	348e
benzotellurophene·picric acid	147-8	--	348e

4) Dibenzotellurophene and its derivatives

The dibenzotellurophene or 2,2'-biphenylylene telluride ring system *(95)* was first synthesized by Courtot and Bastani[84] from biphenyl and tellurium tetrachloride or tetrabromide. They obtained the 5,5-dihalides in small yields. Later dibenzotellurophene was prepared by heating the sulfones *(96)*[87] and *(97)*[330] with elemental tellurium.

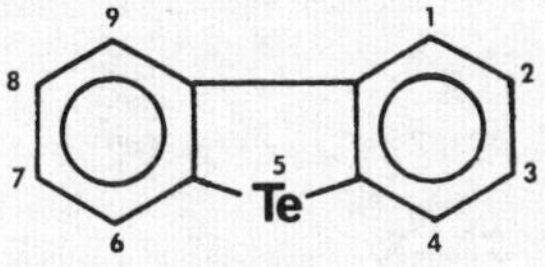

(95)

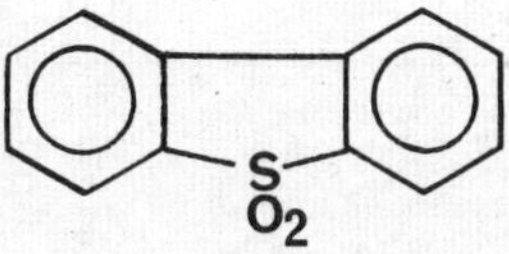

(96)

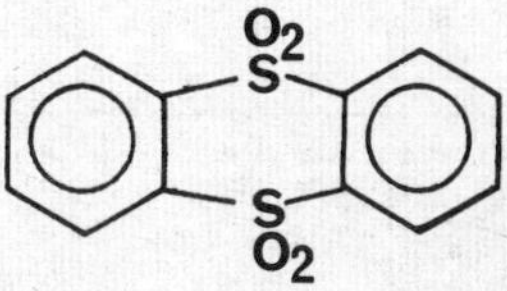

(97)

2,2'-Diiodoperfluorobiphenyl and elemental tellurium kept of 325° for 20 hours gave octafluorodibenzotellurophene[82,83]. Hellwinkel and Fahrbach found that the mercury compound *(98)* and elemental tellurium gave an excellent yield of dibenzotellurophene[171,173] and its 3,7-dimethyl derivative [173](eq. 186).

(186)

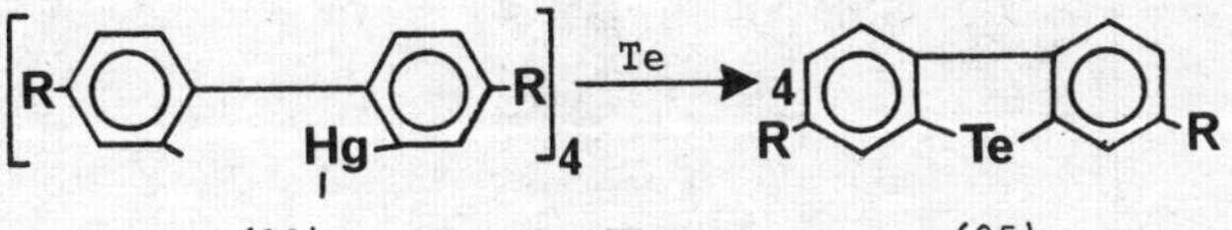

(98) R = H, CH_3 *(95)*

The other modes of formation of dibenzotellurophene listed in Table XII-4 are of no preparative importance.

Dibenzotellurophene reacted with halogens to give 5,5-dihalides, which can be reduced to the parent telluride by potassium disulfite and sodium sulfide. Mercuric chloride, picric acid and 1,3,5-trinitrobenzene adducts of dibenzotellurophene are known[173]. Dibenzotellurophene, unlike tellurophene, formed with methyl iodide the rather unstable telluronium iodide, which decomposed when heated in ethanol[172]. Most of the

Table XII-4

Dibenzotellurophene and Its Derivatives

Compound	Method	mp. °C	Yield %	Reference
dibenzotellurophene *(95)*	Te + *(96)*	93	--	87
	Te + *(97)*	96-7	--	330
	Te + *(98)* (R = H)	93-4	82	171,173
	$TeCl_2$ + 2,2'-dilithiobiphenyl	93-4	54	171,173
	$TeCl_4$ + 2,2'-dilithiobiphenyl	93-4	28	173
	(99) + $K_2S_2O_5$	91-2	--	84
	(99) + Na_2S	--	75	171,173
	(100) + Na_2S	--	96	171,173
	(101) + heat	--	96	171
	(101) + C_4H_9Li	--	--	172
	$(CH_3)_3TeI$ + 2,2'-dilithiobiphenyl	--	37	172
dibenzotellurophene 5,5-dichloride *(99)*	$TeCl_4$ + biphenyl	200 (dec)	2	84
	(95) + Cl_2	333-5	84	171,173
dibenzotellurophene 5,5-dibromide	$TeBr_4$ + biphenyl	210-20 (dec)	small	84
dibenzotellurophene 5,5-diiodide *(100)*	*(95)* + I_2	335-40 (dec)	100	171,173

Table XII-4 (cont'd)

dibenzotellurophene 5-oxide	*(99)* + chloramine T in DMF	214-5	34	173
	(99) + H_2O	230-40	--	84
dibenzotellurophene·picric acid	*(95)* + picric acid	118-9	82	173
dibenzotellurophene·1,3,5-trinitrobenzene	*(95)* + 1,3,5-trinitrobenzene	136-7	85	173
dibenzotellurophene·$HgCl_2$	*(95)* + $HgCl_2$	248-50	53	173
4-nitrobenzotellurophene	*(95)* + HNO_3	184	--	330
1-nitrobenzotellurophene*	*(95)* + HNO_3	158-61	--	330
3,7-dimethyldibenzotellurophene	Te + *(98)*(R = CH_3)	157-8	79	173
3,7-dimethyldibenzotellurophene·1,3,5-trinitrobenzene	*(95)* + 1,3,5-trinitrobenzene	151-2	--	173
octafluorodibenzotellurophene	Te + 2,2'-dilithiooctafluorobiphenyl	116-9	66	82,83

* The position of the nitro-group is uncertain.

telluronium salts listed in Table XII-5 were, however, obtained through cleavage of one carbon-tellurium bond in the tetravalent tellurium compounds *(101)* (eq. 187).

(187)

(101)

$\xrightarrow{YX}$

R = H, CH_3; YX = H-OH, HCl, Br_2, I_2

The anions in these telluronium salts can be exchanged as described in section VIII-1c. Dibenzotellurophene 5,5-dichloride hydrolyzed to the oxide, which regenerated the dihalides with hydrohalic acids[84]. The oxide was also obtained as a hydrolysis product of 2,2'-biphenylylene tellurtosylimine[173]. The dibromide decomposed on heating to 4,4'-dibromobiphenyl[84]. Dibenzotellurophene and butyl lithium exchanged their metal atoms and yielded 2,2'-dilithiobiphenyl and dibutyl telluride[172]. Nitration with nitric acid (d = 1.42g/ml) at elevated temperatures produced mononitro derivatives of dibenzotellurophene[330]. Dibenzothiophene was formed when the tellurium derivative was heated with sulfur[84]. Tables XII-4 and XII-5 list the known dibenzotellurophene derivatives and summarize the methods of their preparation.

The reactions of dibenzotellurophene compounds leading to tetraorganyl tellurium derivatives have been discussed in chapter IX.

Table XII-5

2,2'-Biphenylylene Organyl Telluronium Compounds

$$\left[R\text{-}C_6H_3\text{—}C_6H_3\text{-}R \;(2,2'\text{-}Te\text{-}R') \right]^+ X^-$$

R'	R	X	Method	mp. °C	Yield %	Reference
methyl	H	I	*(95)* + CH_3I	102-4 (dec)	87	172
2-biphenylyl	H	Cl	*(101)* + HCl	269-70	81	171,173
		I	$R_3TeOH + I^-$	253-5	--	171,173
			(101) + CH_3I	253-5	--	173
			R_3TeCl + KI	--	--	173
		OH	*(101)* + H_2O	186-8	23	171,173
		$(C_6H_5)_4B$	*(101)* + $H_2O/Na[B(C_6H_5)_4]$	185	--	173
			$R_3TeCl + Na[B(C_6H_5)_4]$	185	--	173
2-bromo-2'-biphenylyl	H	Br	*(101)* + Br_2	264	79	171,173
2-iodo-2'-biphenylyl	H	I	*(101)* + I_2	265	65	173
4,4'-dimethyl-2-biphenylyl	CH_3	Cl	*(101)* + HCl	280-1	94	173
		$(C_6H_5)_4B$	*(101)* + $H_2O/Na[B(C_6H_5)_4]$	207-9	--	173
4,4'-dimethyl-2-bromo-2'-biphenylyl	CH_3	Br	*(101)* + Br_2	277-8	85	173

B) Six membered, Heterocyclic Tellurium Compounds with Tellurium as the only Heteroatom

This section is dealing with six-membered ring systems, which contain one tellurium atom as the only heteroatom. Telluracyclohexane *(102)* and telluroisochroman *(103)* are the known representatives of this class of compounds. 1-Tellura-3,5-cyclohexanedione and a large number of its substituted derivatives were prepared from tellurium tetrachloride and 1,3-diketones.

1) Telluracyclohexane

Tellurium and pentamethylene dihalides[121] and sodium telluride[12], magnesium telluride[294] or aluminum telluride[294] and pentamethylene dihalides formed the telluracyclohexane ring system. The reaction of aluminum telluride with pentamethylene dichloride, dibromide and diiodide did not produce directly telluracyclohexane *(102)*.

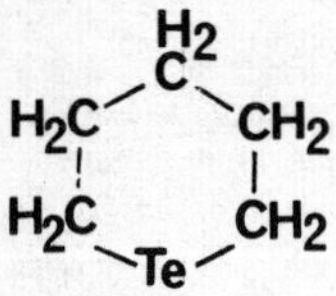

(102) 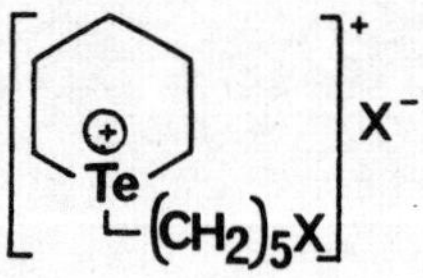 *(107)-(109)* 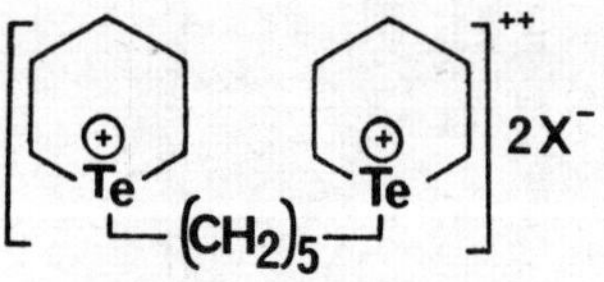 *(110)-(113)*

X = Cl, Br, I

The primary products were the telluronium salts *(107)-(113)*, which were formed by addition of excess halide to *(102)*. These telluronium salts lost a molecule of dihalide upon heating in vacuum forming telluracyclohexane *(102)*[294]. Telluracyclohexane added elemental halogens to yield the respective telluracyclohexane 1,1-dihalides. Telluracyclohexane 1,1-dichloride and 1,1-dibromide were also obtained from the telluronium salts *(107)* and *(108)*, respectively (see Table XII-6). The dihalides were reduced to the cyclic tellurides with potassium disulfite[294]. The reactions of telluracyclohexane and its derivatives are summarized in Table XII-6.

Gilbert and Lowry[143] studied the ultraviolet and visible spectra of the tellurocyclohexane dihalides. The results of conductivity measurements in aqueous solutions suggested that the dihalides exist as hydroxy telluronium halides *(114)*. The telluronium halides *(107)*, *(108)* and *(109)* (Table XII-6) were found to be binary electrolytes, while *(110)*, *(111)* and *(112)* behaved as ternary electrolytes.

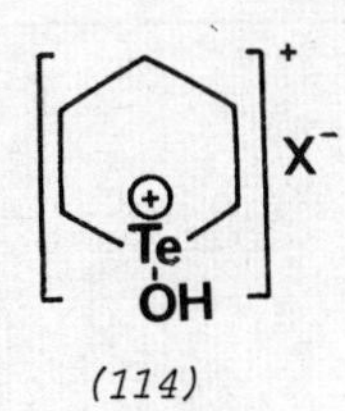

(114)

2) <u>1-Tellura-3,5-cyclohexanedione and its derivatives</u>

Morgan and coworkers prepared a large number of compounds containing the telluracyclohexane ring system by condensing tellurium tetrachloride with 1,3-diketones. The diketone must have at least two hydrogen atoms each on the two carbon atoms in α-position to the carbonyl groups (see section IV-B-1a). The condensation products, 1-telluracyclohexane-3,5-dione 1,1-dichlorides *(115)*, were in the earlier publications formulated as 1-tellura-2-oxacyclohexene-5-one derivatives *(116)*.

(115) *(116)*

Table XII-6

Telluracyclohexane and Its Derivatives

Compound	Method	mp. °C bp. °C(torr)	Yield %	Reference
(102) telluracyclohexane	*(104)* + $K_2S_2O_5$	82-3(12); 45(1-2)	--	294,P-3
	Na_2Te + $(CH_2)_5Cl_2$	--	33	121
	(108), *(111)* at 160-90°/ 30 torr	--	90	294
(103) telluracyclohexane 1,1-dichloride	*(102)* + Cl_2	105-6	37	121,294,P-3
	(107) + Cl_2	--	--	294
(104) telluracyclohexane 1,1-dibromide	*(102)* + Br_2	106-7	90	121,294,P-3
	(108) + Br_2	--	--	294
(105) telluracyclohexane 1,1-diiodide	*(102)* + I_2	135-6	--	294
	Te + $I(CH_2)_5I$	136	63	121,294
telluracyclohexane 1-iodide 1-triodide	*(105)* + I_2	82-4	--	143
telluracyclohexane 1,1-dihydroxide (or 1-oxide)	*(104)* + Ag_2O/H_2O	--	--	143
(106) telluracyclohexane 1,1-dioxide	*(102)* + H_2O_2	195 (dec)	--	294,P-3
(107) *c*-pentamethylene-5'-chloropentyl telluronium chloride	Al_2Te_3 + $Cl(CH_2)_5Cl$	149-51	--	294,P-3
(108) *c*-pentamethylene 5'-bromopentyl	Al_2Te_3 + $Br(CH_2)_5Br$	143-5	17†	294,P-3

Table XII-6 (Cont'd)

(109) *c*-pentamethylene 5'-iodopentyl telluronium iodide	$Al_2Te_3 + I(CH_2)_5I$	135-6*	17	294
(110) pentamethylene bis(*c*-pentamethylene telluronium chloride)	$Al_2Te_3 + Cl(CH_2)_5Cl$	224-5	--	294,P-3
(111) pentamethylene bis(*c*-pentamethylene telluronium bromide)	$Al_2Te_3 + Br(CH_2)_5Br$	223-4	19†	294,P-3
(112) pentamethylene bis(*c*-pentamethylene telluronium tribromide)	*(111)* + Br_2	102-4	--	294,P-3
(113) pentamethylene bis(*c*-pentamethylene telluronium iodide)	$Al_2Te_3 + I(CH_2)_5I$	216-7 (dec)	12	294
pentamethylene bis(*c*-pentamethylene telluronium hydroxide)	*(111)* + Ag_2O/H_2O	--	--	143
pentamethylene bis(*c*-pentamethylene telluronium triodide)	*(113)* + I_2	127-9	--	143
pentamethylene bis(*c*-pentamethylene telluronium) dichromate	*(110)* + $Na_2Cr_2O_7$	193 (dec)	--	294,P-3
bis(telluracyclohexane) 1,1'-oxide 1,1'-diiodide	*(105)* + K_2CO_3	195 (dec)	--	143

* solidified above 135° and remelted at 197-8°

† yield varied

Chemical[285,287,289], spectroscopic[97] and structural results[327] prove unequivocally that stucture *(115)* is correct.

The dichlorides were reduced to telluracyclohexanediones by potassium disulfite or alkali metal hydrogen sulfite in aqueous solution. The cyclic telluride added iodine and bromine to give the dihalides. The individual compounds are listed in Tables XII-7 and XII-8.

1-Tellura-3,5-cyclohexanedione decomposed when treated with concentrated hydrochloric acid or potassium hydroxide[275]. Its dichloride[275] *(115)* and the corresponding 4-ethyl derivative[277] eliminated tellurium when reacted with sulfur dioxide and hydrochloric acid, respectively. 1-Tellura-3,5-cyclohexanedione 1,1-dichloride refluxed in chloroform solution in presence of hydrogen chloride and ethyl chloride gave the linear tellurium trichloride *(117)*[276,285] (eq. 188).

(188)

O, O, Te, Cl_2 ⟵ HO, O, Te, Cl_2 —RCl⟶ RO, O, Te, Cl_2

—HCl⟶

$$H_3C-\overset{OR}{\overset{|}{C}}=CH-\overset{O}{\overset{\|}{C}}-CH_2-TeCl_3$$

(117)

The cyclic diketones yielded dioximes with hydroxylamine sulfate in boiling aqueous solutions. 2,4-Substituted telluracyclohexanediones and the 4,4-dimethyl derivative formed monoximes in dilute acidic acid solutions, while dioximes were obtained in alkaline solutions. The oximes thus far prepared are given in Table XII-8. The monoximes and dioximes should exist in various isomeric forms depending on the substitution of the heterocyclic ring and the orientation of the hydroxyl groups in the dioximes.

Table XII-7

1-Tellura-3,5-cyclohexanedione and Its Derivatives

1-tellura-3,5-cyclohexanedione						1-tellura-3,5-cyclohexanedione 1,1-dichloride		
R^1	R^2	R^3	R^4	mp. °C	Yield %	mp. °C	Yield %	Reference
H	H	H	H	182	65	151/173*	62	275-7
CH_3	H	H	H	100	79	170-1(dec)	65	279
C_2H_5	H	H	H	110-2	100	153-4	58	282
C_3H_7	H	H	H	80	70	125	50	284
C_4H_9	H	H	H	86	--	102	--	284
C_5H_{11}	H	H	H	86	--	87	15	282
C_6H_{13}	H	H	H	74-5	--	80	23	287
C_7H_{15}	H	H	H	89	--	89	25	287
C_8H_{17}	H	H	H	64	--	49	--	284
$C_{10}H_{21}$	H	H	H	98-9(dec)	--	89	21	289

Table XII-7 (cont'd)

$C_6H_5CH_2$	H	H	H	159(dec)	--	not cryst.	--	289
H	CH_3	H	H	170(dec)	76	180-90(dec)	67	277
H	CH_3	CH_3	H	125	100	194	60	285
H	CH_3	C_2H_5	H	--	--	--	--	289†
H	C_2H_5	H	H	141-2	96	185-90(dec)	88	277,291
H	C_2H_5	C_2H_5	H	85-6	100	178-80	23	282
H	C_3H_7	H	H	106-7	--	~180	--	283
H	*i*-C_3H_7	H	H	152-3	--	150(dec)	--	283
H	C_4H_9	H	H	129	--	155	--	284
H	*i*-C_4H_9	H	H	150	--	142	36	289
H	*sec*-C_4H_9	H	H	145	--	168-9	37	289
H	$(2-CH_3)C_4H_8$	H	H	138-9	--	162	35	289
H	$C_6H_5CH_2$	H	H	153	--	180(dec)	52	287
H	$C_6H_5CH_2$	$C_6H_5CH_2$	H	128	--	189-90(dec)	23	287
H	Cl	H	H	153-4(dec)	51	161-2	5	277
CH_3	CH_3	H	H	124-5	100	166-7	80	282,285
CH_3	C_2H_5	H	H	109	--	167	63	283

Table XII-7 (cont'd)

CH_3	C_3H_7	H	H	102(dec)	--	145-50(dec)	62	289
CH_3	i-C_3H_7	H	H	127	--	173(dec)	--	289
CH_3	C_4H_9	H	H	93	--	103	30	284
CH_3	$C_6H_5CH_2$	H	H	124(dec)	--	168	63	289
C_2H_5	C_2H_5	H	H	113	--	140(dec)	54	282
CH_3	H	H	CH_3	151	--	162	78	282,287
CH_3	H	H	C_2H_5	101-2	--	156(dec)	19	282
C_2H_5	H	H	C_2H_5	97	--	120/138/148*	--	283
CH_3	CH_3	H	CH_3	135(dec)	100	180(dec)	68	287
CH_3	C_2H_5	H	CH_3	137-8(dec)	--	182(dec)	76	282
CH_3	$(2\text{-}CH_3)C_4H_8$	H	CH_3	--	--	oily	--	289

* polymorphic

† not isolated

Table XII-8

Oximes, 1,1-Dibromides and 1,1-Diiodides of 1-Tellura-3,5-cyclohexanediones

R^1	R^2	R^3	R^4	Monoxime mp. °C(dec)	Dioxime mp. °C(dec)	Ref.	Dibromide mp. °C(dec)	Diiodide mp. °C(dec)	Ref.
H	H	H	H	--	190-207	285	124/160†	121/141†	276,277
H	CH_3	H	H	--	--	--	153	176	277
H	C_2H_5	H	H	--	192	285	161-70	176	277
H	$C_6H_5CH_2$	H	H	--	168-70	287	--	--	--
H	CH_3	CH_3	H	184	235	285	--	--	--
CH_3	H	H	H	--	161.5	289	156	190	279
C_2H_5	H	H	H	--	149/183*	285	--	--	--
CH_3	CH_3	H	H	164	198	285	--	--	--
CH_3	H	H	CH_3	--	168-70	287	--	--	--
CH_3	CH_3	H	CH_3	--	170	287	--	--	--
CH_3	C_2H_5	H	H	157	182	289	--	--	--

* [illegible]

3) Telluroisochroman

Holliman and Mann[180,259] prepared telluroisochroman *(118)* in 50 per cent yield from 2-(β-bromoethyl)benzyl bromide and sodium telluride. Studies indicate that this cyclic telluride undergoes the same reactions given by the other diorganyl tellurides. Figure XII-1 summarizes the reactions of telluroisochroman.

4-Chlorophenacyl bromide produced the racemic telluronium salt *(119)*, which was partially resolved with silver *d*-bromocamphorsulfonate. The *ℓ*-*d* salt precipitated while the *d*-*d* compound was obtained from the mother liquor. Treatment of the active sulfonates with sodium picrate gave the active picrates. All the optically active compounds showed a slow mutarotation. The molecular rotation of these compounds varied indicating that the samples were optically impure.

C) Tellurium Containing, Six-Membered Heterocyclic Ring Systems with Oxygen, Sulfur or Tellurium as Additional Heteroatoms

Six-membered ring systems containing tellurium and oxygen or sulfur in 1,4 position to each other have been prepared. Among these compounds phenoxtellurine is the most extensively investigated substance. There is only one literature report on telluranthrene. The trimer of telluroformaldehyde, $(CH_2Te)_3$, 1,3,5-tritelluracyclohexane, is the only ring system with three tellurium atoms.

1) 1-Oxa-4-telluracyclohexane

1-Oxa-4-telluracyclohexane was prepared by addition of bis(2-chloroethyl) ether dissolved in ethanol to an aqueous solution of sodium telluride. To facilitate the isolatiion of the heterocyclic compound it was converted to the 4,4-dichloride or dibromide[121]. The

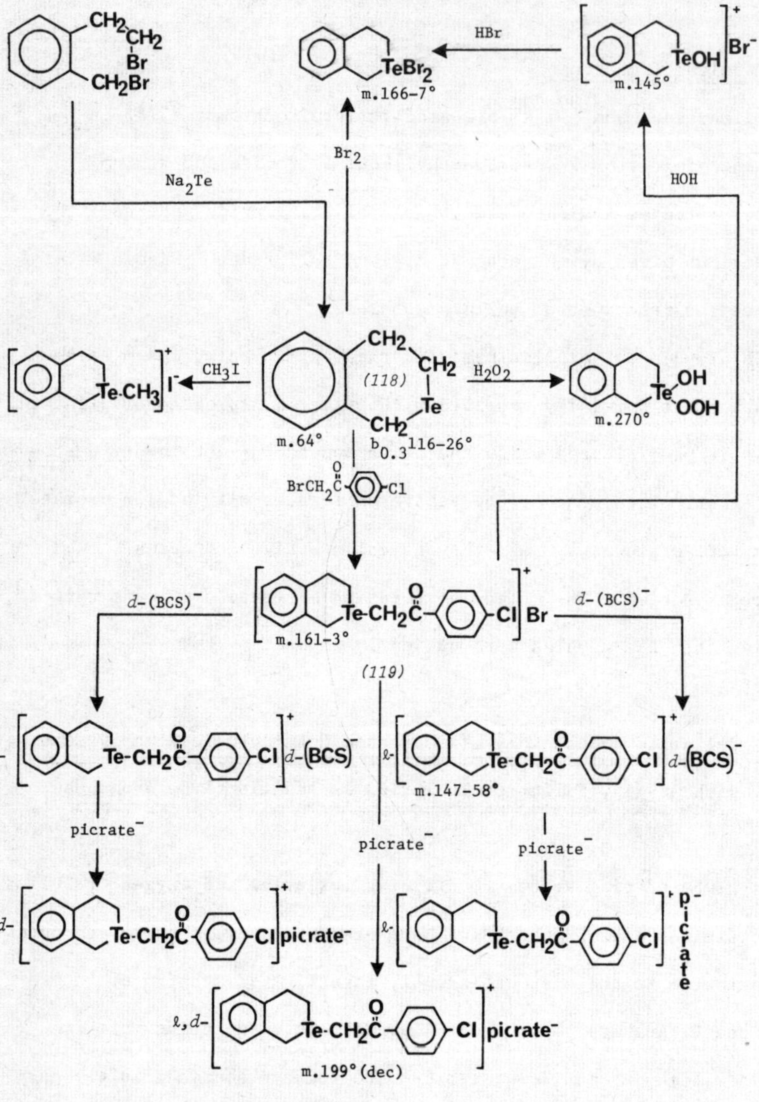

CH_2CH_2Br / CH_2Br — Na_2Te → (118) m.64° $b_{0.3}$ 116–26°

(118) — Br_2 → TeBr$_2$ m.166–7°

[TeOH]$^+$ Br$^-$ m.145° — HBr → TeBr$_2$

(118) — CH_3I → [Te-CH$_3$]$^+$ I$^-$

(118) — H_2O_2 → Te(OH)(OOH) m.270°

(118) — $BrCH_2C(O)C_6H_4Cl$ → [Te-CH$_2$C(O)-C$_6$H$_4$-Cl]$^+$ Br$^-$ m.161–3° (119)

(119) — HOH → [TeOH]$^+$ Br$^-$

(119) — d-(BCS)$^-$ → d-[Te-CH$_2$C(O)-C$_6$H$_4$-Cl]$^+$ d-(BCS)$^-$ — picrate$^-$ → d-[Te-CH$_2$C(O)-C$_6$H$_4$-Cl]$^+$ picrate$^-$

(119) — d-(BCS)$^-$ → ℓ-[Te-CH$_2$C(O)-C$_6$H$_4$-Cl]$^+$ d-(BCS)$^-$ m.147–58° — picrate$^-$ → ℓ-[Te-CH$_2$C(O)-C$_6$H$_4$-Cl]$^+$ picrate$^-$

(119) — picrate$^-$ → ℓ,d-[Te-CH$_2$C(O)-C$_6$H$_4$-Cl]$^+$ picrate$^-$ m.199° (dec)

Fig. XII-1: Syntheses and Reactions of Telluroisochroman

reaction between elemental tellurium and bis(2-iodoethyl) ether gave only a 10 per cent yield of 1-oxa-4-telluracyclohexane 4,4-diiodide, due to the instability of the ether under the reaction conditions. The dichloride was reduced by potassium disulfite in aqueous solution in the presence of carbon tetrachloride. The pure 1-oxa-4-telluracyclohexane was stable. Extensive decomposition occurred, however, when the compound was dissolved in carbon tetrachloride[121]. The reactions of 1-oxa-4-telluracyclohexane are presented in Fig. XII-2.

2) Phenoxtellurine

Phenoxtellurine 10,10-dichloride *(121)* was obtained, when tellurium tetrachloride was heated with an equimolar amount of diphenyl ether at 200°[103,105]. Hydrogen chloride was evolved in two stages, first at 120° and then near 200°. Under milder conditions 4-phenoxyphenyl tellurium trichloride was formed. In order to make intramolecular condensation possible, the trichlorotelluro group must migrate into the 2-position. When the trichloride was heated for 2 hours at 150-60° a trichloride melting at 125° was isolated, which could have been the 2-substitution product. Potassium disulfite reduced it to a ditelluride. When 4-phenoxyphenyl tellurium trichloride was heated for 4.75 hours at temperatures rising from 150° to 210° phenoxtellurine 10,10-dichloride was formed[103] (eq. 189).

(189) $TeCl_4$ + C_6H_5–O–C_6H_5 $\xrightarrow[-HCl]{120°}$ C_6H_5–O–C_6H_4–$TeCl_3$ (m.156°) $\xrightarrow{160°-200°}$ 2-(C_6H_5O)$C_6H_4$$TeCl_3$ (m.125°) $\xrightarrow[-HCl]{200°}$ phenoxtellurine 10,10-dichloride (Te Cl_2) *(121)*

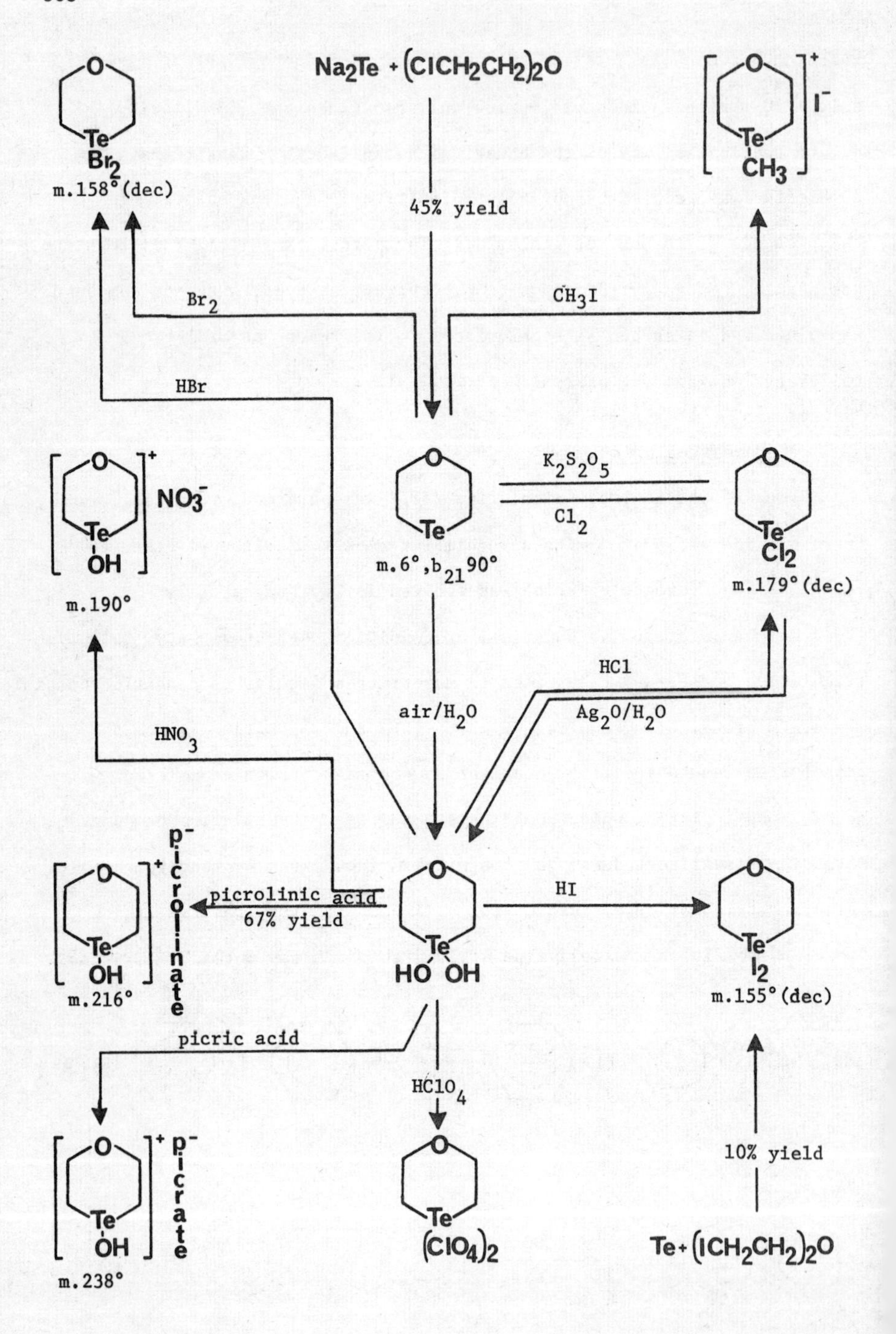

Fig. XII-2: 1-Oxa-4-telluracyclohexane and Its Derivatives

Campbell and Turner[55], Gioaba[146] and Vasiliu[422,422a] prepared 2,8-disubstituted phenoxtellurine derivatives by direct condensation of tellurium tetrachloride and the appropriate diphenyl ethers.

Campbell and Turner[55] found a convenient synthesis for substituted phenoxtellurines, which is described by equation (190).

(190) 2-NH_2-diphenyl ether (4'-R) $\xrightarrow{NaNO_2/HCl}$ 2-$N_2^+Cl^-$ (4'-R) $\xrightarrow{HgCl_2}$ 2-HgCl (4'-R) $\xrightarrow{TeCl_4}$ 2-$TeCl_3$ (4'-R) $\xrightarrow{heat}$ phenoxtellurine Cl_2 (R)

2-Methylphenoxtellurine 10,10-dichloride was obtained in an overall yield of 36 per cent based on 2-amino-4'-methyldiphenyl ether[55]. For other derivatives see Table XII-9.

Attempts to condense nitro substituted diphenyl ethers with tellurium tetrachloride led to extensive decomposition. Nitrophenoxtellurines were, however, obtained by nitration of phenoxtellurine with fuming nitric acid under the conditions outlined in scheme (191). Mono- and dinitrated derivatives were the only products. Trinitro or higher substituted compounds were never isolated[105]. The positions

(191)

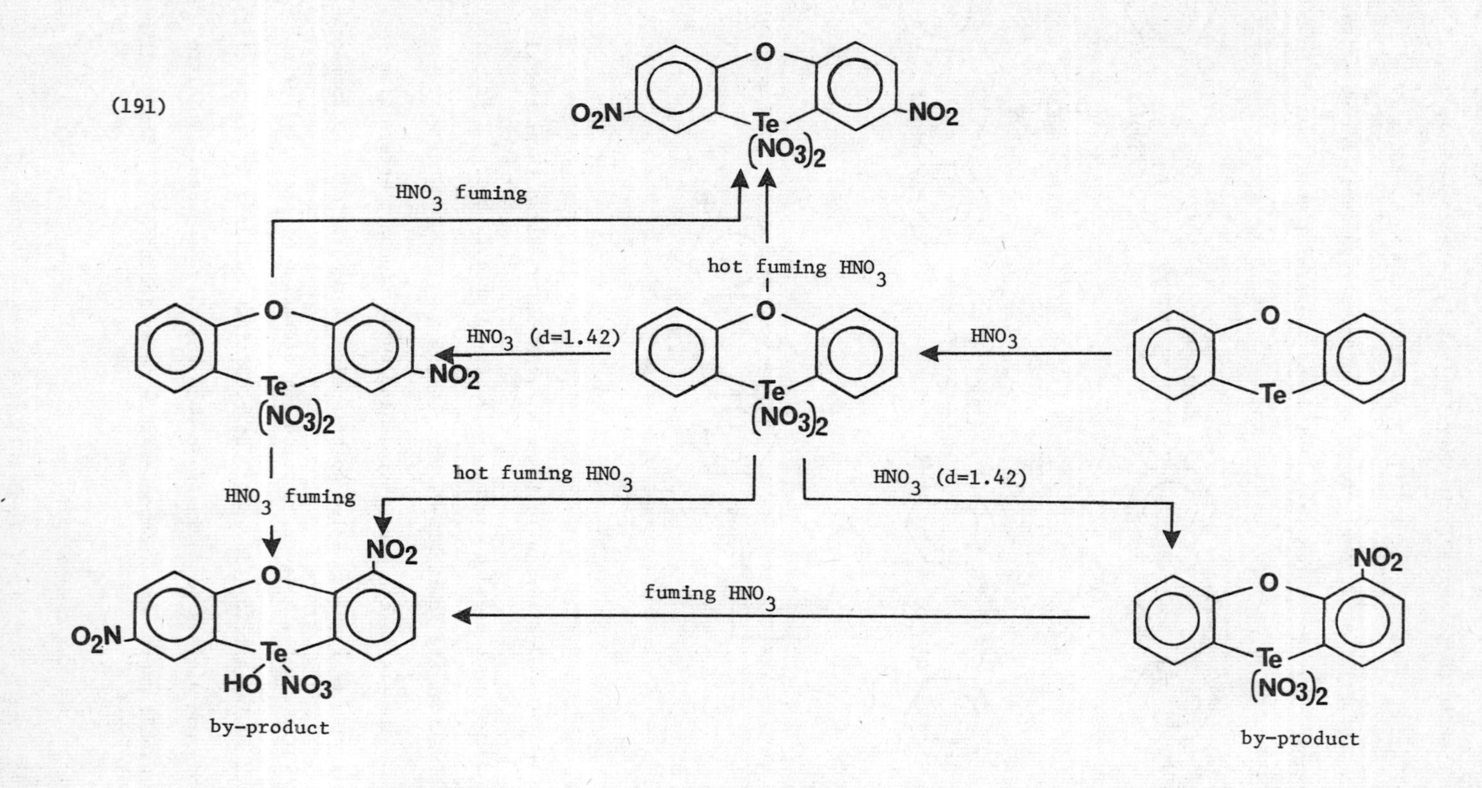
O
Te
O_2N
NO_2
$(NO_3)_2$
HNO_3 fuming
hot fuming HNO_3
HNO_3 (d=1.42)
HNO_3
hot fuming HNO_3
HNO_3 (d=1.42)
HNO_3 fuming
fuming HNO_3
HO
NO_3
by-product
by-product

Table XII-9

Phenoxtellurine and Its Derivatives

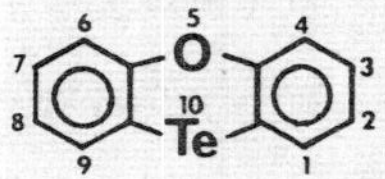

Compound or Substitutents	Method	mp. °C	Yield %	Ref.
(120) phenoxtellurine	*(121)* + $K_2S_2O_5$	78-9	100	103,105,109,146a
	(122) + $NaHSO_3$	--	--	104
	heating *(125)* in quinoline	--	--	55
	(121) + Na_2S	79	100	359
(121) 10,10-dichloride	$RTeCl_3$ + heat; $TeCl_4$ + $(C_6H_5)_2O$	265	50	103,105
	(121a) + Cl_2 in $CHCl_3$	--	--	146a
	(121b) + Cl_2 in $CHCl_3$	--	82.7	146a
	(121a) + $SOCl_2$	--	--	146a
	(120) + 1,2-$C_6H_4(O_2CCl_2)^*$/$TiCl_4$	263-4	--	421a
(121a) 10,10-dibromide	*(120)* + Br_2	290 (dec)	--	103
	(121) + KBr in CH_3OH	--	--	146a
	(121b) + PBr_3	--	85	146a
	(121b) + Br_2	--	94	146a

Table XII-9 (cont'd)

(121b) 10,10-diiodide	*(120)* + I_2	270 (dec)	--	103
	(121) + KI in CH_3OH	--	--	146a
	(121a) + KI in CH_3OH	--	--	146a
10,10-dinitrate	*(120)* + HNO_3	258 (dec)	--	103,104
(122) 10,10-bis(hydrogen sulfate)	*(123)* + H_2SO_4	--	--	104,109
(123) 10,10-diacetate	*(120)* + H_2O_2 + glac.CH_3COOH	205-7	90	104,109
10-oxide	*(121)* + Ag_2O	--	--	109
10-tetrachloroplatinate(II)	*(122)* + K_2PtCl_4	--	--	109
2-methyl	*(124)* + $K_2S_2O_5$	50-2	93	55
(124) 2-methyl 10,10-dichloride	$RTeCl_3$ + heat	274-5	80	55
(125) 2-carboxy	*(126)* + $K_2S_2O_5$	231-3	--	55
(126) 2-carboxy 10,10-dichloride	$RTeCl_3$ + heat	319	62	55
d-2-carbo-ℓ-menthoxy	*(125)* + ℓ-menthol	123-5	--	55
2,8-dimethyl	*(127)* + $KHSO_3$	48-9	--	146
(127) 2,8-dimethyl 10,10-dichloride	$TeCl_4$ + $(CH_3C_6H_4)_2O$	285-6	--	146
(127a) 2,8-difluoro	*(127b)* + $K_2S_2O_5$ in acetone	156-7	94	422a
(127b) 2,8-difluoro 10,10-dichloride	$TeCl_4$ + $(FC_6H_4)_2O$	287-8	37	422a

Table XII-9 (cont'd)

	2,8-difluoro 10,10-dibromide	*(127a)* + Br_2 in CCl_4	312-3 (dec)	91	422a
	2,8-difluoro 10,10-diiodide	*(127a)* + I_2 in CCl_4	266-7 (dec)	85	422a
(128)	2,8-dichloro	*(129)* + $K_2S_2O_5$	111-2	100	146a,422
(129)	2,8-dichloro 10,10-dichloride	$TeCl_4$ + $(ClC_6H_4)_2O$	308-9	42.8	422
		(121b) + Cl_2 in $CHCl_3$	--	93	146a
	2,8-dichloro 10,10-dibromide	*(128)* + Br_2	327-9 (dec)	93	422
	2,8-dichloro 10,10-diiodide	*(128)* + I_2	267-71 (dec)	88	422
(130)	2-chloro-8-methyl	*(131)* + $K_2S_2O_5$	67-8	--	55,106
(131)	2-chloro-8-methyl 10,10-dichloride	$TeCl_4$ + $ClC_6H_4OC_6H_4CH_3$	284	--	55,106
	2-chloro-8-methyl 10,10-dibromide	*(130)* + Br_2	315 (dec)	--	106
	2-chloro-8-methyl 10,10-diacetate	*(130)* + O_2 + CH_3COOH	230-2	--	106
(132)	2-nitro	*(133)* + $K_2S_2O_5$	128-9	--	55,105
(133)	2-nitro 10,10-dinitrate	*(120)* + HNO_3 (d = 1.42)	--	20	55,105
		(132) + 11M HNO_3	196-7	--	104
(134)	4-nitro	*(135)* + $K_2S_2O_5$	104	--	105
(135)	4-nitro 10-hydroxide 10-nitrate·$1H_2O$	*(120)* + HNO_3 (d = 1.42)	--	--	105
		(134) + 11M HNO_3	243	--	104,105

Table XII-9 (cont'd)

	4-nitro 10,10-dibromide	*(134)* + Br_2	302	--	105
(136)	2,8-dinitro	*(137)* + $K_2S_2O_5$	228	--	105
(137)	2,8-dinitro 10,10 dinitrate	*(120)* + HNO_3(d = 1.50)	259	--	105
		(136) + 11 <u>M</u> HNO_3	259	--	104
(138)	2,8-dinitro 10,10 dinitrate	*(132)* + HNO_3 fuming	--	--	105
(139)	4,8-dinitro	*(140)* + $K_2S_2O_5$	197-8	--	105
(140)	4,8-dinitro 10-hydroxide 10-nitrate	*(120)* + HNO_3(d = 1.50)	--	--	105
		(139) + 11 <u>M</u> HNO_3	237-9	--	104
		(134) + HNO_3 fuming	--	60	105
		(132) + HNO_3 fuming	--	small	105
	2-amino	*(132)* + Sn/HCl	157	65	55,105
	2,8-diamino	*(136)* + Sn/HCl	198	--	105
	4,8-diamino	*(139)* + Sn/HCl	156	--	105

* pyrocatechol dichloromethylene ether

of the nitro groups were determined by alkali cleavage of the ring system (eq. 192). The position taken by the incoming nitro group seems to be determined by the oxygen atom. The influence of the tellurium dinitrate groups seems to be negligible.

(192) O_2N–[O, $TeCl_2$ tricycle]–NO_2 $\xrightarrow{+KOH}$ O_2N–C_6H_4–O–C_6H_4–NO_2 $+ K_2TeO_3$ $+ KCl$

In these nitration reactions the divalent tellurium atom was oxidized to the tetravalent state. All the nitrophenoxtellurines were therefore isolated as the 10,10-dinitrates. When phenoxtellurines were treated with only 11M nitric acid nitration of the aromatic rings did not occur. The products isolated in these reactions were 10,10-dinitrates or hydroxide nitrates[104]. The 10,10-dinitrates and 10-hydroxide 10-nitrates are reducible with potassium disulfite.

The cyclic tellurides react with bromine and iodine to give 10,10-dihalides. Phenoxtellurine 10,10-dichloride was also obtained from phenoxtellurine and pyrocatechol dichloromethylene ether in dichloromethane solution in the presence of titanium tetrachloride[421a]. The halogen exchange reactions carried out with phenoxtellurine 10,10-dihalides are summarized in equation (193). The 10,10-dihalides of phenoxtellurine[146a], 2,8-difluorophenoxtellurine[422a] and 2,8-dichlorophenoxtellurine[146a] were reduced by potassium disulfite to the cyclic tellurides.

(193)

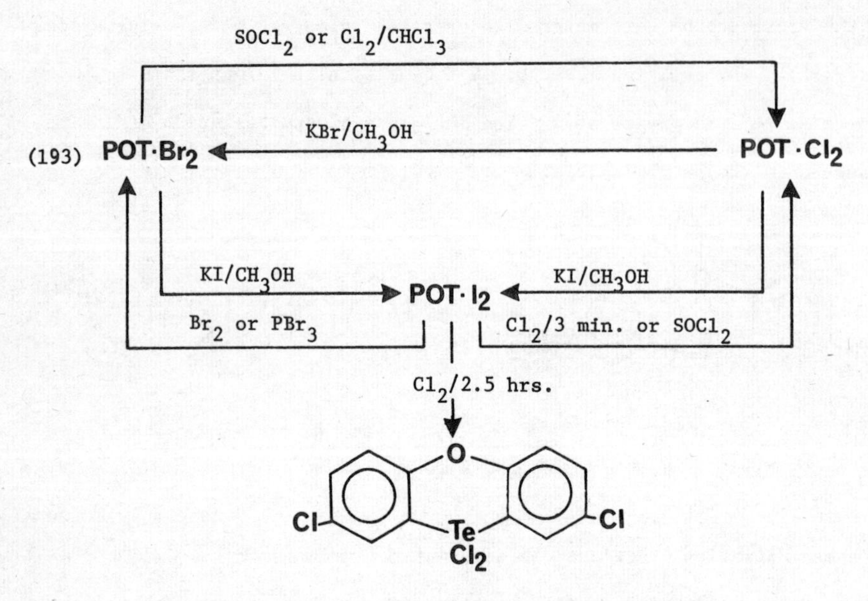

POT = Phenoxtellurine

The nitrophenoxtellurines are readily reduced by tin and hydrochloric acid to the corresponding amines, which can be converted to diazonium salts[105]. The telluroxide and tellurone of phenoxtellurine are discussed in sections VII-C and VII-D. The reactions with pentacarbonylmanganese chloride is treated in section XI-E. All these reactions of phenoxtellurine and its derivatives are listed in Table XII-9.

Phenoxtellurine (POT) and its derivatives were found to be capable of forming colored complexes. Upon reduction of the crude nitration products of phenoxtellurine the adducts $[2\text{-}(NO_2)POT]\cdot[2,8\text{-}(NO_2)_2POT]$ and $[2\text{-}(NO_2)POT]\cdot[2,8\text{-}(NO_2)_2POT]$ were isolated as recrystallizable entities. The complex $POT\cdot[2\text{-}Cl,8\text{-}CH_3POT]$ was obtained as a by-product in the synthesis of phenoxtellurine from diphenyl ether and tellurium tetrachloride Phenyl 4-tolyl ether, present as impurity in diphenyl ether, gave 8-methylphenoxtellurine 10,10-dichloride. Tellurium tetrachloride then chlorinated the heterocyclic compound or the starting ether[105]. Upon reduction the complex with phenoxtellurine was formed.

Hetnarski and Hofman[174,174a,174b] studied the charge transfer complexes between phenoxtellurine and 1,3,5-trinitrobenzene, picric acid and picryl chloride. According to spectroscopic measurements in dichloroethane the reactants combined in a 1:1 ratio to form the charge-transfer complexes. Only the complex between phenoxtellurine and 1,3,5-trinitrobenzene was obtained as a solid, which melted at 73-5°. The ir spectrum of this complex showed, that the bands caused by the symmetric and antisymmetric stretching vibrations of the nitro groups were shifted by 20 cm^{-1} towards lower wavenumbers, while the phenoxtellurine bands remained unchanged. The authors believed that in these complexes the tellurium atom transfers electrons to the aromatic rings. Further data are listed in Table XII-10.

Drew[104] noted the appearance of a bluish-violet intermediate, when phenoxtellurine was oxidized to its 10,10-dinitrate by nitric acid. The reduction of the dinitrate with sulfur dioxide produced the same intermediate. Such highly colored substances were also formed when the dry compounds listed in Table XII-11 were rubbed together between glass plates. The color disappeared upon treatment of these complexes with water, ethanol, ether, benzene and chloroform. The components forming the complexes were recovered[104]. The reactants given in Table XII-11 combined also in glacial acetic acid. The isolation of the complexes in analytically pure form from these solutions was not possible, since decomposition occurred in the presence of the oxidizing nitrate ions. When phenoxtellurine was dissolved in concentrated sulfuric acid, sulfur dioxide was evolved indicating that phenoxtellurine had been oxidized and converted into a sulfate. During these reactions the solutions became red. Deeply colored substances containing the phenoxtellurine ring system, sulfate, hydrogen sulfate, and hydroxide ions, sulfuric acid and water were isolated[104,106]. The structure of

Table XII-10

Phenoxtellurine Complexes

Component 1	Component 2	Molar Ratio	mp. °C	Remarks	Ref.
Phenoxtellurine	2-chloro-8-methyl-phenoxtellurine	1:1	59	---	105,106
	2-nitrophenoxtellurine	1:1	145-85	---	105
	2,8-dinitrophenoxtellurine	1:2	145-220	---	105
	1,3,5-trinitrobenzene	1:1	73-5	$K = 13.2 \pm 2.0^{*}$; $\Delta H^{\circ\dagger} = -1.0 \pm 0.8$; $\Delta G^{\circ\dagger} = -1.52 \pm 0.09$	174,174a,174b
	picric acid	1:1	--	$K = 8.4 \pm 1.8^{*}$	174
	picryl chloride	1:1	--	$K = 8.0 \pm 2.1^{*}$	174

* $(\text{mole fraction})^{-1}$, 23° in CH_2Cl_2

† Kcal/mole

Table XII-11

Complexes formed by Phenoxtellurine and Its Derivatives in the Solid State

Phenoxtellurine	Phenoxtellurine Dinitrate or Hydroxide Nitrate	Color
phenoxtellurine	phenoxtellurine	intense violet
	2,8-dinitro	feeble violet
	4,8-dinitro	intense violet
2-nitro	phenoxtellurine	intense violet
	2-nitro	feeble violet
2,8-dinitro	phenoxtellurine	none
	2,8-dinitro	none
4,8-dinitro	4,8-dinitro	none

these complexes is unknown. Farrar[122] suggested, that the color of these adducts is caused by the presence of the radical cation *(141)*. The analytical results reported by Drew[104] would be consistent with such a formulation.

(141)　　　　*(142)*

Evidence for the existence of such a radical cation was found in electrochemical oxidation experiments with the heterocyclic compounds *(142)* (X = S, Se, Te)[20].

Campbell and Turner[55] attempted to resolve 2-carboxy- and 2-aminophenoxtellurine into their optically active antipodes by reacting these compounds with ℓ-menthol, optically active bases and potassium *d*-tartrate, respectively. The 2-substituted phenoxtellurine derivatives are asymmetric provided that the molecules are folded along the oxygen-tellurium axis. The failure to resolve these compounds was taken as an indication that rapid interconversion is taking place.

The triplet-triplet energy transfer in phenoxtellurine at 25° in cyclohexane solution has been used to measure the triplet state lifetime and the quantum yield of triplet formation[38b].

The phenoxtellurine ring system is very stable. In hot fuming nitric acid and in concentrated sulfuric acid the ring system remained intact. In basic solution, however, the tellurium atom was eliminated with formation of diphenyl ether[105]. The tellurium atom was replaced by a sulfur atom when 2,8-dimethylphenoxtellurine[146], 2,8-difluoro-phenoxtellurine or its 10,10-dichloride[422a], was fused with sulfur.

3) <u>1-Thia-4-telluracyclohexane</u>

Sodium telluride and bis(2-chloroethyl) sulfide in aqueous-ethanolic solution produced 1-thia-4-telluracyclohexane *(143)* in a 6.6 per cent yield[269].

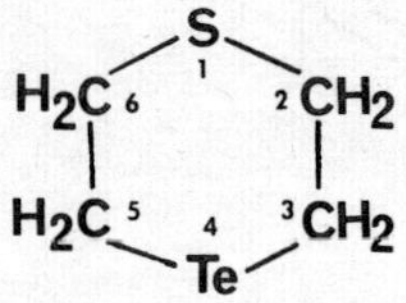

(143)

The white solid melted at 69.5°, and gave with the elemental halogens the 4,4-dichloride [41 per cent, m. 201-2°(dec)], the dibromide [68 per cent, m. 191-2°(dec)], and the diiodide [97 per cent, m. 150-1°

(dec)]. The structure of the diiodide has been determined by X-ray methods[194a]. With methyl iodide in acetone an 82 per cent yield of the telluronium salt, which sublimed under partial decomposition between 200-230°, was obtained.

4) <u>Thiophenoxtellurine</u>

Petragnani[339] prepared thiophenoxtellurine 10,10-dichloride *(144)* by heating 2-thiophenoxyphenyl tellurium trichloride at 240-50° for 30 minutes in a glass tube. The compound, obtained in 42 per cent yield, melted at 265-70°. At 230° a change in the crystalline form was noted. Sodium sulfide reduced the dichloride to thiophenoxtellurine (96 per cent yield, m. 122-3.5°). Equation (194) presents the various steps in the preparation of thiophenoxtellurine.

(194)

S, $NH_2 \cdot HCl$ —$NaNO_2/HCl$→ [S, $N_2^{\oplus}$]$^+$ Cl^-

—$HgCl_2/Cu$, 54.5%→ S, HgCl

—$TeCl_4$, dioxane, 80.6%→ S, $TeCl_3$

—heat 240-50°, 42%→ S 5, 10 Te Cl_2 (positions 1, 2, 3, 4, 6, 7, 8, 9) *(144)*

—Na_2S, 96%→ S, Te

5) Telluranthrene

Telluranthrene *(145)*, melting at 188-90°, was claimed as a product of the reaction between tellurium and tetraphenyl tin in an evacuated glass tube. When 0.05 moles of tetraphenyl tin and 0.1 mole of tellurium were heated for eight hours at 310°, eight grams of telluranthrene, four grams of diphenyl telluride and ten grams of diphenyl were isolated. The telluranthrene was probably formed by thermal decomposition of diphenyl telluride[379].

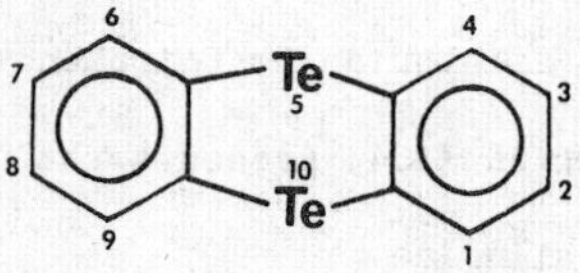

(145)

6) 1,3,5-Tritelluracyclohexane

1,3,5-Tritelluracyclohexane was formed when methylene generated by thermal decomposition of diazomethane reacted with tellurium mirrors[441]. The same cyclic compound might have been obtained by earlier investigators[23,335,365] (see chapter X). The chemistry of this compound remains unexplored.

XIII. TELLURIUM CONTAINING POLYMERS

Almost no efforts have been made to synthesize well defined polymeric organic tellurium compounds. Polymeric materials were observed as unwanted by-products. In the reaction between sodium telluride and bis(2-chloroethyl) sulfide a polymeric condensation product was formed, which was not further characterized[269]. A similar substance was obtained from sodium telluride and bis(2-chloroethyl) ether[121]. Diiodomethane heated with tellurium gave as the main product a red polymer[121,153]. The by-product, bis(iodomethyl) tellurium diiodide, when reduced gave the telluride, which polymerized upon heating and in solution[121]. Gaseous telluroformaldehyde, obtained from the action of methylene on tellurium mirrors, polymerized on condensation[23,335,365]. The degree of polymerization of these samples is unknown. Morgan and Drew[286] reduced methylene bis(tellurium trichloride) with potassium disulfite and isolated ditelluromethane, CH_2Te_2, a red amorphous, polymeric substance, which was insoluble in water and organic solvents.

Hellwinkel and Fahrbach[173] identified polymeric tellurides as products of the thermal decomposition of bis(2,2'-biphenylylene) tellurium. These insoluble substances did not melt up to 550°. The structure *(146)* was tentatively assigned to these compounds.

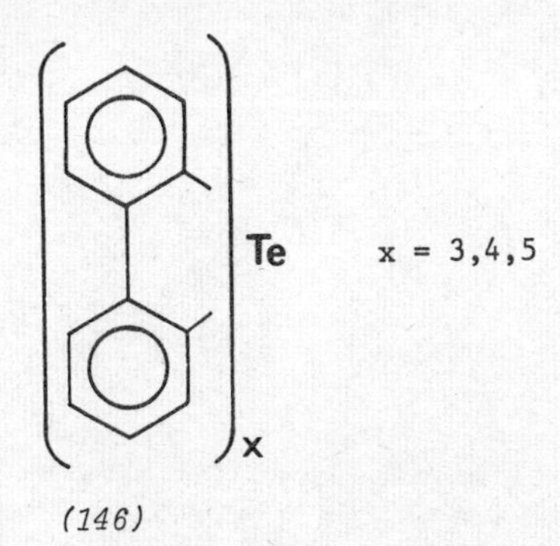

(146)

Akobjanoff[4] prepared crosslinked polymers by reacting tellurium halides with unsaturated organic compounds. The viscosity of the products depended on the ratio of tellurium halide to unsaturated compound employed in the reaction. The "transgel condensates" obtained in this manner are claimed to be useful as lacquers and adhesives for natural and synthetic rubbers.

Diphenyl tellurium dichloride reacted with the silver salts of dicarboxylic acids according to equation (195).

(195)

$$n(C_6H_5)_2TeCl_2 + n\,AgOOC(CH_2)_pCOOAg \xrightarrow{-2n\,AgCl} \left[-OOC(CH_2)_pCOO-Te(C_6H_5)_2- \right]_n \qquad (195)$$

p, solvent, mp.%yield: 2, dioxane, 255-7°,-; 8, CH_3CN, 164-6°, 35%.

The molecular mass of the polymer with n = 8, as determined in benzene solution by vapor pressure osmometry, was 3860[245a].

XIV. PHYSICOCHEMICAL INVESTIGATIONS OF ORGANIC TELLURIUM COMPOUNDS

Infrared and Raman spectroscopy, ultraviolet and visible spectroscopy, nuclear magnetic resonance, nuclear quadrupole resonance, and X-ray and mass spectrometric techniques have been employed to characterize organic tellurium compounds.

Most of the infrared investigations cover the range 3100 cm^{-1} to 300 cm^{-1}. A few compounds have been examined beyond this region where the absorptions caused by Te–Te vibrations and most bending modes involving the tellurium atom occur. Normal coordinate analyses for a few methyl tellurium compounds have been performed. Ultraviolet and visible spectral characteristics of a few organic tellurium compounds have been determined, but assignments of the observed bands to certain electronic transition have not been made.

The first nmr investigation of an organic tellurium compounds seems to have been carried out by Simonnin[399] in 1963 on phenyl propargyl telluride. Detailed analyses of the spectra of the simpler molecules, dimethyl and diethyl telluride, appeared in the literature a few years later. The only nuclear quadrupole resonance investigation in this area was performed on dimethyl tellurium dichloride probing the chlorine nucleus.[195]

The structural investigations have been carried out mainly by McCullough and Foss. Little is known about the mass spectral behavior of organic tellurium compounds.

A) Infrared, Ultraviolet and Visible Spectroscopy

A detailed investigation of the ir and/or Raman spectra of methanetellurol[401], dimethyl telluride, its dichloride, dibromide and diiodide[168], of dimethyl ditelluride and dimethyl ditelluride-d_6[401a] have been carried out. A special 10 cm gas cell was designed to take the spectrum of methanetellurol[167a]. The assignments of the observed bands are given in Table XIV-1. The tellurium-carbon stretching frequency, found for the dimethyl derivatives in the range 472-544 cm^{-1} [136,168,401,401a], was located in the region 460-506 cm^{-1} for the diethyl compounds[136] listed in Table XIV-2. The carbon(phenyl)-tellurium stretching frequencies of eight compounds containing phenyl groups were tentatively assigned to weak bands between 455 and 487 cm^{-1} appearing as shoulders on the intense deformation vibrations of the phenyl rings[136]. Values between 2.10 and 2.28 mdyn/Å were calculated as the force constants for the carbon(alkyl)-tellurium stretching modes[133,136,401a]. Other pertinent force constants derived from the spectra of dimethyl telluride and dimethyl ditelluride were tabulated by Freeman and Henshall[133] and Sink and Harvey[401a], respectively. A value of 4 mdyn/Å was suggested as the force constant for the carbon(phenyl)-tellurium stretching vibration[136].

The far infrared spectrum in the range 350 cm^{-1} to 33 cm^{-1} of solid dimethyl telluride at -190° consisted of bands arising from a skeletal bending mode (198 cm^{-1}), a torsional mode (185 cm^{-1}) and lattice modes (122, 103, 90, 45 cm^{-1}). A barrier to internal rotation of 2.27 kcal/mole was calculated using the torsional frequency[113a]. The C-Te-C bending vibration in the complexes *trans*-$[(CH_3)_2Te]_2PtX_2$ caused absorptions at 210 cm^{-1} (m) (X = Br) and 214 cm^{-1} (m) (X = I).

Cyclic and linear condensation products between tellurium tetrachloride and 2,4-pentanedione and 3-chloro-2,4-pentanedione were investigated with infrared techniques to elucidate the structures of these compounds[97]. The formulation of the cyclic condensation products as 1-tellura-3,5-cyclohexanediones was confirmed. The linear tellurium dichlorides *(147)*, by-products in these reactions, were found to exist

Table XIV-1

Infrared Absorptions of Methyl Tellurium Compounds*

Compound / Assignment	CH_3TeH	$(CH_3)_2Te$	$(CH_3)_2TeCl_2$	$(CH_3)_2TeBr_2$	$(CH_3)_2TeI_2$	$(CH_3)_2Te_2$	$(CD_3)_2Te_2$	CH_3TeCl_3	CH_3TeBr_3	CH_3TeI_3	$(CH_3)_3TeBr$
	I	II	III	IV	V	VI	VII	VIII	IX	X	XI
ν_{C-Te}	*516*vs	528s (A_1,B_1) *526*	544 (A_1,B_1) *544*s	540vs *536*s	527wm *524*wm	507ms *509*s	472vs *475*		524vw *470*wm†	480w	534w
ν_{Te-H}	*216*s	--	--	--	--	--	--	--	--	--	--
ν_{Te-X}	--	--	281s *276*s	154ms *158*ms	113s *116*s	--	--	338s(br)	202s 208m,sh†	172s(br)	--
$\nu_{as,Te-X}$	--	--	248s *251*vw	183s	144s *148*w	--	--	315s(br)	226vs 234s†	--	--
ν_{Te-Te}	--	--	--	--	--	189mw *188*vs	193vw *189*s	--	--	--	--
δ_{C-Te-H}	*608*w	--	--	--	--	--	--	--	--	--	--
δ_{C-Te-C}	--	*198*	213s *218*m	*191*s	190wm	--	--	--	--	--	--
$\delta_{as,Te-Te-C}$	--	--	--	--	--	170	--	--	--	--	--

Table XIV-1 (cont'd)

	I	II	III	IV	V	VI	VII	VIII	IX	X	XI
δ_{X-Te-X}	--	--	--	--	--	--	--	200ms 140m 103m	133s(br) 105m(br) 143ms† 109m† 61w†	139m 98m	--
$\nu_{s,C-H}$	*2931*s	*2948*,2918 *2923*	2926wm	2922vw	2920w	2920vs	2118vs *2118*w	--	--	--	--
$\nu_{as,C-H}$	*3019*vw	*3018*, 2998s *3000*	3022w	3020w	--	3007wm	2258wm	--	--	--	--
$\delta_{s,H-C-H}$	*1220*m	1227m(A_1) *1225*,1208m(B_1) *1228*	1229wm *1230*wm	1220m *1227*w	1212s	1206vvs *1218*w	929vs *921*w	--	--	--	1212w
$\delta_{as,H-C-H}$	*1417*vvw	*1414*(A_2) *1440*(B_2) *1426*,1415vs(B_1) *1420*	1403s *1404*vw	1400s *1400*vw	1390ms	1405ms 1405vvw	1038vs	--	--	--	1233w
CH_3 rock	*870*vww	*872*,875s(A_1) *805*,790vw(B_2) *843*,837s(B_2)	916s(A_1) 869s(B_2)	911ms(A_1) 862s(B_2)	901ms(A_1) 852ms(B_2)	824vs	620vs	--	858s	--	967wm 905m

Table XIV-1 (cont'd)

	I	II	III	IV	V	VI	VII	VIII	IX	X	XI
Reference	401	8,77,136, 168	168	77,168	136,168	76,401a	401a	447a,b,c	77,447c[†]	447c	77

* Underlined frequencies in italics are taken from infrared gas phase spectra. Frequencies in italics were obtained from Raman spectra.

† These frequencies are taken from reference 447c.

Table XIV-2

Infrared Absorption Bands of Alkyl Tellurium Compounds

Range cm^{-1} / Compound	$Te(CN)_2$	$[i\text{-}Pr_2NNHC(S)Te]_2$	$(CH_3)_2Te$	$(CH_3)_2TeI_2$	$(CH_3)_3TeI$	$C_2H_5TeCl_3$	$C_2H_5TeBr_3$	$C_2H_5TeI_3$	$(C_2H_5)_2Te$	$(C_2H_5)_2TeI_2$	$(C_2H_5)_3TeI$	$CH_3(C_2H_5)_2TeI$
	I	II	III	IV	V	VI	VII	VIII	IX	X	XI	XII
3100–3000		3070s	3002ms		3014m							
3000–2100	2181m,sh 2179m	2965s 2920s 2867s 2800s	2922s 2810m 2420w	2930m 2908w	2923w				2930s 2902s 2856s	2971m 2940m 2897vw,sh	2955ms 2908m 2805m	2931m 2900m 2845w,sh
1900 1800												
1700			1736w		1745w							
1600					1631w							
1500												

Table XIV-2 (cont'd)

	I	II	III	IV	V	VI	VII	VIII	IX	X	XI	XII
1300	1316w	1388m 1368m 1324m 1308s		1396s						1375m	1399w 1376m	1379m
1200		1208w	1225m,sh 1208vw	1226m 1209vs	1248m 1209ms				1294m	1203vs		1203vs
1100		1183w 1170s 1138s 1124m 1100w							1183vs 1117w	1182vs	1199vs	
1000	1086	1007s							1035w	1041w	1037m	1038w
900		992m 940w 923w							961m	965w,sh 948m	945m	985w 957w
800		897m 850w	876s 840vs	898s 849vs	894vs 826m							855m
700		770s 704m	782s 756vs						794m	717m		
600		611m	676w	608w					671m			671m

Table XIV-2 (cont'd)

	I	II	III	IV	V	VI	VII	VIII	IX	X	XI	XII
500 500		570w 520m 510m	525vs 511w,sh	525s 519m,sh	530m,sh 528vs 520s				513sh 503m		591w 552vw 506m	529m
400	460vw 403s	475w 460w				486m	469w	460w		498m		477m
300						385m 350ms 316ms						
200							213m,sh 212s	230w				
100						140s (br)	118s,98s 57w	183ms 153s 105m 87s				
Cond-itions	KI	KBr	neat	KBr, Nujol	KBr, Nujol	Nujol	Nujol	Nujol	neat	KBr, Nujol	KBr, Nujol	KBr, Nujol
Ref.	136	12	136	136	136	447c	447c	447c	136	136	136	136

in the enol form with an intramolecular hydrogen bond[97] (see sections XII-B-2, VII-B-1, IV-B-1a for more information on these compounds).

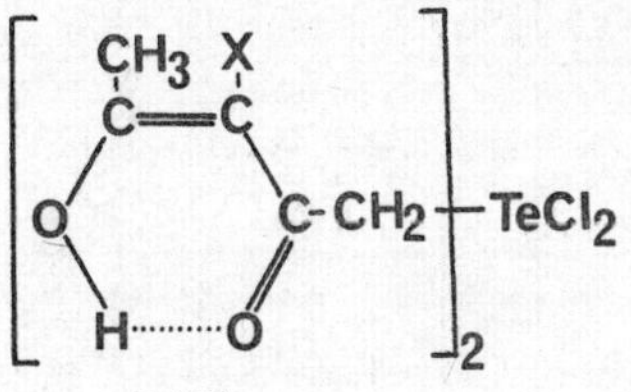

$$\left[\begin{array}{c} CH_3 \;\; X \\ C{=}C \\ O \qquad\qquad C{-}CH_2 \\ H \cdots\cdots O \end{array} \right]_2 TeCl_2 \qquad X = H,Cl$$

(147)

Fringuelli[133b] reported infrared bands arising from carbon-oxygen modes in 2-R-tellurophenese [R = -COOH, $-COOCH_3$, -C(O)H, $-C(O)CH_3$] and 2-acetyl-5-R-tellurophenes [R = -COOH, $-COOCH_3$] and listed the hydroxyl frequencies for 2-hydroxymethyl- and 2-(1'-hydroxyethyl)tellurophene.

Fritz and Keller[136] recorded the infrared spectra of several tellurium compounds. The reported absorption bands are listed in Tables XIV-2 and XIV-3. In addition to the data collected in Tables XIV-1, 2 and 3, the following infrared bands (cm^{-1}) for $[(4\text{-}RC_6H_4)_2C{=}CH]_2Te$ have been reported[119a]: R = H: 3050m, 3010m, 1600m, 1590m, 1490m, 1440s; R = CH_3: 3020m, 2920m, 2850w, 1600m, 1550m, 1500s, 1450m, 1400m. Radchenko and coworkers published the infrared spectra of 5-methyltelluroethynyl- and 5-ethyltelluroethynyl-1,3-diphenyl-2-pyrazoline[354] and of 1-methyltelluro-3-methyl-3-buten-1-yne[355]. Detailed assignments of the observed bands to certain vibrations have not been made for these tellurides and the compounds given in Tables XIV-2 and XIV-3.

Egorochkin and coworkers[115,116,117] recorded the infrared spectra of the compounds $(C_2H_5)_3M\text{-}Te\text{-}M'(C_2H_5)_3$ (M, M' = Si, Ge, Sn). The frequencies characteristic of the ethyl groups were found to be independent of the metal M. The metal-carbon stretching frequencies were tabulated. The tellurium-metal modes expected to occur below 250 cm^{-1} were not detected.

The Si-H stretching frequencies for the compounds $(C_2H_5Te)_n(C_2H_5)_{3-n}SiH$

Table XIV-3

Infrared Absorption Bands of Phenyl Tellurium Compounds

Range, cm^{-1} / Compound	$(C_6H_5)_2Te$	$(C_6F_5)_2Te$	$(C_6H_5)_2TeCl_2$	$(C_6H_5)_3TeI$	$(C_6H_5)_4Te$	$CH_3TeC_6H_5$	$C_6H_5C(O)TeC_6H_5$	$CH_3(C_6H_5)_2TeI$	$C_6H_5TeCl_3$	$(C_6H_5)_2Te_2$	tellurophene	tetrachloro tellurophene
	I	II	III	IV	V	VI	VII	VIII	IX	X	XI	XII
3100–3000	3064 w		3025w	3059w	3020w	3030m		3044w		3040s	3090* 3080m† 3080m†	
3000–2100				2934w	2995w 2950vw	2920m		2935vw	2900w 2840w	2903m,sh		
1900	1950vw					1943w				1945w		
1800	1873vw 1806vw					1863w				1866w 1800w		
1700	1742vw					1799w 1734w	1725s			1748w		
1600		1637s		1624w			1682vs			1637w		
1500	1574m	1508vs	1570m	1578m	1570w 1565w,sh 1560w,sh	1595sh 1577s	1580m 1570m	1570m		1570s		
1400	1475m	1484vs	1475ms [illegible]	1479ms [illegible]	1473s 1435m,sh	1475s 1433s	1475m 1450m	1478m 1436m	1467m 1432m	1468s 1430s	1431* 1425s†	1490m

Table XIV-3 (cont'd)

	I	II	III	IV	V	VI	VII	VIII	IX	X	XI	XII
1300	1324w	1383m	1387w 1327m 1304w	1379vw 1330vw 1305vw	1376s	1325w	1310w 1300w	1399vw	1366m	1377w 1322w	1316* 1310m†	
1200	1297w		1270w 1207m	1297vw	1290w	1294w 1217m	1200vs	1214w	1285m 1249s	1294w 1260vw	1245* 1227* 1240s† 1220s†	1230s
1100	1154w		1159m,sh 1150w	1157w	1175w 1152w	1174w 1154w	1170s 1100w	1196w 1179w 1164w 1156w	1159m 1108s	1175w 1152w	1100m†	
1000	1063w 1017s	1083vs	1095w 1066m 1057m,sh 1052m 1015m	1061m,sh 1054w 1015w	1047m 1015w	1061m 1017s	1070w 1020m 1000m	1094w 1060w 1017w	1093s 1063s	1089vw 1059ms 1014s	1078*	1000m
900	998m	971vs	994s 974w 923w	996m	994m	997s	925w	997s 917w	993ms	995w 963w 908w	983* 975m†	
800			824w		845w	832w	855vs	858m 839ms 824ms	888ms 842s	850w 842w		805s
700	726vs	793s	747s 737vs	733vs	733s 724s	779w 727vs	735m	753s 742vs 730vs	737vs	732vs	796* 785m†	

Table XIV-3 (cont'd)

	I	II	III	IV	V	VI	VII	VIII	IX	X	XI	XII
600	688vs 662s 645w 611w		688s 682s 656w 606vw	684s	695s 649w 630w	698w,sh 689s 651w 606vw	690s 660s 655s 620w	687s 680w	680s	685s 652ms 614w	672* 655s†	695m
500						542vw 520w		529m 523m		585vw		
400	469w 461s 445s		464m 455s	464m,sh 455s 450w,sh	480m 474m 462s 453s 442w	484vw 413s		465s 458w 450s 424vw	474vw 463m	449s		
300									311m 305m			
200			287m 271m									
Conditions	neat	Nujol	KBr, Nujol	KBr, Nujol	KBr, Nujol	neat	CCl_4	KBr, Nujol	KBr, Nujol	neat	neat*	--
Ref.	136	95a	136	136	136	136	348d	136	136	136	*133b †257	256

were located at 2117 cm^{-1} (n = 1), 2122 cm^{-1} (n = 2) and 2138 cm^{-1} (n = 3)[117a]. The shift toward higher wavenumbers with increasing number of ethyltelluro groups bonded to the silicon atom was explained on the basis of a small contribution from $(Te)_{p\pi}-(Si)_{d\pi}$ bonding to the Te-Si bond. The Si-H stretching frequency for $(C_2H_5Te)(CH_3)_2SiH$ was found at 2131 cm^{-1} [117a].

The tellurium atom in tellurides of the type $[(C_2H_5)_3M]_nTeR_{2-n}$ [n = 2, M = Si, Ge, Sn; n = 1, M = Si, R = $(C_2H_5)_3Sn$, $(C_2H_5)_3Ge$; n = 1, M = Ge, R = $(C_2H_5)_3Sn$] acts as deuterium acceptor towards $CDCl_3$. An investigation of the C-D stretching vibration revealed that the basicity of the tellurium atom increases slightly in the order Si<Ge<Sn[117b]. Diethyl telluride in carbon tetrachloride solution forms similarly hydrogen bonds with phenol and indole[77a].

Schumann and Schmidt[386] investigated the compounds $(C_6H_5)_3M-Te-M'(C_6H_5)_3$ (M, M' = Ge, Sn, Pb) in the region 5000 to 250 cm^{-1}. The tellurium-metal frequencies were not detected. However, absorption bands at 162 cm^{-1} (164 cm^{-1}) and 125 cm^{-1} (122 cm^{-1}) in the complex $[(CH_3)_3Sn]_2Te\cdot W(CO)_5$ $\{[(CH_3)_3Sn]_2Te\cdot Cr(CO)_5\}$ were assigned to the asymmetric and symmetric Sn-Te stretching frequencies, respectively[390a]. Hooton and Allred[181] reported the silicon-carbon and carbon-hydrogen modes for bis(trimethylsilyl) telluride. Metal-tellurium stretching and bending frequencies were located in the following compounds (R = Raman):

$[(CH_3)_3Si]_2Te$:	ν_{as} 323 cm^{-1} ν_s 330 cm^{-1}(ir)	(Ref. 49)
$(H_3Si)_2Te$:	ν_{as}, ν_s 335 cm^{-1}(ir), 334 cm^{-1}(R)	(Ref. 49)
	δ_{Si_2Te} 85 cm^{-1}(R)	(Ref. 49)
$(H_3Ge)_2Te$:	ν_{as}, ν_s 228 cm^{-1}(ir, R)	(Ref. 86)
	δ_{Ge_2Te} 73 cm^{-1}(R)	(Ref. 86)

The infrared spectra of the germanium compounds were recorded using the non-annealed solid and the crystal at -196°. The Raman spectrum was taken with the liquid compound at room temperature[86].

The infrared absorption frequencies for Te=C=X (X = O, S, Se) are listed in Table V-1 (chapter V).

Infrared spectra of many transition metal complexes with organic tellurium compounds as ligands have been published. In most of these papers only frequencies arising from carbonyl modes in carbonyl complexes or from metal-halogen modes in halogen containing complexes have been reported.

The few transition metal-tellurium stretching frequencies, which have been assigned are listed below:

ν_{Pt-Te} 185 cm^{-1} (wm) in *trans*-$[(C_3H_7)_2Te]_2Pt_2Cl_4$ [2a]

ν_{Pt-Te} 197 cm^{-1} (ms) (or 177 cm^{-1}) in *trans*-$[(C_2H_5)_2Te]_2Pt_2Cl_4$ [2a]

ν_{Pt-Te} 169 cm^{-1} (m) in *trans*-$[(CH_3)_2Te]_2PtBr_2$ [9]

ν_{Pt-Te} 187,156 cm^{-1} in *cis*-$[(CH_3)_2Te]_2PtCl_2$ [9]

ν_{Pd-Te} 183 cm^{-1} in $[(CH_3)_2Te]_2Pd_2Cl_4$ [9]

ν_{Pd-Te} 228 or 158 cm^{-1} in $[(CH_3)_2Te]_2Pd_2Br_4$ [9]

Infrared data for the following complexes have been reported:

$[(CH_3)_3Sn]_2Te \cdot Cr(CO)_5$ — ν_{CO} [390a]

$[C_6H_5TeMo(CO)_2\pi\text{-}cp]_2$ — ν_{CO} [419]

$[C_6H_5Te \cdot Mo(CO)_3\pi\text{-}cp]$ — ν_{CO} [419]

$[(CH_3)_3Sn]_2Te \cdot W(CO)_5$ — ν_{CO} [390a]

$(C_6H_5)_2Te \cdot Mn(NO)_3$ — ν_{NO} [176]

$[(C_6H_5)_2Te]_2Mn(CO)_3X$ — ν_{CO} (X = Cl, Br, I) [175,176]

$(POT)_2Mn(CO)_3Cl$ — ν_{CO} (POT = phenoxtellurine) [176]

$[C_6H_5TeMn(CO)_4]_2$ — ν_{CO} [176]

$(C_6H_5)_2TeMn(CO)_3(AAB)Br$ — ν_{CO} (AAB = 4-aminoazobenzene) [178]

fac-$[(C_2H_5)_2Te]_2ClRe(CO)_3$ — ν_{CO} [120a]

cis-$[(C_2H_5)_2Te]ClRe(CO)_4$ — ν_{CO} [120a]

$(C_6H_5)_2TeFe(CO)(NO)_2$	ν_{CO}, ν_{NO} [176]	
$(C_6H_5)_2TeFe(CO)_4$		ν_{CO} [176]
$(C_6H_5)_2TeFe(CO)_3I_2$	ν_{CO} [176]	
$[4\text{-}CH_3OC_6H_4TeFe(CO)_3]_2$		ν_{CO} [176]
$[C_6H_5TeFe(CO)_3]_2$	ν_{CO} [197]	
$[C_6F_5TeFe(CO)_3]_2$		ν_{CO} [197]
$[RTeFe(CO)_3]_2$	ν_{CO} [357a]	($R = C_6H_5$, $4\text{-}CH_3OC_6H_4$, C_6F_5)
$[C_6H_5TeFe(CO)_3]_2$		ν_{CO} [378b]
$[C_6H_5TeFe(CO)_2\pi\text{-cp}]$	ν_{CO} [378a]	
$[(C_6H_5)_2Te]_3Ru(CO)X_2$	ν_{CO} [178a]	(X = Cl, I)
$[(C_6H_5)_2Te]_2Ru(CO)_2Cl_2 \cdot 1/2\ CH_2Cl_2$		ν_{CO} [178a]
$[(C_4H_9)_2Te]_3Ru(CO)I_2$	ν_{CO} [178a]	
$[C_6H_5TeRu(CO)_3]_2$		ν_{CO} [378b]
$[(C_6H_5Te)_2Ru(CO)_2]_n$	ν_{CO} [378b]	
$[(R_2Te)_2Ru(CO)_2X_2$		ν_{CO} [189a] ($R = C_6H_5$, C_4H_9; X = Cl, Br, I)
$[CO_2Te(CO)_5]_n$	ν_{CO} [176]	
trans-$\{[(C_2H_5)_2Te]_2Rh(CO)Cl$		ν_{CO}, ν_{Rh-Cl} [120b]
$\{[(C_2H_5)_2Te]_2Rh(CO)R(Cl)X\}$	ν_{CO}, ν_{Rh-Cl} [120b]	(R,X: CH_3CO, I; $C_6H_5SO_2$, Cl; CH_3, I)
$\{[(C_2H_5)_2Te]_2Rh(CO)ClX_2\}$		ν_{CO}, ν_{Rh-Cl} [120b] (X = Cl, Br, I)
$[(CH_3)_2Te]_2Pd_2X_4$	range: 400-100 cm^{-1} [9]	
trans-$\{[(CH_3)_2Te]_2PdI_2\}$		range: 400-100 cm^{-1} [9]
trans-$\{[(C_2H_5)_2Te]_2Pd(R)X\}$	ν_{Pd-C}, ν_{Pd-X}	(R,X: see Table XI-3)
cis-$\{[(CH_3)_2Te]_2PtCl_2\}$		range: 400-100 cm^{-1} [9]
trans-$\{[(CH_3)_2Te]_2PtX_2\}$	range 400-100 cm^{-1} [9]	(X = Br, I)
trans-$\{[(C_2H_5)_2Te]_2PtCl$ (mesityl)$\}$		ν_{Pt-Cl} [392b]
trans-$\{[(C_2H_5)_2Te]_2PtCl(C_6H_5)\}$	ν_{Pt-Cl} [392b]	
$[R_2Te(amine)PtCl_2]$		ν_{N-H} [67,71]

Chen and George[77] examined the infrared spectra of the adducts $(CH_3)_3TeBr \cdot BBr_3$ and $[(CH_3)_2TeBr_2]_2 \cdot BBr_3$. Wynne and coworkers reported the spectra of adducts of the type $RTeX_3 \cdot SbCl_5$ [447b] and $RTeX_3 \cdot$tetramethylthiourea[447c]. Their assignments for modes involving the tellurium atom are summarized in Table XIV-4.

Table XIV-4

Infrared Absorption Bands of Compounds $R_nTeX_{4-n} \cdot L$ (cm^{-1})

$RnTeX_{4-n} \cdot L$	ν_{Te-C}	ν_{Te-X}	Te-X and/or Te-S modes	δ_{X-Te-X}	Conditions	Ref.
$CH_3TeCl_3 \cdot$tmtu*	473w	251s	210s, 155wm	--	Nujol	447c,d
$CH_3TeBr_3 \cdot$tmtu	475w	?	195s(br),165s(br),115m	--	Nujol	447c
$CH_3TeI_3 \cdot$tmtu	--	170s,135m,98m	210m,sh	--	Nujol	447c
$CH_3TeCl_3 \cdot SbCl_5$	526w	375s	--	283w,265	Nujol	447b
$(CH_3)_3TeBr \cdot BBr_3$	538m	--	--	--	Nujol	77
$[(CH_3)_2TeBr_2]_2 \cdot BBr_3$	535w	188s,149s	--	102s,98s	Nujol	77
$C_2H_5TeCl_3 \cdot$tmtu	460w	248ms 248ms	159w,55w(br) 162s	--	Nujol CH_2Cl_2	447c
$C_2H_5TeBr_3 \cdot$tmtu	472wm	--	190s(br),168s,sh(br),140m,sh	--	Nujol	447c
$C_2H_5TeI_3 \cdot$ tmtu	478wm	165s,130s,60w	215wm,sh	--	Nujol	447c
$C_2H_5TeCl_3 \cdot SbCl_5$	--	370s	--	275	Nujol	447b
4-$CH_3OC_6H_4TeCl_3 \cdot$tmtu	585wm,520ms 533w	 265ms	225s,150w,90w(all br) 251s,205w,sh,142w	--	Nujol CH_2Cl_2	447c
4-$CH_3OC_6H_4TeCl_3 \cdot SbCl_5$	418w	373s	--	265	Nujol	447b

As the result of infrared investigations of phosphine tellurides, $R_3P{=}Te$, the P-Te stretching vibration was assigned to bands falling into the range 400-518 cm^{-1} for R = C_nH_{2n+1} (n = 2,3,4,5,8) and cyclohexyl. The corresponding vibration in $[(CH_3)_2N]_3PTe$[352a] was found at 519 cm^{-1}, from which the rather large force constant of 5.89 mdyne/Å was obtained. The calculation of the P-Te bond order according to Siebert's method gave the anomalous value of 2.83[352a].

The organic tellurium compounds, whose ultraviolet and visible spectra have been recorded, are listed in Table XIV-5. The wavelength(s) of maximum absorption and the logarithm(s) of the molar absorbance(s) are also tabulated. The absorption of the methyl tellurium compounds was drastically reduced, when wet ethanol was used as a solvent[248] in these investigations.

B) Nuclear Magnetic Resonance Spectroscopy

The organic tellurium compounds which have been investigated with nmr techniques, are compiled in Table XIV-6 together with pertinent chemical shifts and coupling constants. A review on coupling constants for M-C-H and M-C-C-H, where M represents a variety of metal atoms and a few non-metal atoms including tellurium, has been published by Frischleder and co-workers[134].

Lambert and Keske[209, 210] determined the conformation of the telluracyclohexane ring system employing nmr techniques. The conformation of 1-telluracyclohexane 1,1-dibromide as shown in *(148)* was deduced from the values of the H-H coupling constants[203].

Table XIV-5

The Ultraviolet and Visible Absorption Spectra of Organic Tellurium Compounds

COMPOUND	$\lambda_{max} x10^9 m (\log \varepsilon)$	Solvent	Reference
2-naphthyl tellurium iodide	580(2.4), 320(3.9), 260(4.2)	cyclohexane	427
methyl tellurium triiodide	350(4.14), 290(4.37)	ethanol	249
$CH_3TeCl_3 \cdot (CH_3)_3TeCl$	log ε increases monotonously: 340(1.0), 290(2.0), 275(3.0)	ethanol	248
$CH_3TeBr_3 \cdot (CH_3)_3TeBr$	log ε = 4.0 from 285 to 265	ethanol	248,249
$CH_3TeI_3 \cdot (CH_3)_3TeBr$	348(3.64), 289(3.90)	ethanol	249
$CH_3TeI_3 \cdot (CH_3)_3TeI$	336(4.49), 264(4.70) 366(3.8), 290(4.25)	cyclohexane ethanol	144 248,249
$C_2H_5TeBr_3 \cdot (C_2H_5)_3TeBr$	290(4.0), 269(4.2)	cyclohexane	144
$C_2H_5TeI_3 \cdot (C_2H_5)_3TeI$	335(4.51), 267(4.80)	cyclohexane	144
dibutyl ditelluride	399(3.8)	cyclohexane	26
bis(carboxymethyl) ditelluride	---	---	26
diphenyl ditelluride	407(2.97), 250(4.40)	chloroform	122
$(C_6H_5)_2Te_2 \cdot UCl_5$	---	---	392a
dimethyl telluride	257(-), 250(-), 242(-), 200(-)	--	83a

Table XIV-5 (cont'd)

bis(2,2-diphenylvinyl) telluride	344(4.23)	$CHCl_3$	119a
	340(4.31)	CH_3CN	119a
bis[2,2-*p*-tolylvinyl) telluride	258(4.34), 356(4.51)	$CHCl_3$	119a
	256(4.38), 350(4.53)	CH_3CN	119a
ethyl phenyl telluride	265(3.42), 251(3.60), 224(4.35)	ethanol	39
	330(2.83), 270(3.60), 251(3.52), 225(4.16)	cyclohexane	39
2-naphthyl phenyl telluride	strong absorption from 400-200	cyclohexane	427
trans-[$(C_2H_5)_2$Te,piperdine $PtCl_2$]	340(2.54), 245(4.3)	---	75
dimethyl tellurium dichloride	log ε increases monotonously: 300(1.0), 280(2.0), 260(3.0)	ethanol	248
dimethyl tellurium dibromide	log ε = 3.7 from 280 to 260	ethanol	248
dimethyl tellurium diiodide	357(3.7), 284(4.05)	ethanol	248
	344(3.72), 272(4.00)	ethanol	249
	340(4.33) 272(4.40)	chloroform	249
	340(4.27), 268(4.47)	cyclohexane	144
	333(2.40)	water	249
dimethyl tellurium dicyanide	log ε increases monotonously: 290(1.8), 258(3.20)	ethanol	249
dimethyl tellurium diperchlorate	log ε increases monotonously: 280(1.36), 250(2.20)	ethanol	249
dimethyl tellurium iodide perchlorate	344(3.94), 256(4.12)	ethanol	249

Table XIV-5 (cont'd)

dimethyl tellurium iodide cyanide	342(3.54), 274(3.70)	ethanol	249
dimethyl tellurium iodide triiodide	361(4.3), 290(4.5)	ethanol	248
diethyl tellurium dibromide	290(3.8), 266(4.0)	cyclohexane	144
diethyl tellurium diiodide	335(4.36), 270(4.65)	cyclohexane	144
diethyl tellurium iodide triiodide	335(4.82), 263(4.95)	cyclohexane	144
bis(2,4-dioxopentyl) tellurium dichloride	290(4.09)	chloroform	97
bis(3-chloro-2,4-dioxopentyl) tellurium dichloride	315(4.36)	chloroform	97
trimethyl telluronium iodide	log ε increases monotonously: 270(1.0), 260(2.0), 250(3.0)	ethanol	248
diethyl phenyl telluronium iodide	260(3.30), 220(4.47)	ethanol	39
diphenyl telluronium tetraphenylcyclopentadienylide	347(-), 280(-)	benzene	132a
	280(-), 252(-)	methanol	132a
1-telluracyclopropane	247(-), 241(-), 235(-)	---	83a
1-telluracyclobutane	248(-), 246(-)	---	83a
1-tellura-3,5-cyclohexanedione	306(4.66), 243(4.98)	chloroform	97

Table XIV-5 (cont'd)

1-tellura-3,5-cyclohexanedione 1,1-dichloride	308(4.15), 243(4.80)	chloroform	97
1-telluracyclopentane 1,1-dibromide	log ε = 3.85 from 285-260	ethanol	143
1-telluracyclopentane 1,1-diiodide	338(3.92), 277(4.05)	ethanol	143
1-telluracyclopentane 1-iodide 1-triiodide	356(4.28), 288(4.45)	ethanol	143
pentamethylene bis(cyclopentamethylene telluronium triiodide)	356(4.25), 290(4.44)	methanol	143
tellurophene	280(3.62) 279(3.93), 241(3.36), 209(3.56)	cyclohexane hexane	257 133b
tetraphenyltellurophene	310(4.2), 235(4.6)	cyclohexane	273
tetrachlorotellurophene	295(4.05)	cyclohexane	256
2-acetyltellurophene	346(3.58), 282(3.87), 211(3.93)	hexane	133b
benzotellurophene	320(4.17), 252(4.43), 212(4.43)		348e
phenoxtellurine·1,3,5-trinitrobenzene	440(3.01)	$(ClCH_2)_2$	174
phenoxtellurine·picric acid	420(2.96)	$(ClCH_2)_2$	174

Table XIV-5 (cont'd)

phenoxtellurine·picryl chloride	420(2.92)	$(ClCH_2)_2$	174
5,10-epitelluro-5,10-dihydroarsanthrene	360(2.89), 350(2.92), 291(3.49), 260(4.04)	$CHCl_3$	7a

Table XIV-6

Nuclear Magnetic Resonance Studies of Organic Tellurium Compounds

Compound	Solvent	Chemical Shifts, ppm* TMS = 0ppm	Coupling Constants† Hz	Reference
CH_3TeCl_3	benzene	2.83		447a
	CH_2Cl_2	3.70		447a
CH_3TeCl_3	CH_2Cl_2	3.70		447c
$CH_3TeCl_3\cdot$tmtu	CH_2Cl_2	CH_3Te 3.50, CH_3(urea)3.20		447c
CH_3TeBr_3	benzene	3.10		447c
$CH_3TeBr_3\cdot$tmtu	CH_2Cl_2	CH_3Te 3.63, CH_3(urea)3.20		447c
CH_3TeI_3	acetone	3.60		447c
	benzene	3.53		397b
$CH_3TeI_3\cdot$tmtu	$CDCl_3$	CH_3Te 3.61, CH_3(urea)3.20		447c
$(CH_3)_2Te_2$	CCl_4	2.71		76
	benzene	2.63	$^1J_{C-H}$ + 142.4±0.2	348a
	14.1 mole %		$^2J_{Te-H}$ + 11.9±0.2	
			$^3J_{Te-H}$ + 2.4±0.2	
			$^1J_{Te-C}$ − 74±3	
			$^2J_{Te-C}$ + 75±3.	

Table XIV-6 (cont'd)

$(CH_3)_2Te_2$	benzene 83.3 mole %	2.51		348a
$CH_3TeSeCH_3$	benzene	CH_3Te 2.53, CH_3Se 2.66	$^2J_{Te-H}$ − 23.9±0.2 $^3J_{Se-H}$ ± 1.2±0.2 $^3J_{Te-H}$ ± 2.0±0.2 $^1J_{Se-Te}$ − 169±2	348a
$(CH_3)_2Te$	benzene	1.0	$^1J_{Te-C}$ + 158.5±0.5 $^1J_{C-H}$ + 140.7±0.2 $^2J_{Te-H}$ − 20.8±0.2 $^2J_{C-C}$ ± 0.2±0.1 $^3J_{C-H}$ + 3.1±0.2 $^4J_{H-H}$ ± 0.1	102,193a, 271
$(CH_3)_2Te$	benzene 16.5 mole %	1.83	$^1J_{C-H}$ + 140.7±0.2 $^3J_{C-H}$ + 3.1±0.2 $^4J_{H-H}$ 0.0±0.1 $^2J_{Te-H}$ −20.8±0.2	348a

Table XIV-6 (cont'd)

$(CH_3)_2Te$ (cont'd)			$^1J_{Te-H}$ + 158.5±0.5 $^3J_{C-C}$ ± 0.2±0.1	348a
$(CH_3)_2Te$	-	2.17	$^2J_{Te-H}$ 21.2	379a
$(CH_3)_2Te$	CCl_4	1.88		77
$(CH_3)_2Te \cdot BCl_3$	-	2.35	$^2J_{Te-H}$ 19.5	379a
$(CH_3)_2Te \cdot BBr_3$	-	2.34	$^2J_{Te-H}$ 20.0	379a
$(CH_3)_2Te \cdot BI_3$	-	2.32		379a
$(CH_3)_2TeBr_2$	CCl_4	3.33		77
$(CH_3)_2TeI_2$	benzene	3.26		397a
$(CH_3)_3TeI$	H_2O	-	J_{C-H} + 146.0, $^2J_{Te-H}$ - 20.9	271
$C_2H_5TeCl_3$	CH_2Cl_2	CH_3 2.15, CH_2 4.05	$^3J_{H-H}$ 8	447b,c
$C_2H_5TeCl_3 \cdot$tmtu	$CDCl_3$	CH_3 2.10, CH_2 3.85 CH_3(urea) 3.28		447c
$C_2H_5TeBr_3$	CH_2Cl_2	CH_3 2.30, CH_2 - ††		447c
$C_2H_5TeI_3 \cdot$tmtu	$CDCl_3$	CH_3 2.05, CH_3(urea)3.20		447c
$(C_2H_5Te)Si(CH_3)_2H$	$(CH_3)_6Si_2O$	CH_3 1.56, CH_2 2.44 (Te) CH_3 0.61, H 4.87 (Si)		117a

Table XIV-6 (cont'd)

$(C_2H_5Te)Si(C_2H_5)_2H$	$(CH_3)_6Si_2O$	CH_3 1.55, CH_2 2.43, (Te)		117a
		C_2H_5 1.03**, H 4.73 (Si)		
$(C_2H_5Te)_2Si(C_2H_5)H$	$(CH_3)_6Si_2O$	CH_3 1.60, CH_2 2.54 (Te)		117a
		CH_3 1.08, CH_2 1.48, H 5.18 (Si)		
$(C_2H_5)_2Te_2$	-	CH_3 1.60, CH_2 3.00	$^3J_{H-H}$ 9	447b
$(C_2H_5)_2Te$	neat	second order spectrum	$^1J_{C-H}(CH_3)$ 127.1±0.4	46,193a
			$^1J_{C-H}(CH_2)$ 141.0±0.4	
			$^3J_{H-H}$ 7.6±0.1	
			$^2J_{Te-H}$ + 24.8±0.2	
			$^3J_{Te-H}$ + 22.2±0.3	
i-C_3H_7-Te-C_6H_5	CCl_4	CH 3.18-3.55(m) (1H)		323d
		CH_3 1.30(d) (6H)	$^3J_{H-H}$ 13	
		C_6H_5 7.41-7.61(m) (2H)		
		7.05-7.27 (m) (3H)		
$C_4H_9TeC_6H_5$	CCl_4	2.85(t) (2H)	$^3J_{H-H}$ 15	323d
		1.13-2.05 (m) (4H)		
		0.92(t) (3H)	J_{H-H} 10	

Table XIV-6 (cont'd)

$C_4H_9TeC_6H_5$ (cont'd)		C_6H_5 7.58–7.92(m) (2H)		323d
		7.08–7.24(m) (3H)		
$HC{\equiv}C{-}CH_2{-}Te{-}C_6H_5$	CCl_4	CH_2 3.32, CH 1.98	$^4J_{H-H}$ 2.75	399
$(4{-}CH_3OC_6H_4Te)_2CH_2$	$CDCl_3$	CH_2 3.77		347b
$(4{-}C_2H_5OC_6H_4Te)_2CH_2$	$CDCl_3$	CH_2 3.81		347b
$[(C_6H_5)_2C{=}CH]_2Te$	CCl_4	CH 7.49, C_6H_5 7.26		119a
$[(4{-}CH_3C_6H_4)_2C{=}CH]_2Te$	CCl_4	CH 7.26, C_6H_4 7.08, CH_3 2.32		119a
bis(2-oxo-3-chloro-4-hydroxy-3-penten-1-yl) tellurium dichloride	$CDCl_3$	CH_3 2.33, CH_2 4.87		97
		OH 1.46		
$4{-}CH_3C_6H_4TeCl_3$	CH_3CN, 3%	CH_3 1.79, C_6H_4 5.90		323a
$4{-}CH_3OC_6H_4TeCl_3$	CD_3CN	CH_3 3.90, C_6H_4 7.24*(o)*, 8.44*(m)*	J_{23} 9.0	447c
$4{-}CH_3OC_6H_4TeCl_3{\cdot}$tmtu	CH_2Cl_2	CH_3 3.80, CH_3(urea) 3.22,		447c
		C_6H_4 6.90*(o)*, 8.34*(m)*	J_{23} 9.0	
$1,2{-}(4'{-}C_2H_5OC_6H_4Te)_2C_6H_4$	$CDCl_3$, 1<u>M</u>	C_6H_4 7.255, 6.902	J_{34} 7.63, J_{35} 1.45, J_{36} 6.40	306a
			J_{45} 7.29, J_{46} 1.45, J_{56} 7.63	
		TeC_6H_4OR 7.741, 6.779	J_{23} 8.43, J_{25} 0.37, J_{26} 2.05	
			J_{35} 2.84, J_{36} 0.37, J_{56} 8.43	

Table XIV-6 (cont'd)

1-telluracyclopentane	CCl_4	CH_2(2,5) 3.118		409
		CH_2(3,4) 2.041		
	C_6D_6	CH_2(2,5) 2.838		409
		CH_2(3,4) 1.615		
1-telluracyclopentane	CCl_4	CH_2(2,5) 3.871		409
1,1-dibromide		CH_2(3,4) 2.921		
	C_6D_6	CH_2(2,5) 2.887		409
		CH_2(3,4) 2.020		
tellurophene	$CDCl_3$	H(3,4) 7.83		257
		H(2,5) 8.94		
	$CDCl_3$	H(3.4) 7.78	J_{23} 6.70, J_{24} 1.30, J_{25} 2.60	133b
		H(2,5) 8.87	J_{34} 4.00	
2-carboxytellurophene	$CDCl_3$	H(3) 8.59, H(4) 7.91, H(5) 9.34,	J_{34} 4.00, J_{45} 6.80, J_{35} 1.20	133b
		COOH 11.78		
2-carbomethoxytellurophene	$CDCl_3$	H(3) 8.31, H(4) 7.96, H(5) 9.32,	J_{34} 4.10, J_{45} 6.60, J_{35} 1.30	133b
		CH_3 2.55		

Table XIV-6 (cont'd)

2-(1'-hydroxyethyl)tellurophene	$CDCl_3$	H(3) 7.42, H(4) 7.67, H(5) 8.77, OH 2.43, CH_3 1.52, CH 4.95	J_{34} 4.00, J_{45} 6.80, J_{35} 1.30, J_{CH-CH_3} 6.3, $J_{CH_3-H(3)}$ 1.0	133b
2-carboxy-5-methyltellurophene	$CDCl_3$	H(3) 8.36, H(4) 7.38, CO_2H 10.68, CH_3 2.64,	J_{34} 4.10, $J_{CH_3-H(4)}$ 1.0	133b
2-carbomethoxy-5-acetyltellurophene	$CDCl_3$	H(3) 8.59, H(4) 8.32, $COOCH_3$ 3.82, CH_3CO 2.53	J_{34} 4.00	133b
2-carboxy-5-acetyltellurophene	$CDCl_3$	H(3) 8.69, H(4) 8.34, COOH 9.24, CH_3 2.55	J_{34} 4.10	133b
2,5-bis(hydroxymethyl)-tellurophene	$CDCl_3$	H(3,4) 7.43, CH_2 4.85, OH 2.16		257
2,5-bis(1'-hydroxy-1'-methylethyl)tellurophene	$CDCl_3$	H(3,4) 7.29, CH_3 1.62, OH 2.74		257
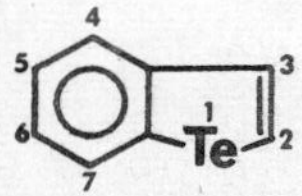		H(2) 8.56, H(3) 7.85, H(4,7) 7.77, H(5,6) 7.15	J_{23} 7.1,	348e
1-telluracyclohexane-4,4-d_2	$HOSO_2F/SO_2$		J_{trans} 8.62, J_{cis} 3.12	210

Table XIV-6 (cont'd)

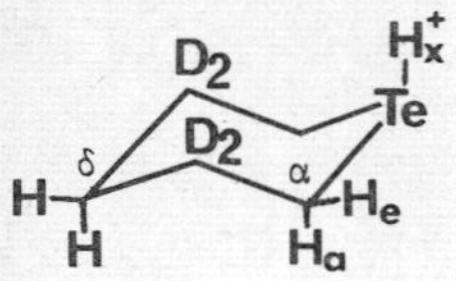	$HOSO_2F/SO_2$		J_{ae}(α) 13.5, J_{ae}(δ) 15.0 $^3J_{ax}$ 11.2, $^3J_{ex}$ 2.3	210
1-telluracyclohexane 1,1-dibromide	--	--	$J_{13} = J_{24}$ 5.2, $J_{14} = J_{23}$ 7.8, $J_{35} = J_{46}$ 2.56, $J_{36} = J_{45}$ 9.21	209
1-tellura-3,5-cyclohexanedione	$CDCl_3$	CH_2(4) 3.45 CH_2(2,6) 3.90		97
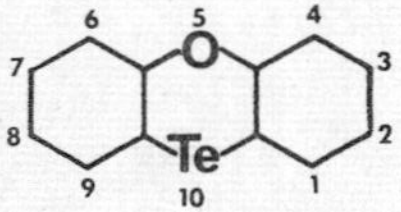	acetone	H(1) 7.25, H(2) 7.30, H(3) 7.07, H(4) 7.64	J_{12} 8.20, J_{13} 1.17, J_{14} 0.41	54a
	acetone/$CHCl_3$(1:1)	7.21 7.22 7.20 7.54	J_{23} 7.54, J_{24} 1.57, J_{34} 7.62	
	$CDCl_3$	7.23, 7.21 7.01 7.49		
	CCl_4	7.14, 7.13 6.96 7.39		
disilyl telluride	neat	3.71±0.1	$^1J_{Si-H}$ 224.4±0.5	49
digermyl telluride	cyclopentane	3.663±0.005	$^3J_{Te-H}$ 19.4±0.1	86
	cyclohexane	3.59±0.01		86
bis(trimethylsilyl) telluride	CCl_4	0.6		181

Table XIV-6 (cont'd)

bis(triethylsilyl) telluride		CH_2 0.81, CH_3 1.00		114
bis(triethylgermyl) telluride		CH_2 1.05, CH_3 1.05		114
bis(triethylstannyl) telluride		CH_2 1.09, CH_3 1.20		114
triethylsilyl telluride triethylgermyl		CH_2 0.80, CH_3 0.97(Si) CH_2 1.09, CH_3 1.31(Ge)		114 114
triethylgermyl telluride triethylstannyl		CH_2 1.05, CH_3 1.05(Ge) CH_2 1.08, CH_3 1.20(Sn)		114
$[(CH_3)_3Sn]_2Te\cdot Cr(CO)_5$	benzene, 5%	+7 Hz		390a
$[(CH_3)_3Sn]_2Te\cdot W(CO)_5$	benzene, 5%	+7.5 Hz		390a

† The symbol C, Se, Te represents ^{13}C, ^{77}Se and ^{125}Te, respectively.

†† Peaks overlapped by solvent signal.

* m = multiplett, s = singlet, d = doublet, t = triplet

** center of multiplet

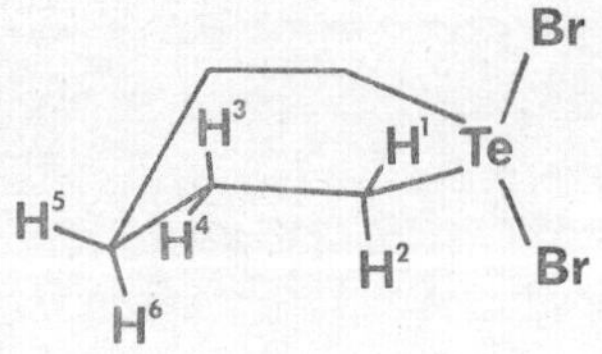

(148)

Telluracyclohexane-4,4-d_2, telluracyclohexane-3,3,5,5-d_4 and protonated telluracyclohexane-3,3,5,5-d_4 were also investigated. A study of the nmr spectrum of telluracyclohexane-3,3,5,5-d_4 at various temperatures showed that ring inversion is still fast at -100°[210]. The nmr spectra of the following complexes are reported in the literature: [π-cpMo(CO)$_3$(TeC$_6$H$_5$)][419], [π-cpMo(CO)$_2$(TeC$_6$H$_5$)]$_2$[419], [π-cpFe(CO)$_2$(TeC$_6$H$_5$)][37] [π-cpFe(CO)(TeC$_6$H$_5$)]$_2$[378a] and [Fe(CO)$_3$(TeC$_6$F$_5$)]$_2$[357]. Chemical shifts are given only for cyclopentadienyl groups. No further details have been published.

Radchenko and Petrov[355] reported the nmr spectrum of 1-methyltelluro-3-methyl-3-buten-1-yne relative to water as standard.

C) Structural Investigations

Dimethyl tellurium diiodide was the first organic tellurium compound to be investigated by X-ray techniques. A preliminary report about the structure of this compound was published in 1941. Earlier attempts to elucidate the structures of organic tellurium compounds made use of a variety of physical properties, which could be measured rather easily. The parachors of aliphatic and aromatic tellurium compounds were determined[50,249,329,400], the magnetic susceptibilities of dimethyl tellurium dihalides and dinitrates were measured[29,30,98] and the dipole moments of dimethyl tellurium diiodide (2.26D)[447], bis(4-methylphenyl) tellurium dichloride (2.98D), dibromide (3.21D) and the anhydride of bis(4-methylphenyl)tellurium chloride hydroxide (6.1D)[188] were calculated. Smyth[404] deduced a value of 0.7D for the carbon-tellurium bond moment. Jensen[188] proposed a trigonal bipyramidal structure for bis(organyl) tellurium dihalides with the two organic groups and the lone electron pair occupying the equatorial positions. Gillespie[145] and Abrahams[2] discussed the stereochemistry of organic tellurium compounds in context of other, similar molecules. Lowry and Gilbert investigated the conductivity of the dimethyl tellurium dihalides[248,249], diethyl tellurium compounds[144], and telluracyclopentane derivatives[143]. The dihalides hydrolyzed in an aqueous medium and behaved then as binary electrolytes. The conductivities of diphenyl tellurium dichloride and diiodide, of triphenyl telluronium chloride and iodide and of methyl diphenyl telluronium iodide were determined in acetonitrile and dimethylformamide. The telluronium salts were found to be ionized in these solvents. The behavior of the dihalides was dependent upon the solvent[83d]. The conductivity of dimethyl telluride and dimethyl tellurium dichloride was measured in liquid hydrogen chloride[333].

Billows[31] reported crystallographic data on diphenyl tellurium dibromide and showed, that it can form tetragonal and triclinic crystals. Knaggs and Vernon[194] described the crystals of dimethyl tellurium diiodide and the crystals of the complex $(CH_3)_3TeI \cdot CH_3TeI_3$.

The structural parameters of the organic tellurium compounds, which have been examined by single crystal X-ray techniques are compiled in Table XIV-7. The carbon-tellurium bond lengths vary between 2.01 and 2.18Å with standard deviations between 0.02 and 0.20Å. The tellurium-halogen bond lengths in diorganyl tellurium dihalides and organyl tellurium trihalides are not equal. The inequality is caused by the interaction of one halogen atom with the tellurium atom of the neighboring molecule. Carbon-tellurium-carbon bond angles between 101° and 91° have been observed. Bis(4-chlorophenyl) ditelluride has a conformation

Table XIV-7

Bond Distances and Bond Angles in Organic Tellurium Compounds

Compound	$(CH_3)_2TeCl_2$	$(CH_3)_3Te^+$	$CH_3TeI_4^-$	$ClCH_2CH_2TeCl_3$	1-thia-4-tellura-cyclohexane 4,4-diiodide
Te-C	2.08±0.03	2.01±0.08	2.15±0.08	2.164	2.18
Te-C'	2.10±0.03	2.08±0.08	—	—	2.13
Te-C"	—	2.13±0.08	—	—	—
Te-Te'	—	—	—	—	—
Te'-C'	—	—	—	—	—
Te-X	2.480±0.010	—	2.948±0.007	2.386†	2.884
Te-X'	2.541±0.01	—	2.984±0.007	2.717††	2.939
Te-X"	—	—	2.891±0.007	—	—
Te-X"'	—	—	2.840±0.007	—	—
C-Te-C'	98.2±0.3	91,97,99±3.2	—	—	100
C-Te-X	87.6±0.7	—	87,88,90, 92±1.7	92.7	88.3
C-Te-X'	88.0±0.7	—	—	82.5	91.6
C'-Te-X	86.4±0.8	—	—	—	87.4
C'-Te-X'	88.0±0.8	—	—	—	91.8
X-Te-X'	172.3±0.3	—	88.6,98.3,90.6 91.8±0.2	—	175
Te-Te'-Cl	—	—	—	—	—
Te'-Te-Cl	—	—	—	—	—
dihedral angle	—	—	—	—	—
Te-Te-Te	—	—	—	—	—
Reference	79	118	118	194b	194a

† terminal Cl-Te distance

†† bridging Cl-Te distance

Table XIV-7 (cont'd)

Compound	$(C_6H_5)_2TeBr_2$	$(4\text{-}ClC_6H_4)_2TeI_2$	$(4\text{-}CH_3C_6H_4)_2Te$	$[(C_2H_5)_2P(S)Se]_2Te$	Te
Te-C	2.14	2.13±0.02	2.05	-	-
Te-C'	2.14	2.10±0.02	--	-	-
Te-C"	--	--	--	-	-
Te-Te'	--	--	--	-	2.36±0.02
Te'-C'	--	--	--	-	-
Te-X	2.682±0.005	2.922±0.002	--	2.50	-
Te-X'	--	2.947±0.002	--	-	-
Te-X"	--	--	--	-	-
Te-X"'	--	--	--	-	-
C-Te-C'	96.3	101.1±1.0	101	-	-
C-Te-X	90.7±0.6	87.3±0.7	--	-	-
C-Te-X'	90.8±0.6	89.1±0.7	--	-	-
C'-Te-X	--	88.1±0.7	--	-	-
C'-Te-X'	--	87.3±0.7	--	-	-
X-Te-X'	--	173.5±0.1	--	100.7	-
Te-Te'-Cl	--	--	--	-	-
Te'-Te-Cl	--	--	--	-	-
dihedral angle	90.0	--	--	-	-
Te-Te-Te	--	--	--	-	102
Reference	78	66	36, 377	185a	447

Table XIV-7 (cont'd)

Compound	$(4\text{-}ClC_6H_4)_2Te_2$	$C_6H_5TeCl \cdot tu$*	$C_6H_5TeBr \cdot tu$*	$C_6H_5TeCl \cdot tu_2$*	5,10-epitelluro-5,10-dihyroarsanthrene
Te-C	2.16±0.02	2.12	2.12	2.12	As-Te 2.571
Te-C'	--	--	--	--	As'-Te 2.575
Te-C"	--	--	--	--	--
Te-Te'	2.702±0.010	--	--	--	As-Te-As 80.8°
Te'-C'	2.10±0.20	--	--	--	--
Te-X	--	3.00	3.11	3.00	--
Te-X'	--	--	--	--	--
Te-X"	--	--	--	--	--
Te-X"'	--	--	--	--	--
C-Te-C'	--	--	--	--	--
C-Te-X	--	84	86	84	--
C-Te-X'	--	--	--	--	--
C'-Te-X	--	--	--	--	--
C'-Te-X'	--	--	--	--	--
X-Te-X'	--	--	--	--	--
Te-Te'-Cl	95.2±1.0	--	--	--	--
Te'-Te-Cl	93.5±1.0	--	--	--	--
dihedral angle	72.0±1.5	--	--	--	--
Te-Te-Te	--	--	--	--	--
Reference	208	128, 130	128, 130	128; 129	192b

* tu = thiourea

similar to that of hydrogen peroxide with a dihedral angle of 72°[208]. The molecular structures of the phenyl tellurium halide thiourea adducts, which are characterized by distorted square planar coordination of the divalent tellurium atom, are depicted in Fig. IV-2.

Crystals of 2-chloroethyl tellurium trichloride contain polymer chains held together by bridging chlorine atoms with the five coordinated tellurium atoms situated very nearly in the centers of the bases of square pyramids. The organic group occupies the apical position. The authors suggest that this compound possesses a predominantly ionic lattice with $ClCH_2CH_2TeCl_2^+$ and Cl^- ions alternating along the polymer chain[194b].

The diorganyl tellurium dihalides form distorted trigonal bipyramids, in which the organic groups and the lone electron pair occupy the equatorial positions. The axial chlorine atoms in dimethyl tellurium dichloride are slightly displaced towards the methyl groups[79] while the bromine atoms in diphenyl tellurium dibromide point away from the phenyl groups[78]. The trimethyl telluronium cation has the shape of a tetrahedron with the lone electron pair occupying the fourth position. The tellurium atom in the anion $[CH_3TeI_4]^-$ can be considered to be octahedrally ligated. The methyl group and the lone electron pair take up the apical positions, while the four iodine atoms form the basis of the bipyramid[118].

A chair conformation as pictured in *(149)* has been found for 1-thia-4-telluracyclohexane 4,4-diiodide.

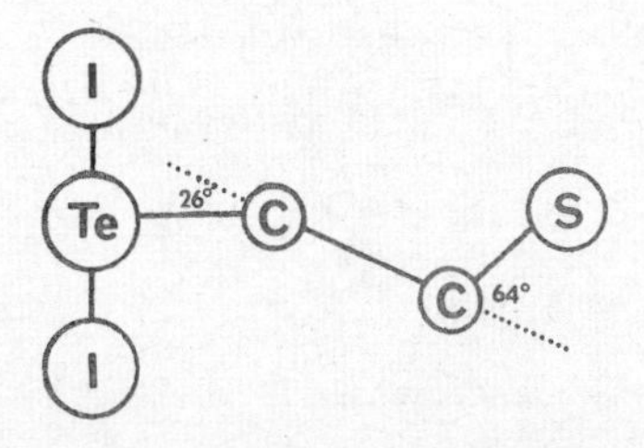

The configuration around the tellurium atom approximates a trigonal bipyramid with two carbon atoms and a lone electron pair in the equatorial positions and the iodine atoms in the axial positions[194a].

1-Tellura-3,5-cyclohexanedione is claimed to have a chair structure[327]. The structure was derived from 100 independently observed intensity data.

D) Mass Spectrometry of Organic Tellurium Compounds

Tellurium has eight naturally occurring isotopes, which give rise to a series of mass spectral peaks for each ion containing one or more atoms of this element. Such isotope clusters are of great help in identifying tellurium containing ions. Calculated relative intensities of the mass spectral peaks expected from Te, Te_2, Te_3, TeSe, Te_2Se, $TeSe_2$[185c], $TeCl_2$[119a], $TeBr_2$, Te_2Br_2 and Te_2Br[185c] have been published.

Mass spectrometry has been employed to characterize bis(biphenylylene) tellurium, bis(4,4'-dimethyl-2,2'-biphenylylene) tellurium[173] and 1,3,5-tritelluracyclohexane[441]. A molecular ion was observed in the spectrum of diphenyl telluronium tetraphenylcyclopentadienylide at low electron energy. At higher energies only diphenyl telluride was detected[132a]. 5,10-Epitelluro-5,10-dihydroarsanthrene produced a base peak corresponding to the 9-arsafluorenyl ion, a prominent molecular ion peak and a strong peak at *m/e* 152, which is certainly due to the biphenylylene ion[7a].

The electron impact fragmentation of the complex $[Fe_2(CO)_6(TeC_6H_5)_2]$ produced a weak molecular ion peak and weak peaks corresponding to $[Fe_2(CO)_n(TeC_6H_5)_2]$ (n = 4, 3, 2, 1). The base peak is caused by Fe_2Te_2. The completely decarbonylated ion $[Fe_2(TeC_6H_5)_2]^+$ possesses an intensity of 43.8% relative to the base peak[378b].

The spectrum of bis(2,2-diphenylvinyl) telluride[119a] contained the following ions: $[(C_6H_5)_2C{=}CH]_2Te$ (50%), $(C_6H_5)_2C_2$ (100%), $(C_6H_5)_2C{=}CH$ (second largest peak), $[(C_6H_5)_2C{=}CH]_2$ (12%) and $[(C_6H_5)_2C{=}CH]_2$-H (35%).

Among the heterocyclic tellurium compounds only telluracyclopentane and benzotellurophene have been investigated by mass spectrometry. The base peak in the spectrum of telluracyclopentane occurs a *m/e* 55 ($C_4H_7^+$). The molecular ion is approximately half as intense as the base peak. A weak group of peaks represents ions formed by loss of C_2H_4 from the parent molecular ion. A second series of peaks originates from loss of cyclopropane. Ions corresponding to Te^+ and TeH^+ have also been observed[112].

The base peak in the mass spectrum of benzotellurophene[51a] at *m/e* 102 is probably due to benzocyclobutadiene. The tellurium containing ions with their intensity relative to the base peak are molecular ion (49.3% for ^{130}Te), 1,2-epitellurobenzene (0.8% for ^{130}Te), tellurium (7.4% for ^{130}Te) and doubly charged molecular ion (2.6% for ^{130}Te).

Bogolyubov[37b] investigated the mass spectral behavior of diethyl telluride and diethyl ditelluride and determined the appearance potentials of the ions formed. The results are summarized in Table XIV-8.

The diaryl ditellurides, $(R\text{-}C_6H_4)_2Te_2$ (R = H, 4-CH_3, 4-F, 3-F, 4-Br, 4-C_6H_5) produced upon electron impact (70 ev) the ions $Ar_2Te^{\cdot +}$, $Ar_2Te^{\cdot +}$, $ArTe_2^+$, $ArTe^+$, TeC_nH (n = 2-5), $Te_2^{\cdot +}$, $Te^{\cdot +}$, Ar^+ and $Ar_2^{\cdot +}$. The hydrocarbon fragments were further broken down by loss of hydrogen and carbon. Low intensity peaks corresponding to the combination ions $ArTe_2H$, ArTeH, Te_2H and TeH have also been observed. Table XIV-9 lists the relative abundances of the more important ions in the mass spectra of diaryl ditellurides[185c].

Table XIV-8

Mass Spectral Data for Diethyl Ditelluride and Diethyl Telluride

m/e	$(C_2H_5)_2Te_2$ Ion	AP,eV	I,%*	$(C_2H_5)_2Te$ Ion	AP,eV	I,%*
318	R_2Te_2	9.0	23.9			
290	RTe_2H	10.4	8.40			
289	RTe_2	10.9	14.60			
275	CH_3Te_2	—	0.40			
274	CH_2Te_2	—	0.20			
262	Te_2H_2	12.0	1.85			
261	Te_2H	12.8	15.10			
260	Te_2	13.0	14.00			
188				R_2Te	9.2	16.40
173				C_3H_7Te	—	0.40
161				C_2H_7Te	—	0.22
160	RTeH	10.3	0.67	RTeH	10.9	11.0
159	RTe	13.0	3.48	RTe	13.4	8.13
158				C_2H_4Te	—	0.42
145				CH_3Te	14.5	0.88
144	CH_2Te	—	0.35	CH_2Te	—	0.68
143	CHTe	—	0.30	CHTe	—	0.50
132	TeH_2	12.8	0.58	TeH_2	13.2	7.48
131	TeH	15.0	4.06	TeH	15.2	11.00
130	Te	14.4	3.48	Te	12.8	17.30
31				C_2H_7	—	1.97
30	C_2H_6	—	0.08	C_2H_6	—	0.15
29	C_2H_5	—	3.50	C_2H_5	—	7.58
28	C_2H_4	—	1.05	C_2H_4	—	2.28
27	C_2H_3	—	2.90	C_2H_3	—	5.08
26	C_2H_2	—	0.40	C_2H_2	—	0.91

* The intensities are expressed in percent relative to the sum of all t peaks in the spectrum containing ^{130}Te.

Table XIV-9

Relative Abundances* of Ions in the Mass Spectra of Diaryl Ditellurides, Ar_2Te_2

Ion	Ar-Te-Te-Ar						Ar_2Te
	C_6H_5	$4-CH_3C_6H_4$	$4-FC_6H_4$	$3-FC_6H_4$	$4-BrC_6H_4$	$4-C_6H_5C_6H_4$	$4-C_6H_5C_6H_4$
Ar_2Te_3	0.9	0.6	trace	-	trace	-	-
Ar_2Te_2	100 (2490)	100 (941)	100 (1101)	100 (1145)	100 (331)	0.6	-
$ArTe_2$	14.1	3.4	9.4	13.5	10.9	-	-
Ar_2Te	30.5	41.5	61.3	42.0	83.4	100 (1576)	100 (476)
ArTe	73.0	58.0	126.0	86.5	187.6	23.2	22.5
Te_2	9.7	4.7	11.4	14.0	18.1	trace	trace
Te	8.8	2.7	11.4	8.1	51.4	3.9	3.8
Ar_2 †	22.0	29.2	53.5	29.0	69.5	116.0	156.0
Ar †	64.0	46.7	68.8	52.9	77.5	39.2**	43.1**
Probe	75°	35°	85°	80°	140°	190°	170°
Source	210°	190°	175°	198°	200°	200°	200°

* The numbers in parentheses give the actual peak heights in millimeters † These values include the isotope peaks.

** Contains a contribution from doubly charged quaterphenyl.

XV. ANALYTICAL TECHNIQUES

In order to determine tellurium in organic compounds containing this element, the organic meterial must be oxidized. During this process the tellurium is converted into tellurite or tellurate. The following reagents were employed as oxidants: fuming nitric acid[108], aqua regia[138], a mixture of concentrated sulfuric acid and 70 per cent perchloric acid[363], a mixture of nitric acid and 70 per cent perchloric acid[207], fused sodium carbonate (2 parts) and sodium nitrate (1 part)[109], and a mixture of sodium peroxide, potassium chlorate and sucrose in a Parr-bomb[417]. The acidic solutions thus obtained were usually evaporated to dryness on the water bath and the residues dissolved in 10 per cent hydrochloric acid. The cooled fusion mixtures were dissolved in water and the solutions acidified. The reduction of tetra- or hexavalent tellurium to the element was accomplished with hydrazine sulfate and sulfur dioxide[363], with hydrazine chloride and sulfur dioxide[108,293,373,417], and with hydrazine chloride[138]. The elemental tellurium was filtered, washed with water and ethanol and dried at 105°. Kruse and coworkers[207] developed a volumetric method for the determination of tellurium. After oxidizing the organic tellurium compound with nitric acid and perchloric acid, the colorless solution was diluted with water. Excess standardized 0.1N potassium dichromate solution was added. The solution was then treated with 0.1N iron(II) ammonium sulfate and the excess iron(II) back-titrated with dichromate using sodium diphenylamine sulfonate as indicator (eq. 195).

$$(195)\quad 3TeO_2 + K_2Cr_2O_7 + 8HCl \longrightarrow 3H_2TeO_4 + 2KCl + 2CrCl_3 + H_2O$$

In the micro-determination of tellurium in organic compounds the substance is oxidized with fuming nitric acid. Hydrazine chloride together with sulfur dioxide was used as reducing agent[108,373].

Belcher and coworkers[22] published in 1952 a review on the determination
metals in organic compounds and included tellurium. The analytical
mistry of tellurium and the separation of this element from other
ments, especially from selenium is the subject of treatises by
en and Turley[156] and by Frommes[139].

In the qualitative determination of tellurium, tetra- or hexavalent
lurium is reduced to the element. The solutions to be reduced are
ated with alkali stannite in alkaline medium, or with sulfur dioxide
highly acidic medium. Tellurium is precipitated as the element.
enium remains in solution under these conditions. Another method fuses
solid residue, obtained by evaporation of the tellurium containing
ution, with potassium cyanide in a hydrogen atmosphere. Potassium
uride and - in the presence of selenium - potassium selenocyanate are
ed. The aqueous solution from the melt deposits only tellurium
ontact with air[372].

The chromatographic behavior of organic tellurium compounds on paper
on thin aluminum oxide layers was investigated in order to find
conditions, which would allow the separation and identification of
e compounds and their corresponding polonium derivatives. A paper-
matographic method was employed for the determination of the
ility of hexavalent tellurium derivatives[325]. The following compounds
investigated: $C_6H_5TeCl_3$[319b], $(C_6H_5)_2TeCl_2$[319b,429], $(C_6H_5)_2TeI_2$[319c],
$)_2Te$[319b], $(C_6H_5)_3TeCl$[319b], $(4\text{-}CH_3C_6H_4)TeCl_3$[311,313*], $(4\text{-}CH_3C_6H_4)_2TeCl_2$[311,
318*,428a], $(4\text{-}CH_3C_6H_4)_2TeBr_2$[318*], $(4\text{-}CH_3C_6H_4)_2TeI_2$[318*], $(4\text{-}CH_3C_6H_4)_2Te$
[13*,428b], $(4\text{-}CH_3C_6H_4)_3TeCl$[311,313*], $(3\text{-}CH_3C_6H_4)_2TeCl_2$[428a,429], $(3\text{-}CH_3C_6H_4)_2Te$
$(2\text{-}CH_3C_6H_4)_2TeCl_2$[428a,429], $(2\text{-}CH_3C_6H_4)_2Te$[428b], $[2,5\text{-}(CH_3)_2C_6H_3]_2Te$[428b],
$_3OC_6H_4)_2TeCl_2$[317*], $(4\text{-}CH_3OC_6H_4)_2Te$[428b], $(1\text{-}C_{10}H_7)_2TeI_2$[316*] and
$_0H_7)_2Te$[428b]. The starred references refer to papers which

contain data on polonium compounds. Vapor phase chromatography was employed in the purification of telluracyclopentane[112].

Organic tellurium compounds have received some attention as reagents for the determination of metal ions. Alimarin and coworkers[5,6,7] reported that benzenetellurinic acid forms precipitates with a variety of metal ions in neutral or weakly acidic medium. The corresponding salts of benzetelluronic acid were all soluble. Pitombo[349] employed diphenyl telluride to detect palladium in a 0.1M hydrochloric acid solution. The polarographic reduction of triphenyl telluronium chloride in 0.1M deairated potassium chloride solution proceeded in two steps[263]. A correlation was sought between the half wave potential of the telluronium salt and the turbidity produced upon addition of complex metal ions[264]. The telluronium chloride gave precipitates with various anions, which were, however, too solubl to be of use in analytical chemistry[263]. The half wave potential of triethyl telluronium chloride was found to be -1.42V independent of pH. In the presence of iodide ions the potential was found to be sensitive to the concentration of Bi^{+3}[394]. Portratz and Rosen[350] employed triphenyl telluronium iodide as a reagent for the detection of bismuth and cobalt. Shinagawa and Matsuo[395] reviewed in 1956 the analytical applications of onium compounds and included telluronium salts.

XVI BIOLOGY OF ORGANIC TELLURIUM COMPOUNDS

Morgan and coworkers prepared a large number of compounds containing the 1-tellura-3,5-cyclohexanedione ring system (see section XII-B-2). The first paper describing the inhibitory action of these compounds on the growth of bacteria appeared in 1922[277]. In the following years the mean bactericidal concentrations of 1-tellura-3,5-cyclohexanedione and many of its derivatives with respect to a variety of organisms were determined. The data collected in these investigations are presented in Table XVI-1. The experiments were preformed in sterile aqueous medium employing lemco-peptone broth at 37°. B. *coli communi* treated under these conditions with the appropriate tellurium compound, did not grow when subcultured to fresh broth indicating bactericidal and not only inhibitory action of the tellurium compound. The coliform organisms were more affected than the cocci organisms[278]. Appreciable bactericidal action was exerted after one hour, but maximum efficacy was reached after four hours[281]. Protozoa[281] *(paramoecia)* required comparatively high concentrations of $1{:}4x10^5$. The cyclic tellurium compounds, especially 2,6-dimethyl-1-tellura-3,5-cyclohexanedione, were extremely active as germicides in the presence of urine[280,281]. In serum, however, concentrations of $1{:}4x10^5$ to $1{:}2x10^5$ were required for complete disinfection. Serum from ox blood reduced the bactericidal power even after dialysis, showing that the incompatible substance is collodial in nature[281]. Addition of cholesterol and nucleic acids to the broth reduced the germicidal power slightly. Substances like oleic acid, olive oil, coconut oil and lecithine caused a remarkable collapse of germicidal power[292]. Gulland and Farrar[159,160] pointed out the structural similarity between the cyclic tellurium compounds and pyridoxine. Their hypothesis, that an enolic hydroxyl group in the 3-position of the 1-tellura-3,5-cyclohexanedione ring is necessary for bactericidal activity, is now

Table XVI-1

The Bactericidal Action of Organic Tellurium Compounds

Compound or Substituent	Organism	Mean Bactericidal Concentration	Ref.
1-tellura-3,5-cyclohexanedione	*E. coli*	$1:5x10^5$	159, 160
	B. coli communis	$1:8.16x10^5$ $1:5x10^5$	292 278, 280, 281
	*B. typhosus i**	$1:9x10^5$	280
	Staphylococcus pyogenes aureus i	$1:1.85x10^4$	280
1-tellura-3,5-cyclohexanedione dioxime	*E. coli, typhoid bacilli, staphylococci, hemolytic streptococci*	$1:1.66x10^5$	159, 160
	B. coli	$1:1.66x10^5$	292
2-methyl-1-tellura-3,5-cyclohexanedione	*E. coli, typhoid bacilli, staphylococci, hemolytic streptococci*	$1:3x10^6$	159, 160
	B. coli communis	$1:3x10^6$	277, 278, 280, 281
	B. typhosus i	$1:2.3x10^6$	280
	Staphylococcus pyogenes aureus i	$1:4x10^5$	280
4-methyl-	*E.coli, typhoid, bacilli, staphylococci, hemolytic streptococci*	$1:9x10^5$	159, 160
	B. coli communis	$1:9x10^5$	278, 280

Table XVI-1 (cont'd)

4-methyl- (cont'd)	*B. typhosus i*	$1:1.5x10^6$	280
	Staphylococcus pyogenes aureus i	$1:4.2x10^4$	280
2,4-dimethyl-	*E. coli, typhoid bacilli, staphylococci, hemolytic streptococci*	$1:5x10^6$	159, 160
	treatment of cystitis and eye infection	-	281
	B. coli communis	$1:5x10^6$ $1:3x10^6$ $1:2x10^7$	280 292 281
	B. acidi lactici	$1:7x10^6$	292
	B. lactis aerogenes	$1:4.5x10^6$	292
	B. prodigiosus	$1:3x10^6$	292
	B. mesentericus	$1:3x10^5$	292
	B. typhosus i	$1:5x10^6$	280
	*B. typhosus ii**	$1:8x10^6$	280
2,6-dimethyl-	*E. coli, typhoid bacilli, staphylococci, hemolytic streptococci*	$1:9x10^6$	159, 160, 281
	B. coli communis	$1:2x10^7$ $1:9x10^6$	292 253, 278, 280
	B. acidi lactici	$1:1.2x10^7$	292

Table XVI-1 (cont'd)

2,6-dimethyl- (cont'd)	*B. lacti aerogenes*	$1:3x10^6$	292
	B. prodigiosus	$1:4x10^7$	292
	B. mesentericus	$1:9x10^7$	292
	B. typhosus i	$1:7.5x10^6$	280
	B. typhosus ii	$1:2x10^7$	280
	Staphylococcus pyogenes aureus i	$1:9x10^6$	280
	Staphylococcus pyogenes aureus ii	$1:6x10^6$	280
	B. paratyphosus A	$1:1.7x10^7$	280
	B. paratyphosus B	$1:4x10^6$	280
	B. pyocyaneus	$1:2.5x10^5$	280
	B. phlei	$1:7.5x10^5$	280
	Streptococcus haemolyticus a	$1:5.4x10^5$	280
	Streptococcus haemolyticus b	$1:2.8x10^6$	280
2-methyl-6-ethyl-	*B. coli communis*	$1:3x10^6$	278, 280
	B. typhosus i	$1:3x10^6$	280
	Staphylococcus pyogenes aureus i	$1:6.2 \times 10^5$	280

Table XVI-1 (cont'd)

2,4,6-trimethyl-	*powerful bactericide*		287
2-methyl-4-ethyl-	*B. coli communis*	$1:1.3x10^7$ $1:2x10^6$	281 280
	B. typhosus ii	$1:2.5x10^6$	281
2-methyl-4-propyl-	*B. coli communis*	$1:8x10^6$	281
2-methyl-4-butyl-	*B. coli communis*	$1:7.5x10^6$	281
2,6-dimethyl-4-ethyl	*B. coli communis*	$1:2.8x10^6$	280
	Staphylococcus pyogenes aureus ii	$1:5x10^6$	280
2-ethyl-	*E. coli, typhoid bacilli, hemolytic streptococci*	$1:3x10^6$	159, 160
	B. coli communis	$1:6.25x10^6$	292
	B. acidi lactici	$1:1.2x10^7$	292
	B. lactis aerogenes	$1:1.7x10^6$	292
	B. prodigiosus	$1:4.5x10^6$	292
	B. mesentericus	$1:7x10^6$	292
	B. typhosus i	$1:2.4x10^6$	280
	Staphylococcus pyogenes aureus i	$1:4.5x10^5$	280
	Staphylococcus pyogenes aureus ii	$1:4x10^6$	280

Table XVI-1 (cont'd)

2-ethyl- dioxime	*B. coli communis*	$1:1.66x10^5$	292
4-ethyl-	*E. coli, typhoid bacilli, staphylococci hemolytic streptococci*	$1:2.5x10^6$	159, 160
	B. coli communis	$1:4.33x10^5$ $1:2.5x10^6$	292 278, 280
	B. acidi lactici	$1:3x10^6$	292
	B. lactis aerogenes	$1:1.7x10^6$	292
	B. prodigiosus, B. mesentericus	$1:1.7x10^6$	292
	B. cloacae	$<1:2.6x10^5$	292
4-ethyl- dioxime	*B. coli communis*	$1:1.66x10^5$	292
	B. typhosus i	$1:2.2x10^6$	280
	Staphylococcus pyogenes aureus i	$1:1.22x10^5$	280
4,4-diethyl-	*E. coli, typhoid bacilli, staphylococci hemolytic streptococci*	$1:9x10^5$	159, 160
	B. coli communis	$1:9x10^5$	280
	B. typhosus i	$1:5x10^5$	280
2,6-diethyl-	*E. coli, typhoid bacilli, staphylococci, hemolytic streptococci*	$1:5x10^6$	159, 160
	B. coli communis	$1:5x10^6$	280
	B. typhosus i	$1:3x10^6$	280

Table XVI-1 (cont'd)

2-propyl-	*E. coli, typhoid bacilli, staphylococci, hemolytic streptococci*	1:1.2x10^{6}	159, 160
	B. coli communis	1:1.2x10^{6}	280
2-*i*-propyl	*B. coli*	1:1.218x10^{6}	292
4-*i*-propyl	*B. coli*		292
2-butyl-	*E. coli, typhoid bacilli, staphylococci hemolytic streptococci*	1:7x10^{5}	159, 160
	B. coli communis	1:7x10^{5}	280
4-butyl-	*E. coli, typhoid bacilli, staphylococci, hemolytic streptococci*	1:2.8x10^{6}	159, 160
	B. coli communis	1:3.5x10^{6}	292
4-*i*-butyl-	*B. coli*	1:6.66x10^{5}	292
4-*sec*-butyl-	*B. coli*	1;1.197x10^{6}	292
2-pentyl-	*E. coli, typhoid bacilli, staphylococci hemolytic streptococci*	1:5x10^{5}	159, 160
	B. coli	1:5x10^{5}	280
4-*i*-pentyl	*B. coli*	1:1.25x10^{5}	292
4-chloro-	*E. coli, typhoid bacilli, staphylococci, hemolytic streptococci*	1:1.25x10^{4}	159, 160

Table XVI-1 (cont'd)

1-telluracyclopentane 1,1-dichloride	*B. coli*	--	294
1-telluracyclopentane 1,1-dibromide	*B. coli*	--	294
bis(4-acetamidophenyl) ditelluride	*B. paramelitensis*	$1{:}2x10^6$	155
	Staphylococcus aureus	$1{:}1\text{-}5x10^4$	155
	B. coli	$1{:}2x10^4$	155
bis(carboxymethyl) ditelluride	*B. coli*	$1{:}1x10^5$	292
bis(carboxymethyl) telluride	*B. coli*	$1{:}1x10^5$	292
bis(carbethoxymethyl) telluride	*B. coli*	$1{:}1.8x10^4$	292
bis(carboxymethyl) telluride, NH_4 salt	*B. coli*	$1{:}1.8x10^4$	292
guaicyl telluride	*B. coli*	$1{:}5x10^4$	292
phenoxtellurine	*B. coli*	no effect	292
bis(4-methoxyphenyl) tellurium dichloride	*Mycobacterium tuberculosis*	$1{:}1x10^6$	415
bis(4-methylphenyl) tellurium dichloride	weak inhibitory effects on the growth of Ehrlich tumors	--	416
bis(4-methoxyphenyl) tellurium dibromide		--	416
bis(4-methoxyphenyl) tellurium diiodide		--	416

Table XVI-1 (cont'd)

bis(4-ethoxyphenyl) tellurium dichloride	weak inhibitory effects on the growth of Ehrlich tumors	--	416

* strain i: somewhat old culture ** strain ii: isolated shortly before experiment was performed

untenable. Dewar and coworkers[97] have shown, that the compounds exist in the dione-form. The 1-tellura-3,5-cyclohexanedione 1,1-dihalides were found to be ineffective[159].

Taniyama and coworkers[415] found that diaryl tellurium dihalides possess a strong antibacterial action. The unsymmetric diorganyl tellurides prepared by Rogoz[370] (section VII-A-2b) had a much weaker bacteriostatic action than chloramphenicol against 21 strains of various microorganisms.

The cyclic tellurium compounds are highly toxic. 2,6-Dimethyl-, 2,4-dimethyl- and 4-ethyl-1-tellura-3,5-cyclohexanedione in aqueous solution injected subcutaneously into the back legs of mice cause death in doses of 0.005, 0.014 and 0.013 g/kg body weight, respectively. No symptoms were observed with doses well below the lethal dose. At higher doses a slight malaise was observed. Blood passed through urine a few hours after injection. The animals recovered in two days. With a lethal dose the signs of malaise soon appeared, hematuria followed and the mouse soon died. Dissections showed enlargement of kidneys, liver and spleen. Death was probably cause by hematuria. The tellurium compounds were found to be cumulative in large but not in small doses[280].

Levaditi claimed weak therapeutic properties for 4-ethyl-1-tellura-3,5-cyclohexanedione[243]. The 2,4-dimethyl derivative performed very satisfactori in treatment of infective conditions of the eye like conjunctivitis, blepharitis and corneal ulceration[281]. Bis(2-chloroethyl) tellurium dichloride was found to be a mild irritant of the human and dog skin indicated by simple hyperemia without vesication, mild urticarial rash, moderate swelling and edema and very little or no necrosis[166]. Glazebrook and Pearson[150] reported, that continued inhalation of dimethyl telluride and other volatile tellurium compounds induced an unpleasant halitosis and debility and caused headaches.

Trimethyl telluronium methyl trichlorotellurate(IV) acts as a powerful medullary stimulant and specifically excites the suprarenal glands.

Large doses paralyze nerve structures in the following order: symphatic ganglion cells, other autonomic ganglion cells and medulla and motor nerve endings[85]. Dimethyl tellurium dichloride is relatively inactive toward nerve cells[85]. Diphenyl telluride did not have a curative effect on syphilis in rabbits[243]. The LD_{50} values in mg/kg with mice as experimental animals were found to be 382, 70.3 and 92.4 for sodium 2-methyltellurobenzoate[51], methyl 4-(1'-hydroxy-2'-hydroxymethyl-2'-dichloroacetamidoethyl)-phenyl and biphenylyl telluride[370], respectively. Oefele[322] published in 1912 a review on therapeutic agents containing tellurium. A recent review deals with the toxicology of tellurium and its inorganic compounds[59].

The ability of living systems to convert tellurium and inorganic tellurium compounds into gaseous tellurium derivatives has been noticed very early. Upon administration of potassium tellurite to dogs and men, the odorous compounds exhaled were easily detected as bad breath. Literature references to these early investigations are part of reviews on this subject prepared by Challenger[62,63].

Bird and Challenger[35] identified the gaseous product formed when molds were allowed to grow on moistened bread crumbs in the presence of tellurites. The mold was allowed to form a good mycelium before potassium tellurite was added in an amount sufficient to reach a concentration of 0.5 g per 200 ml of medium. The mold gases, which were very sensitive towards air, were passed into a mercuric chloride solution. The precipitate was identified as $(CH_3)_2Te \cdot HgCl_2$, proving that the inorganic tellurium compound was reduced and methylated by the molds to dimethyl telluride. *Scopulariopsis brevicaulis*[35,64], green *Penicillium (P. notatum)*[35] and *Penicillium chrysogenum*[35] produced varying amounts of dimethyl telluride. Challenger[65] and Challenger and North[61] summarized the biological methylation of compounds of metalloids. It was also found[378] that sodium tellurite in aqueous solution was reduced by tissues of various plants, especially by *Allium cepa* to an extent differing with the various parts of each plant. Dimethyl telluride was formed only by molds but not by higher plants.

APPENDIX I: PATENTS COVERING ORGANIC TELLURIUM COMPOUNDS

Reference	Patent Number	Chem. Abstr.	Compound	Claim
			BELGIAN PATENTS	
P- 1	639 358	62, 9013	organic Te compounds	as constituents of catalysts for the preparation of unsaturated aldehydes from olefins
			BRITISH PATENTS	
P- 2	978	7, 2286	phthalein dyes containing Te	therapeutic agents
P- 3	292 222	23, 1747	telluracyclohexane and derivatives	preparation; therapeutic agents
P- 4	359 328	27, 449	Te dialkyldithiocarbamates	rubber vulcanization accelerators
P- 5	385 980	27, 4246	aliphatic Te compounds	preparation
P- 6	498 315	33, 3816	hydroxyaryl tellurides	oxidation and corrosion inhibitors
P- 7	536 514	36, 1806	thiazolyl tellurium halide	vulcanization accelerator
P- 8	576 740	42, 2069	$(RCTe)_2Te$	additive to improve extreme-pressure properties of lubricating oils
P- 9	585 803	46, 250	tellurophenolsulfonic acid	additive to crank-case oil
P- 10	599 729	42, 7334	Te derivatives of alkyl-phenols	detergents in lubricating oils

P- 11	790 281	52, 17695	organic telluride	oxidation inhibitor in lubricating and hydraulic fluids
P- 12	842 873	55, 4050	Te compound	vulcanization agent
P- 13	1 107 698	69, 77110	2,5-disubstituted telluro-phenes	preparation
P- 14	1 157 588	71, 62176	Te diethyldithiocarbamate	curing accelerator for ethylene-propylene terpolymer
P- 15	1 160 486	71, 82420	Te diethyldithiocarbamate	curing agent for ethylene-propylene terpolymer
			FRENCH PATENTS	
P- 16	443 554	7, 1617	triphenylmethane dyes con-taining Te	preparation
P- 17	758 359	28, 3079	halo(hydroxy)alkyl tellurides	preparation
P- 18	805 666	31, 4345	organic tellurides with functional groups	antiknock agents, antioxidants
P- 19	842 509	34, 5466	organic tellurides	photochemical preparation
P- 20	1 186 992	54, 25974	Te tetramethylthiuram disulfide	additive to polymers to improve their thermal and mechanical properties
P- 21	1 556 085	71, 82421	Te O,O-dibutyl phosphoro-dithioate	curing accelerator for ethylene-propylene-diene rubber
			GERMAN PATENTS	
P- 22	261 556	7, 3422	triphenylmethane dyes containing Te	preparation

P- 23	269 699	8, 2221	tellurols	as reagents in the preparation of As-Te compounds
P- 24	269 700	8, 2221	tellurols	as reagents in the preparation of As-Te compounds
P- 25	1 200 817	63, 15896	telluronium compounds with complex metal anions	to refine metals for semiconductor use or electroplating
			SOUTH AFRICAN PATENT	
P- 25a	70 00 519	75, 13567	diaryl tellurides	photoconductors
			US PATENTS	
P- 26	1 578 731	20, 1631	dialkyl tellurides	preparation from Na_2Te and alkyl sulfates
P- 27	2 030 035	30, 2019	Bz-1,Bz-1'-dibenzanthronyl ditelluride	dye intermediate
P- 28	2 174 110	34, 554	organotellurium compound	preparation from hydrocarbon, chlorine and TeO_2 and oxidation of the product by air; detergent
P- 29	2 195 539	34, 5093	$R(OH)C_6H_3$-Te-$C_6H_3(OH)R'$	oxidation inhibitor, stabilization of mineral and fatty oils
P- 30	2 216 751	35, 1627	R-Te-$(CH_2)_nCOO^-M^{+m}/m$	inhibitors of oxidation and sludge formation in greases and extreme-pressure lubricants
P- 31	2 217 611	35, 1156	bornyl-fenchyl tellurocyanoacetate	wetting agents
P- 32	2 217 612	35, 1156	hydroterpinyl tellurocyanoacylate	preparation

P- 33	2 217 613	35, 1156	terpinyl tellurocyano-acylate	preparation
P- 34	2 217 614	35, 1156	terpinyl tellurocyano-acylate	preparation
P- 35	2 217 615	35, 1156	tellurocyanate of a reduced alloocimene-crotonaldehyde condensate	preparation
P- 36	2 227 058	35, 2646	terpene tellurocyanoacylate	insecticide, flotation and wetting agent
P- 37	2 227 059	35, 2646	bornyl-fenchyl telluro-cyanoricineolates	insecticide, flotation and wetting agent
P- 38	2 227 060	35, 2646	bornyl-fenchyl telluro-cyanonaphthenate	insecticide, flotation and wetting agent
P- 39	2 227 061	35, 2646	terpene tellurocyanoacylate	insecticide, flotation and wetting agent
P- 40	2 239 495	35, 4781	terpene tellurocyanoacylate	insecticide, flotation agent
P- 41	2 239 496	35, 4781	terpene tellurocyanoacylate	insecticide, flotation agent
P- 42	2 257 974	36, 930	thiazolyl tellurium halide	vulcanization accelerator
P- 43	2 263 716	36, 1730	terpene ether tellurocyanate	insecticide, fungicide
P- 44	2 275 606	36, 4240	terpene tellurocyanoacylate	insecticide, fungicide, bactericide, flotation agent
P- 45	2 359 331	39, 1754	aromatic tellurides	mineral oil additives
P- 46	2 366 873	39, 3154	aromatic tellurides	lubricating oil additive
P- 47	2 375 007	40, 2293	$[RTe\text{-}Si(O)\text{-}TeR']_n$	antifoaming additive to lubricating oil

P- 48	2 385 301	40, 2294	aromatic tellurides	antioxidant for lubricating oils
P- 49	2 398 202	40, 3389	$[HOOCCH(R)]_2Te$	corrosion inhibitor for ferrous metals
P- 50	2 398 414	40, 3598	$(C_{16}H_{33})_2Te$	oxidation inhibitor in lubricating oils
P- 51	2 400 095	40, 4747	tellurocyanocarboxylic acid esters	pharmacological agents
P- 52	2 411 583	41, 1093	salts of $[TeC_6H_3(R)SO_3]^{--}$	corrosion inhibitor, wetting agents
P- 53	2 422 484	42, 217	tellurols	stabilizer for xylidine
P- 54	2 438 876	42, 5657	aromatic tellurides	oxidation inhibitor in lubricating oils
P- 55	2 473 511	43, 6820	dialkyl telluride	antioxidant and gum-inhibitor in lubricating oils
P- 56	2 531 427	45, 2120	bentonite telluronium compositions	gel-forming clays
P- 57	2 531 812	45, 3155	clay telluronium compositions	drilling fluids
P- 58	2 622 987	47, 3005	bentonite telluronium composition	coating composition and vehicle
P- 59	2 626 207	47, 3555	alkyl(hydroxy)phenyl tellurides	sludge inhibitor
P- 60	2 805 954	52, 754	bentonite telluronium compositions	coating compositions
P- 61	2 811 502	52, 5871	tellurium compounds	processing of butyl rubber
P- 62	2 813 828	52, 4977	dialkyl telluride	stabilizer for lubricating greases

P- 63	2 911 405	<u>54</u>, 3462	Te compounds of morpholine and piperidine	curing of hot milled rubbery compositions
P- 64	2 980 500	<u>55</u>, 15146	aliphatic tellurols, tell-urides and ditellurides	preparation of photoconductive Cd crystals
P- 65	3 006 829	<u>56</u>, 258	organic Te compounds	bridging agents for linear polyolefins
P- 66	3 050 485	<u>57</u>, 16876	organic Te compounds	curing of organopolysiloxanes
P- 67	3 149 101	<u>63</u>, 1819	tetraphenyltellurophene	preparation
P- 68	3 151 140	<u>61</u>, 16097	tetraphenyltellurophene	preparation
P- 69	3 207 720	<u>63</u>, 18367	tellurium halide	cross-linking of poly(vinyl chlorides)

REFERENCES

A

1. E. W. Abel and D. A. Armitage, Advan. Organometal. Chem., 5, 1 (1967).

1a. R. H. Aborn, Chapter 10 in "Tellurium", W. C. Cooper, ed., Van Nostrand Reinhold Co., New York, 1971, p. 389.

2. S. C. Abrahams, Quart. Rev. (London), 10, 407 (1956).

2a. D. M. Adams and P. J. Chandler, J. Chem. Soc., A, 1969, 588.

3. M. Adloff and J. P. Adloff, Bull. Soc. Chim. Fr., 1966, 3304.

4. L. Akobjanoff, Kunststoffe, 48, 373 (1958); Chem. Abstr., 52, 21226f.

5. I. P. Alimarin, Dokl. Akad. Nauk SSSR, 113, 105 (1957).

6. I. P. Alimarin and V. S. Sotnikov, Chem. Anal. (Warsaw), 2, 222 (1957); Chem. Abstr., 52, 6065a.

7. I. P. Alimarin and V. S. Sotnikov, Vestn. Mosk. Univ., Ser. Mat., Mekh., Astron., Fiz. Khim., 12 (6), 137 (1957); Chem. Abstr., 52, 19721h.

7a. D. W. Allen, J. C. Coppola, O. Kennard, F. G. Mann, W. D. S. Motherwell and D. G. Watson, J. Chem. Soc., C, 1970, 810

8. J. R. Allkins and P. J. Hendra, Spectrochim. Acta, 22, 2075 (1966).

8a. J. R. Allkins and P. J. Hendra, J. Chem. Soc., A, 1967, 1325.

9. J. R. Allkins and P. J. Hendra, Spectrochim. Acta, A, 24, 1305 (1968).

10. E. Amberger and E. Gut, Chem. Ber., 101, 1200 (1968).

11. H. H. Anderson, J. Chem. Eng. Data, 9, 272 (1964).

12. U. Anthoni, C. Larsen and P. H. Nielsen, Acta. Chem. Scand., 21, 2571 (1967).

12a. H. J. Arpe and H. Kuckertz, Angew. Chem., Int. Ed. Eng., 10, 73 (1971).

12b. T. Austad, J. Songstad and K. Ase, Acta Chem. Scand., 25, 33, (1971).

13. E. E. Aynsley, J. Chem. Soc., 1953, 3016.

14. E. E. Aynsley and R. H. Watson, J. Chem. Soc., 1955, 576.

15. E. E. Aynsley and R. H. Watson, J. Chem. Soc., 1955, 2603.

B

16. K. W. Bagnal;, "The Chemistry of Selenium, Tellurium and Polonium", Elsevier, Amsterdam, 1966.

17. M. P. Balfe, C. A. Chaplin and H. Phillips, J. Chem. Soc., 1938, 341.

18. M. P. Balfe and K. N. Nandi, J. Chem. Soc., 1941, 70.

19. A. Baroni, Atti Accad. Naz. Lincei, Rend., Cl. Sci. Fis. Mat. Nat., 27, 238 (1938): Chem. Abstr., 33, 163.

20. C. Barry, G. Cauquis and M. Maurey, Bull. Soc. Chim. Fr., 1966, 2510.

21. F. Becker, Ann. Chem., 180, 257 (1875).

21a. W. M. Becker, V. A. Johnson and W. Nussbaum, Chapter 3 in "Tellurium", W. C. Cooper, ed., Van Nostrand Reinhold Co., New York, 1971, p. 54.

22. R. Belcher, D. Gibbons and A. Sykes, Mikrochim. Acta, 40, 76 (1952).

23. L. Belchetz and E. K. Rideal, J. Amer. Chem. Soc., 57, 1168 (1935).

24. T. N. Bell, B. J. Pullman and B. O. West, Aust. J. Chem., 87, 722 (1963).

25. E. Bergmann, L. Engel, and S. Sandor, Z. Phys. Chem. (Leipzig), B 10, 397 (1930).

26. G. Bergson, Acta Chem. Scand., 11, 571 (1957).

27. F. W. Bergstrom, J. Amer. Chem. Soc., 48, 2319 (1926).

28. J. J. Berzelius, Schweiggers Journal fuer Chemie und Physik, 31, 62 (1821).

29. S. S. Bhatnagar and T. K. Lahiri, Curr. Sci., 1, 380, (1933).

30. S. S. Bhatnagar and T. K. Lahiri, Z. Phys., 84, 671 (1933).

31. E. Billows, Z. Kristallogr., 40, 290 (1902).

32. L. Birckenbach and K. Kellermann, Chem. Ber., 58, 786 (1925).

33. L. Birckenbach and K. Huttner, Z. Anorg. Allg. Chem., 190, 1, (1930).

34. L. Birckenbach, Forschungen und Fortschr., 18, 232 (1942).

35. M. L. Bird and F. Challenger, J. Chem. Soc., 1939, 163.

35a. J. Blachot and L. C. Carraz, Radiochim. Acta, 11, 45 (1969).

36. W. R. Blackmore and S. C. Abrahms, Acta Crystallogr., 8, 317 (1955).

37. M. N. Bochkarev, L. P. Sanina and N. S. Vyazankin, Zh. Obshch. Khim., 39, 135 (1969).

37a. G. M. Bogolyubov, Yu. N. Shlyk and A. A. Petrov, Zh. Obshch. Khim., 39, 1804 (1969).

37b. G. M. Bogolyubov, N. N. Grishin and A. A. Petrov, Zh. Obshch. Khim., 39, 2244 (1969).

38. Yu. A. Boiko, B. S. Kupin and A. A. Petrov, Zh. Org. Khim., 4, 1355 (1968).

38a. Yu. A. Boiko, B. S. Kupin and A. A. Petrov, Zh. Org. Khim., 5, 1553 (1969).

38b. J. M. Bonnier and P. Jardon, J. Chim. Phys. Physicochim. Biol., 68, 432 (1971).

38c. G. N. Bortnikov, M. N. Bochkarev, N. S. Vyazankin, S. Kh. Ratushnaya and Ya. I. Yashin, Izv. Akad. Nauk SSSR, Ser. Khim., 1971, 851.

38d. Yu. Ya. Borovikov, E. V. Rl'tsev, I. E. Boldeskul, N. G. Feshchenko, Yu. P. Makovetskii and Yu. P. Egorov, Zh. Obshch. Khim., 40, 1957 (1970).

38e. A. J. Bowd, D. T. Burns and A. G. Fogg, Talanta, 16, 719 (1969).

39. K. Bowden and A. E. Braude, J. Chem. Soc., 1952, 1068.

40. A. J. Bradley, Phil. Mag., (6) 48, 477 (1924).

41. L. Brandsma, H. E. Wijers and J. F. Arens, Rec. Trav. Chim. Pays-Bas, 81, 583 (1962).

42. L. Brandsma and H. E. Wijers, Rec. Trav. Chim. Pays-Bas, 82, 68 (1963).

43. L. Brandsma, H. E. Wijers and C. Jonker, Rec. Trav. Chim. Pays-Bas, 83, 208 (1964).

44. B. Brauner, J. Chem. Soc., 1889, 382.

45. E. H. Braye, W. Huebel and I. Caplier, J. Amer. Chem. Soc., 83, 4406 (1961).

45a. G. Bredig, Z. Physik. Chem., 13, 289 (1894).

46. V. Breuninger, H. Dreeskamp and G. Pfisterer, Ber. Bunsenges. Phys. Chem., 70, 613 (1966).

47. E. Buchta and K. Greiner, Naturwissenschaften, 46, 532 (1959).

48. E. Buchta and K. Greiner, Chem. Ber., 94, 1311 (1961).

49. H. Buerger and V. Goetze, Inorg. Nucl. Chem. Lett., 3, 549 (1967).

50. F. H. Burstall and S. Sugden, J. Chem. Soc., 1930, 229.

51. N. P. Buu-Hoi, P. H. Chanh and M. Renson, C. R. Acad. Sci., Paris, Ser. D., 268, 2807 (1969).

51a. N. P. Buu-Hoi, M. Mangane, M. Renson and J. L. Piette, J. Heterocycl. Chem., 7, 219 (1970).

C

52. A. Cahours, Ann. Chem., 135, 352 (1865).

53. A. Cahours, C. R. Acad. Sci., Paris, 60, 620 (1865).

54. A. Cahours, Ann. Chim. (Paris), [5] 10, 13 (1877).

54a. I. C. Calder, R. B. Johns and J. M. Desmarchelier, Aust. J. Chem., 24, 325 (1971).

55. I. B. M. Campbell and E. E. Turner, J. Chem. Soc., 1938, 37.

56. L. Carillo and S. J. Nassiff, Radiochim. Acta, 8, 124 (1967).

57. T. A. Carlson and R. M. White, J. Chem. Phys., 38, 2930 (1963).

58. F. Carr and T. G. Pearson, J. Chem. Soc., 1938, 282.

59. E. D. Cerwenka, Jr. and W. C. Cooper, Arch. Environ. Health, 3, 189 (1961).

60. F. Challenger, A. T. Peters and J. Halevy, J. Chem. Soc., 1926, 1648.

61. F. Challenger and H. E. North, J. Chem. Soc., 1934, 68.

62. F. Challenger, Chem. Ind. (London), 54, 657 (1935).

62a. F. Challenger and C. Higginbottom, Biochem. J., 29, 1757 (1935).

63. F. Challenger, Chem. Rev., 36, 315 (1945).

64. F. Challenger, J. Proc. Roy. Inst. Chem. (Gt. Britian and Ireland), 1945, 105.

65. F. Challenger, Sci. Progr., 35, 396 (1947).

65a. C. H. Champness, Chapter 8 in "Tellurium", W. C. Cooper, ed., Van Nostrand Reinhold Co., New York, 1971, p. 322.

66. G. Y. Chao and J. D. McCullough, Acta Crystallogr., 15, 887 (1962).

67. J. Chatt, L. A. Duncanson and L. M. Venanzi, Chem. and Ind., 1955, 749.

68. J. Chatt and L. M. Venanzi, J. Chem. Soc., 1955, 2787.

69. J. Chatt and L. M. Venanzi, J. Chem. Soc., 1955, 3858.

70. J. Chatt, L. A. Duncanson and L. M. Venanzi, J. Chem. Soc., 1955, 4456.

71. J. Chatt, L. A. Duncanson and L. M. Venanzi, J. Chem. Soc., 1955, 4461.

72. J. Chatt and L. M. Venanzi, J. Chem. Soc., 1957, 2351.

73. J. Chatt and L. M. Venanzi, J. Chem. Soc., 1957, 2445.

74. J. Chatt, L. A. Duncanson, B. L. Shaw and L. M. Venanzi, Discussions Faraday Soc., 26, 131 (1958).

75. J. Chatt, G. A. Gamlen and L. E. Orgel, J. Chem. Soc., 1959, 1047.

76. M. T. Chen and J. W. George, J. Organometal. Chem., 12,401 (1968).

77. M. T. Chen and J. W. George, J. Amer. Chem. Soc., 90, 4580 (1968).

77a. J. Chojnowski and W. W. Brandt, J. Amer. Chem. Soc., 90, 1384 (1968).

77b. G. N. Chremos and R. A. Zingaro, J. Organometal. Chem., 22, 637 (1970).

78. G. D. Christofferson and J. D. McCullough, Acta Crystallogr., 11, 249 (1958).

79. G. D. Christofferson, R. A. Sparks and J. D. McCullough, Acta Crystallogr. 11, 782 (1958).

80. G. E. Coates, J. Chem. Soc., 1951, 2003.

81. H. E. Cocksedge, J. Chem. Soc., 1908, 2175.

82. S. C. Cohen, M. L. N. Reddy and A. G. Massey, J. Chem. Soc., D, 1967, 451.

83. S. C. Cohen, M. L. N. Reddy and A. G. Massey, J. Organometal. Chem., 11, 563 (1968).

83a. J. Connor, G. Greig and O. P. Strausz, J. Amer. Chem. Soc., 91, 5695 (1969).

83b. J. Connor, A. Van Roodselaar, R. W. Fair and O. P. Strausz, J. Amer. Chem. Soc., 93, 560 (1971).

83c. W. C. Cooper, Chapter 11 in "Tellurium", W. C. Cooper, ed., Van Nostrand Reinhold Co., New York, 1971, p. 410.

83d. D. A. Couch, P. S. Elmes, J. E. Ferguson, M. L. Greenfield and C. J. Wilkins, J. Chem. Soc., A, 1967, 1813.

84. C. Courtot and M. G. Bastani, C. R. Acad. Sci, Paris, 203, 197 (1936).

85. D. V. Cow and W. E. Dixon, J. Physiol. (London), 56, 42 (1922).

86. S. Cradock, E. A. V. Ebsworth and D. W. H. Rankin, J. Chem. Soc., A, 1969, 1628.

87. N. M. Cullinane, A. G. Rees and C. A. J. Plummer, J. Chem. Soc., 1939, 151.

88. S. J. Cyvin, Spectrochim. Acta, 15, 958 (1959).

89. S. J. Cyvin, Spectrochim. Acta, 17, 1219 (1961).

D

90. A. Damiens, C. R. Acad. Sci., Paris, 171, 1140 (1920).

91. A. Damiens, C. R. Acad. Sci., Paris, 172, 1105 (1921).

92. A. Damiens, C. R. Acad. Sci., Paris, 173, 300, (1921).

93. A. Damiens, C. R. Acad. Sci., Paris, 173, 583 (1921).

94. A. Damiens, Ann. Chim. (Paris), [9], 19, 44 (1923).

95. J. N. E. Day, Sci. Progr., 23, 211 (1928).

95a. G. B. Deacon and J. C. Parrott, J. Organometal. Chem., 22, 287 (1970).

96. E. Demarcay, Bull. Soc. Chim. Fr., 1883, 99.

97. D. H. Dewar, J. E. Fergusson, P. R. Hentschel, C. J. Wilkins and P. P. Williams, J. Chem. Soc., 1964, 688.

98. S. S. Dharmatti, Proc. Indian Acad. Sci., Sect. A, 12, , 212 (1940).

99. W. v. E. Doering and A. K. Hoffmann, J. Amer. Chem. Soc., 77, 521 (1955).

100. L. G. Dowdell and J. Orndoff, Proc. Indian Acad. Sci., 55, 165 (1945).

101. A. W. Downs, J. Chem. Soc., D., 1968, 1920.

102. H. Dreeskamp and G. Pfisterer, Mol. Phys., 14, 295 (1968).

103. H. D. K. Drew, J. Chem. Soc., 1926, 223.

104. H. D. K. Drew, J. Chem. Soc., 1926, 3054.

105. H. D. K. Drew and R. W. Thomason, J. Chem. Soc., 1927, 116.

106. H. D. K. Drew, J. Chem. Soc., 1928, 506.

107. H. D. K. Drew, J. Chem. Soc., 1929, 560.

108. H. D. K. Drew and C. R. Porter, J. Chem. Soc., 1929, 2091.

109. H. D. K. Drew, J. Chem. Soc., 1934, 1790.

110. J. Drowart, Bull. Soc. Chim. Belges., 73, 451 (1964).

111. M. Dubien, Rev. Gen. Sci., 37, 366 (1926).

112. A. M. Duffield, H. Budzikiewicz and C. Djerassi, J. Amer. Chem. Soc., 87, 2920 (1965).

113. P. Dupuy, C. R. Acad. Sci., Paris, 237, 718 (1953).

113a. J. R. Durig, C. M. Player, Jr., J. Bragin and Y. S. Li, J. Chem. Phys., 55, 2895 (1971).

E

114. A. N. Egorochkin, N. S. Vyazankin, G. A. Razyvaev, O. A. Kruglaya and M. N. Bochkarev, Dokl. Akad. Nauk SSSR, 170, 333 (1966).

115. A. N. Egorochkin, N. S. Vyazankin, M. N. Bochkarev and S. Ya. Khorshev, Zh. Obshch. Khim., 37, 1165 (1967).

116. A. N. Egorochkin, S. Ya. Khorshev, N. S. Vyazankin, M. N. Bochkarev, O. A. Kruglaya and G. S. Semchikova, Zh. Obshch. Khim., 37, 2308 (1967).

117. A. N. Egorochkin, S. Ya. Khorshev, and N. S. Vyazakin, Dokl. Akad. Nauk SSSR, 185, 353 (1969).

117a. A. N. Egorochkin, E. N. Gladyshev, S. Ya. Khorshev, P. Ya. Bayushkin, A. I. Burov and N. S. Vyazankin, Izv. Akad. Nauk SSSR, Ser. Khim., 1971, 639.

117b. A. N. Egorochkin, M. N. Bochkarev, N. S. Vyazankin, S. E. Skobeleva and S. Ya. Khorshev, Dokl. Akad. Nauk SSSR, 198, 96 (1971).

118. F. Einstein, J. Trotter and C. Williston, J. Chem. Soc., A, 1967, 2018.

119. J. Ellermann and D. Schirmacher, Chem. Ber., 100, 2220 (1967).

119a. D. Elmaleh, S. Patai and Z. Rappoport, J. Chem. Soc., C, 1971, 3100.

120. H. J. Emeleus and H. G. Heal. J. Chem. Soc., 1946, 1126.

F

120a. F. Faraone, S. Sergi and R. Pietropaolo, J. Organometal. Chem., 24, 453 (1970).

120b. F. Faraone, R. Pietropaolo and S. Sergi, J. Organometal. Chem., 24, 797 (1970).

120c. F. Faraone, S. Sergi and R. Pietropaolo, Atti Soc. Peloritana Sci. Fis. Mat. Natur., 14, 355 (1968).

121. W. V. Farrar and J. M. Gulland, J. Chem. Soc., 1945, 11.

122. W. V. Farrar, Research (London), 4, 177 (1951).

122a. V. K. Finogenova, B. A. Popovkin and A. V. Novoselova, Zh. Neorg. Khim., 15, 1459 (1970).

123. C. H. Fisher and A. Eisner, J. Org. Chem., 6, 169 (1941).

124. O. Foss, Acta Chem. Scand., 4, 1241, (1950).

125. O. Foss, Acta Chem. Scand., 6, 306 (1952).

126. O. Foss, Acta Chem. Scand., 7, 227 (1953).

127. O. Foss and S. Hauge, Acta Chem. Scand., 13, 2155 (1959).

128. O. Foss, S. Husebye and K. Maroey, Acta Chem. Scand., 17, 1806 (1963).

129. O. Foss and K. Maroey, Acta Chem. Scand., 20, 123 (1966).

130. O. Foss and K. Husebye, Acta Chem. Scand., 20, 132 (1966).

131. W. J. Franklin and R. L. Werner, Tetrahedron Lett., 1965, 3003.

132. V. Franzen and C. Mertz, Ann. Chem., 643, 24 (1961).

132a. B. H. Freeman and L. Douglas, J. Chem. Soc., D, 1970, 924.

133. J. M. Freeman and T. Henshall, J. Mol. Structure, 1, 31 (1967/8).

133a. F. Fringuelli, G. Marino, G. Savelli and A. Taticchi, J. Chem. Soc., D, 1971, 1441.

133b. F. Fringuelli and A. Taticchi, J. Chem. Soc. Perkin Trans., 1, 199 (1972

134. H. Frischleder, G. Klose and J. Ranft, Wiss. Z. Karl-Marx Univ., Leipzig, Math. Naturw. Reihe, 14, 863 (1965).

135. E. Fritsman and V. Krinitskii, J. Applied Chem. (U.S.S.R.), 11, 195 (1938).

136. H. P. Fritz and H. Keller, Chem. Ber., 94, 1524 (1961).

137. E. Fritzmann, J. Russ. Phys. Chem. Soc., 47, 588 (1915).

138. E. Fritzmann, Z. Anorg. Allg. Chem., 133, 119 (1924).

139. M. Frommes, Fresenius' Z. Anal. Chem., 96, 447 (1934).

140. H. Funk and W. Weiss, J. Prakt. Chem., [4], 1, 33 (1954).

G

141. E. E. Galloni and J. Publiese, Acta Crystallogr., 3, 319 (1950).

142. F. L. Gilbert and T. M. Lowry, J. Chem. Soc., 1928, 1997.

143. F. L. Gilbert and T. M. Lowry, J. Chem. Soc., 1928, 2658.

144. F. L. Gilbert and T. M. Lowry, J. Chem. Soc., 1928, 3179.

145. R. J. Gillespie, Can. J. Chem., 39, 318 (1961).

146. A. Gioaba and G. Vasiliu, An. Univ. Bucuresti, Ser. Stiint, Natur., Chim., 15, 89 (1966); Chem. Abstr., 70, 68319.

146a. A. Gioaba and O. Maior, Rev. Chim. (Bucharest), 21, 613 (1970).

147. M. Giua and F. Cherchi, Gazz. Chim. Ital., 50, I, 362 (1920); Chem. Abstr., 15, 521.

148. H. H. Glazebrook and T. G. Pearson, J. Chem. Soc., 1936, 1777.

149. H. H. Glazebrook and T. G. Pearson, J. Chem. Soc., 1937, 567.

150. H. H. Glazebrook and T. G. Pearson, J. Chem. Soc., 1939, 589.

151. "Gmelin's Handbuch der Anorganischen Chemie", 8. Auflage; System Nummer 11 "Tellur", Verlag Chemie, G.M.B.H., Berlin 1940, p. 140.

152. A. E. Goddard, "A Textbook of Inorganic Chemistry", Vol. XI, Part IV, J. N. Friend, Ed., C. Griffin & Co., Ltd., London, 1937, p. 166.

153. A. E. Goddard, " A Textbook of Inorganic Chemistry", Vol. XI, Part IV, J. N. Friend, Ed., C. Griffin & Co., Ltd., London, 1937, p. 186.

154. N. P. Grechkin, I. A. Nuretdinov and N. A. Buina, Izv. Akad. Nauk SSSR, Ser. Khim., 1969, 168.

155. H. N. Green and F. Bielschowsky, Brit. J. Exptl. Path., 23, 13 (1942).

156. T. E. Green and M. Turley, "Treatise on Analytical Chemistry, Part II, Analytical Chemistry of the Elements", Vol. 7, I. M. Kolthoff and P. J. Elving, Ed., Interscience Publishers, New York - London, 1961, p. 137.

157. N. N. Greenwood, R. Little and M. J. Sprague, J. Chem. Soc., 1964, 1292.

158. H. G. Grimm and H. Metzger, Chem. Ber., 69, 1356 (1936).

159. J. M. Gulland and W. V. Farrar, Nature, 154, 88 (1944).

160. J. M. Gulland and W. V. Farrar, Annales Farm. Bioquim. (Buenos Aires), 15, 73 (1944); Chem. Abstr., 40, 169.

161. M. Z. Gurshovich and A. V. El'tsov, Zh. Obshch. Khim., 39, 941 (1969).

162. E. Gusarsky and A. Treinin, J. Phys. Chem., 69, 3176 (1965).

163. A. Gutbier and F. Fury, Z. Anorg. Allg. Chem., 32, 108 (1902).

164. V. Gutmann, Monatsh. Chem., 82, 280 (1951).

H

164a. W. S. Haller and K. J. Irgolic, J. Organometal Chem., 38, 97 (1972).

165. S. Hamada, Nippon Kagaku Zasshi, 82, 1327 (1961); Chem. Abstr., 57, 14

166. P. J. Hanzlik and J. Tarr, J. Pharmacol. Exp. Therap., 14, 221 (1919).

167. W. A. Hardy and G. Silvey, Phys. Rev., 95, 385 (1954).

167a. A. B. Harvey, F. E. Saalfeld and C. W. Sink, Appl. Spectrosc., 24, 466 (1970).

168. G. C. Hayward and P. J. Hendra, J. Chem. Soc., A, 1969, 1760.

169. H. G. Heal, J. Chem. Educ., 35, 192 (1958).

170. M. Heeren, Inauguraldissertation, Goettingen, 1861; Chem. Zentr., 1861 916.

171. D. Hellwinkel and G. Fahrbach, Tetrahedron Lett., 1965, 1823.

172. D. Hellwinkel and G. Fahrbach, Chem. Ber., 101, 574 (1968).

173. D. Hellwinkel and G. Fahrbach, Ann. Chem., 712, 1 (1968).

174. B. Hetnarski and W. Hofman, Bull. Acad. Pol. Sci., Ser. Sci. Chim., 17, 1 (1969); Chem. Abstr., 70, 49877W.

174a. B. Hetnarski and A. Grabowska, Bull Acad. Pol. Sci., Ser. Sci. Chim., 17, 333 (1969).

174b. B. Hetnarski and A. Grabowska, Rocz. Chem., 44, 1713 (1970).

175. W. Hieber and K. Wollmann, Chem. Ber., 95, 1552 (1962).

176. W. Hieber and T. Kruck, Chem. Ber., 95, 2027 (1962).

177. W. Hieber and R. Kramolowsky, Z. Anorg. Allg. Chem., 321, 94 (1963).

178. W. Hieber and F. Stanner, Chem. Ber., 102, 2930 (1969).

178a. W. Hieber and P. John, Chem. Ber., 103, 2161 (1970).

179. J. Hoarau "Nouveau Traite De Chemie Minerale", Vol. XIII, P. Pascal, Ed., Masson et Cie., Paris, 1961, p. 1913.

180. F. G. Holliman and F. G. Mann, J. Chem. Soc., 1945, 37.

181. K. A. Hooton and A. L. Allred, Inorg. Chem., 4, 671 (1965).

182. M. G. ter Horst, Rec. Trav. Chim. Pays-Bas, 55, 697 (1936).

183. B. von Horvath. Z. Anorg. Allg. Chem., 70, 408 (1911).

184. M. L. Huggins, J. Chem. Phys., 5, 201 (1937).

185. S. Husebye, Acta Chem. Scand., 20, 2007 (1966).

185a. S. Husebye, Acta Chem. Scand., 23, 1389 (1969).

I

185b. K. J. Irgolic and R. A. Zingaro, Organometal. React., 2, 117 (1971).

185c. K. J. Irgolic and W. S. Haller, Intern. J. Sulfur Chem., in press.

186. D. J. G. Ives, R. W. Pittman and W. Wardlaw, J. Chem. Soc., 1947, 1080.

J

186a. P. H. Jennings, Chapter 2 in "Tellurium", W. C. Cooper, ed., Van Nostrand Reinhold Co., New York, 1971, p. 14.

187. K. A. Jensen, Z. Anorg. Allg. Chem., 231, 365 (1937).

188. K. A. Jensen, Z. Anorg. Allg. Chem., 250, 245 (1943).

189. K. A. Jensen, Z. Anorg. Allg. Chem., 250, 268 (1943).

189a. P. John, Chem. Ber., 103, 2178 (1970).

190. L. H. Jones, Inorg. Chem., 6, 429 (1967).

K

191. L. V. Kaabak, A. P. Tomilov and S. L. Varshavskii, Zh. Vses. Khim. Obshchest. im D. I. Mendeeleva, 9, 700 (1964); Chem. Abstr., 62, 12751a.

192. T. Kaiwa, Tokohu J. Exptl. Med., 20, 163 (1932); Chem. Abstr., 27, 1402.

192a. L. M. Kataeva, N. S. Podkovyrina, A. N. Sarbash and E. G. Kataev, Zh. Strukt. Khim., 12, 931 (1971).

192b. O. Kennard, D. L. Wampler, J. C. Coppola, W. D. S. Motherwell, F. G. Mann, D. G. Watson, C. H. MacGillavry, C. H. Stam and P. Benci, J. Chem. Soc., C, 1971, 1511.

193. C. Klofutar, F. Krasovec and M. Kusar, Croat. Chem. Acta, 40, 23 (1968).

193a. G. Klose, Z. Naturforsch., 16a, 528 (1961).

194. I. K. Knaggs and R. H. Vernon, J. Chem. Soc., 1921, 105.

194a. C. Knobler, J. D. McCullough and H. Hope, Inorg. Chem., 9, 797 (1970).

194b. D. Kobelt and E. F. Paulus, Angew. Chem., Int. Ed. Engl., 10, 74 (1971).

194c. V. Kollonitsch and C. H. Kline, Hydrocarbon Processing and Petroleum Refiner, 43, 139 (1964).

194d. A. Koma, E. Takimoto and S. Tanaka, Phys. Status Solidi, 40, 239 (1970).

194e. A. Koma, O. Mizuno and S. Tanaka, Phys. Status Solidi, B, 46, 225 (1971).

195. S. Kondo, E. Kakiuchi and T. Shimizu, Bull. Soc. Chim. Jap., 42, 2050 (1969).

196. A. G. Konyaeva, Izv. Vyssh. Ucheb. Zaved., Fiz., 10, 91 (1967).

197. E. S. Kostiner, M. L. N. Reddy, D. S. Urch and A. G. Massey, J. Organometal. Chem., 15, 383 (1968).

198. F. Krafft and W. Vorster, Chem. Ber., 26, 2813 (1893).

199. F. Krafft and R. E. Lyons, Chem. Ber., 27, 1768 (1894).

200. F. Krafft and O. Steiner, Chem. Ber., 34, 560 (1901).

201. C. A. Kraus and C. Y. Chiu, J. Amer. Chem. Soc., 44, 1999 (1922).

202. C. A. Kraus and E. W. Johnson, J. Phys. Chem., 32, 1281 (1928).

203. C. A. Kraus and J. A. Ridderhof, J. Amer. Chem. Soc., 56, 79 (1934).

204. E. Krause and G. Renwanz, Chem. Ber., 62, 1710 (1929).

205. E. Krause and G. Renwanz, Chem. Ber., 65, 777 (1932).

206. T. Kruck and M. Hoefler, Chem. Ber., 96, 3035 (1963).

207. F. H. Kruse, R. W. Sanftner and J. F. Suttle, Anal. Chem., 25, 500 (1953).

208. F. H. Kruse, R. E. Marsh and J. D. McCullough, Acta Crystallogr., 10, 201 (1957).

L

209. J. B. Lambert and R. G. Keske, Tetrahedron Lett., 1967, 4755.

210. J. B. Lambert, R. G. Keske and D. K. Weary, J. Amer. Chem. Soc., 89, 5921 (1967).

211. J. O. M. van Langen and T. v. d. Plas, Research Correspondence, Supplement to Research (London), 7, S12 (1954).

212. K. Lederer, C. R. Acad. Sci., Paris, 151, 611 (1910).

213. K. Lederer, Chem. Ber., 44, 2287 (1911).

214. K. Lederer, Ann. Chem., 391, 326 (1912).

215. K. Lederer, Ann. Chem., 399, 260 (1913).

216. K. Lederer, Chem. Ber., 46, 1358 (1913).

217. K. Lederer, Chem. Ber., 46, 1810 (1913).

218. K. Lederer, Chem. Ber., 47, 277 (1914).

219. K. Lederer, Chem. Ber., 48, 1345 (1915).

220. K. Lederer, Chem. Ber., 48, 1422 (1915).

221. K. Lederer, Chem. Ber., 48, 1944 (1915).

222. K. Lederer, Chem. Ber., 48, 2049 (1915).

223. K. Lederer, Chem. Ber., 49, 334 (1916).

224. K. Lederer, Chem. Ber., 49, 345 (1916).

225. K. Lederer, Chem. Ber., 49, 1071 (1916).

226. K. Lederer, Chem. Ber., 49, 1076 (1916).

227. K. Lederer, Chem. Ber., 49, 1082 (1916).

228. K. Lederer, Chem. Ber., 49, 1385 (1916).

229. K. Lederer, Chem. Ber., 49, 1615 (1916).

230. K. Lederer, Chem. Ber., 49, 2002 (1916).

231. K. Lederer, Chem. Ber., 49, 2529 (1916).

232. K. Lederer, Chem. Ber., 49, 2532 (1916).

233. K. Lederer, Chem. Ber., 49, 2663 (1916).

234. K. Lederer, Chem. Ber., 50, 238 (1917).

235. K. Lederer, Chem. Ber., 52, 1989 (1919).

236. K. Lederer, Chem. Ber., 53, 712 (1920).

237. K. Lederer, Chem. Ber., 53, 1430 (1920).

238. K. Lederer, Chem. Ber., 53, 1674 (1920).

239. K. Lederer, Chem. Ber., 53, 2342 (1920).

240. V. Lenher, J. Amer. Chem. Soc., 22, 136 (1900).

241. V. Lenher, J. Amer. Chem. Soc., 24, 188 (1902).

242. V. Lenher, J. Amer. Chem. Soc., 30, 737 (1908).

243. C. Levaditi, Ann. Inst. Pasteur, 41, 369 (1927).

244. D. T. Lewis, J. Chem. Soc., 1940, 831.

245. M. Lipp, F. Dallacker and I. Meier, Monatsh. Chem., 90, 41 (1959).

245a. H. K. Livingstone and R. Krosec, J. Polym. Sci., Part B, 9, 95 (1971).

246. Y. Llabador and J. P. Adloff, Radiochim. Acta, 6, 49, (1966).

247. Y. Llabador and J. P. Adloff, Radiochim. Acta, 9, 171 (1968).

247a. Y. Llabador, Radiochim. Acta, 14, 100 (1970).

248. T. M. Lowry, R. A. Goldstein and F. L. Gilbert, J. Chem. Soc., 1928, 307.

249. T. M. Lowry and F. L. Gilbert, J. Chem. Soc., 1929, 2076.

250. T. M. Lowry and F. L. Gilbert, J. Chem. Soc., 1929, 2867.

251. T. M. Lowry and F. Huether, Rec. Trav. Chim. Pays-Bas, 55, 688 (1936)

252. A. Lowy and R. F. Dunbrook, J. Amer. Chem. Soc., 44, 614 (1922).

253. H. Luers and F. Weinfurtner, Wochschr. Brau., 43, 25, (1926).

254. R. E. Lyons and G. S. Bush, J. Amer. Chem. Soc., 30, 831, (1908).

255. R. E. Lyons and E. D. Scudder, Chem. Ber., 64, 530 (1931).

M

256. W. Mack, Angew. Chem., 77, 260 (1965).

257. W. Mack, Angew. Chem., 78, 940 (1966).

258. J. W. Mallet, Ann. Chem., 79, 223 (1851).

259. F. G. Mann and F. G. Holliman, Nature, 152, 749 (1943).

260. V. M. Marganian, Dissertation, Clemson University, 1966; Diss. Abstr B28, 84 (1967).

260a. V. M. Marganaian, J. E. Whisenhunt and J. C. Fanning, J. Inorg. Nucl Chem., 31, 3775 (1969).

261. A. Marquardt and A. Michaelis, Chem. Ber., 21, 2042 (1888).

262. H. Marshall, Inorg. Syn., 3, 143 (1950).

263. H. Matsuo, J. Sci. Hiroshima Univ., Ser. A, 22, 51 (1958); Chem. Abs 53, 9857b.

264. H. Matsuo, J. Sci. Hiroshima Univ. , Ser. A, 22, 281 (1958); Chem. Abstr., 54, 4206h.

265. L. R. Maxwell and V. M. Mosley, Phys. Rev., 51, 684 (1937).

266. L. R. Maxwell and V. M. Mosely, Phys. Rev., 57, 21 (1940).

267. F. P. Mazza and E. Melchionna, Rend. Accad. Sci. Napoli, [3], 34, 54 (1928); Chem. Abstr., 23, 2955.

268. C. R. McCrosky, F. W. Bergstrom and G. Waitkins, J. Amer. Chem. Soc., 64, 722 (1942).

269. J. D. McCullough, Inorg. Chem., 4, 862 (1965).

270. D. H. McDaniel, Science, 125, 545 (1957).

271. W. McFarlane, Mol. Phys., 12, 243 (1967).

272. F. A. McMahon, T. G. Pearson and P. L. Robinson, J. Chem. Soc., 1933, 1644.

273. E. Montignie, Z. Anorg. Allg. Chem., 307, 109 (1960).

274. E. Montignie, Z. Anorg. Allg. Chem., 315, 102 (1962).

275. G. T. Morgan and H. D. K. Drew, J. Chem. Soc., 1920, 1456.

276. G. T. Morgan and H. D. K. Drew, J. Chem. Soc., 1921, 610.

277. G. T. Morgan and H. D. K. Drew, J. Chem. Soc., 1922, 922.

278. G. T. Morgan, E. A. Cooper and A. W. Burtt, Biochem. J., 17, 30 (1923).

279. G. T. Morgan and H. G. Reeves, J. Chem. Soc., 1923, 444.

280. G. T. Morgan, E. A. Cooper and A. W. Burtt, Biochem. J., 18, 190 (1924).

281. G. T. Morgan, E. A. Cooper and F. J. Corby, J. Soc. Chem. Ind., 43, 304T (1924).

282. G. T. Morgan and H. D. K. Drew, J. Chem. Soc., 1924, 731.

283. G. T. Morgan and R. W. Thomason, J. Chem. Soc., 1924, 754.

284. G. T. Morgan and E. Holmes, J. Chem. Soc., 1924, 760.

285. G. T. Morgan and H. D. K. Drew, J. Chem. Soc., 1924, 1601.

286. G. T. Morgan and H. D. K. Drew, J. Chem. Soc., 1925, 531.

287. G. T. Morgan and C. J. A. Taylor, J. Chem. Soc., 1925, 797.

288. G. T. Morgan and H. D. K. Drew, J. Chem. Soc., 1925, 2307.

289. G. T. Morgan, F. J. Corby, O. C. Elvins, E. Jones, R. E. Kellett and C. J. A. Taylor, J. Chem. Soc., 1925, 2611.

290. G. T. Morgan and O. C. Elvins, J. Chem. Soc., 1925, 2625.

291. G. T. Morgan and A. E. Rawson, J. Soc. Chem. Ind., 44, 462T (1925).

292. G. T. Morgan, E. A. Cooper and A. E. Rawson, J. Soc. Chem. Ind., 45, 106T (1926).

293. G. T. Morgan and R. E. Kellett, J. Chem. Soc., 1926, 1080.

294. G. T. Morgan and H. Burgess, J. Chem. Soc., 1928, 321.

295. G. T. Morgan and H. Burgess, J. Chem. Soc., 1929, 1103.

296. G. T. Morgan and H. Burgess, J. Chem. Soc., 1929, 2214.

297. G. T. Morgan and F. H. Burstall, J. Chem. Soc., 1930, 2599.

298. G. T. Morgan and F. H. Burstall, J. Chem. Soc., 1931, 180.

299. G. T. Morgan, J. Chem. Soc., 1935, 554

300. M. de Moura Campos and N. Petragnani, Tetrahedron Lett., 1959, 11.

301. M. de Moura Campos, N. Petragnani and C. Thome, Tetrahedron Lett., 1960, 5.

302. M. de Moura Campos and N. Petragnani, Chem. Ber., 93, 317 (1960).

303. M. de Moura Campos, Selecta Chim., 19, 55 (1960).

304. M. de Moura Campos and N. Petragnani, Tetrahedron, 18, 521 (1962).

305. M. de Moura Campos and N. Petragnani, Tetrahedron, 18, 527 (1962).

306. M. de Moura Campos, E. L. Suranyi, J. de Andrade, Jr., and N. Petragnani, Tetrahedron, 20, 2797 (1964).

306a. K. Muellen, Org. Magn. Resonance, 3, 331 (1971).

307. A. N. Murin, V. D. Nefedov and O. V. Larionov, Radiokhimiya, 3, 90 (1961).

308. E. L. Muetterties and R. A. Schunn, Quart. Rev. (London), 20, 245 (1966).

N

308a. E. S. Nachtman, Chapter 9 in "Tellurium", W. C. Cooper, Ed., Van Nostrand Reinhold Co., New York, 1971, p. 373.

309. G. Nagarajan and T. A. Hariharan, Acta Phys. Austriaca, 19, 349 (1965).

310. G. Natta, Giorn. Chim. Ind. Applicata, 8, 367 (1926); Chem. Abstr., 20, 3273.

311. V. D. Nefedov, S. A. Grachev and Z. A. Grant, Zh. Obshch. Khim., 32, 1179 (1962).

312. V. D. Nefedov, I. S. Kirin and V. M. Zaitsev, Radiokhimiya, 4, 351 (1962).

313. V. D. Nefedov, M. A. Toropova, S. A. Grachev and Z. A. Grant, Zh. Obshch. Khim., 33, 15 (1963).

314. V. D. Nefedov, V. E. Zhuravlev and M. A. Toropova, Zh. Obshch. Khim., 34, 3719 (1964).

315. V. D. Nefedov, I. S. Kirin and V. M. Zaitsev, Radiokhimiya, 6, 78 (1964).

316. V. D. Nefedov, M. A. Toropova, V. E. Zhuravlev and A. V. Levchenko, Radiokhimiya, 7, 203 (1965).

317. V. D. Nefedov, V. E. Zhuravlev, M. A. Toropova, L. N. Gracheva and A. V. Levchenko, Radiokhimiya, 7, 245 (1965).

318. V. D. Nefedov, V. E. Zhuravlev, M. A. Toropova, S. A. Grachev and A. V. Levchenko, Zh. Obshch. Khim., 35, 1436 (1965).

319. V. D. Nefedov. Zh. Obshch. Khim., 38, 2191 (1968).

319a. V. D. Nefedov, E. N. Sinotova, A. N. Sarbash and S. A. Timofeev, Radiokhimiya, 11, 254 (1969).

319b. V. D. Nefedov, E. N. Sinotova, S. A. Timofeev and S. A. Kazakov, Zh. Obshch. Khim., 40, 1103 (1970).

319c. V. D. Nefedov, E. N. Sinotova and A. N. Sarbash, Radiokhimiya, 12, 192 (1970).

319d. V. D. Nefedov, E. N. Sinotova, A. N. Sarbash, E. A. Kolobov and V. V. Kapustin, Radiokhimiya, 13, 435 (1971).

320. A. N. Nesmeyanov, Usp. Khim., 14, 261 (1945).

320a. L. A. Niselson, G. P. Ustyugov and V. V. Taraskin, Tsvet Metal., 44, 40 (1971).

321. P. Nylen, Z. Anorg. Allg. Chem., 246, 227 (1941).

O

322. F. V. Oefele, New York Med. Monatschr., 22, 223.

323. M. Ogawa, Bull. Chem. Soc. Jap., 41, 3031 (1968).

323a. M. Ogawa, C. Inoue and R. Ishioka, Kogyo Kagaku Zasshi, 73, 1987 (1970).

323b. M. Ogawa and R. Ishioka, Bull. Chem. Soc. Jap., 43, 496 (1970).

323c. K. Ohkubo, Tetrahedron Lett., 1971, 2571.

323d. Y. Okamoto and T. Yano, J. Organometal. Chem., 29, 99 (1971).

324. M. Ordogh and A. Schneer, Magyar Tudomanyos Akad. Koezponti Fis. Kutato Intezetenek Koezlemenyei, 8, 39 (1960).

325. A. Oppenheim, J. Prakt. Chem., 81, 308 (1860).

326. J. D. Orndoff, Phys. Rev., 65, 348 (1944).

P

327. G. L. Parks, Dissertation, Clemson University, 1968; Diss. Abstr., B 29, 3450 (1969).

328. P. Pascal, Bull. Soc. Chim. Fr., [IV], 11, 1030 (1912).

329. P. Pascal, C. R. Acad. Sci., Paris, 156, 1904 (1913).

330. R. Passerini and G. Purello, Ann. Chim. (Rome), 48, 738 (1958).

331. R. C. Paul, R. Kaushal and S. S. Pahil, J. Indian Chem. Soc., 44, 995 (1967).

332. R. C. Paul, R. Kaushal and S. S. Pahil, J. Indian Chem. Soc., 46, 26 (1969).

333. M. E. Peach, Can. J. Chem., 47, 1675 (1969).

334. T. G. Pearson and R. H. Purcell, J. Chem. Soc., 1936, 253.

335. T. G. Pearson, R. H. Purcell and G. S. Saigh, J. Chem. Soc., 1938, 409.

336. G. Pellini, Chem. Ber., 34, 3807 (1901).

337. G. Pellini, Gazz. Chim. Ital., 32, 131, (1902); Chem. Zentr., 1902 I, 970.

338. N. Petragnani and G. Vicentini, Univ. Sao Paulo, Fac. Filosof. Cienc. Letras, Bol. Quim., No. 5, 75 (1959); Chem. Abstr., 58, 11256a.

339. N. Petragnani, Tetrahedron, 11, 15 (1960).

340. N. Petragnani, Tetrahedron, 12, 219 (1961).

341. N. Petragnani and M. de Moura Campos, Chem. Ber., 94, 1759 (1961).

342. N. Petragnani, Chem. Ber., 96, 247 (1963).

343. N. Petragnani and M. de Moura Campos, Chem. Ber., 96, 249 (1963).

344. N. Petragnani and M. de Moura Campos, Anais da Associacao Brasileira de Quimica, XXII, 3 (1963).

345. N. Petragnani and M. de Moura Campos, Chem. Ind. (London), 1964, 1461.

346. N. Petragnani and M. de Moura Campos, Tetrahedron, 21, 13 (1965).

347. N. Petragnani and M. de Moura Campos, Organometal. Chem. Rev., 2, 61 (1967).

347a. N. Petragnani and V. G. Toscano, Chem. Ber., 103, 1652 (1970).

347b. N. Petragnani and G. Schill, Chem. Ber., 103, 2271 (1970).

348. A. A. Petrov, S. I. Radchenko, K. S. Mingaleva, I. G. Savich and V. B. Lebedev, Zh. Obshch. Khim., 34, 1899 (1964).

348a. G. Pfisterer and H. Dreeskamp, Ber. Bunsenges. Phys. Chem., 73, 654 (1969).

348b. J. L. Piette and M. Renson, Bull. Soc. Chim. Belges. 79, 353 (1970).

348c. J. L. Piette and M. Renson, Bull. Soc. Chim. Belges, 79, 367 (1970).

348d. J. L. Piette and M. Renson, Bull. Soc. Chim. Belges, 79, 383 (1970).

348e. J. L. Piette and M. Renson, Bull. Soc. Chim. Belges, 80, 521 (1971).

349. L. R. M. Pitombo, Anal. Chim. Acta., 46, 158 (1969).

350. H. A. Potratz and J. M. Rosen, Anal. Chem., 21, 1276 (1949).

351. G. Pourcelot, M. LeQuan, M. P. Simonnin and P. Cadiot, Bull. Soc. Chim. Fr., 1962, 1278.

352. G. Pourcelot, C. R. Acad. Sci., Paris, 260, 2847 (1965).

R

352a. F. Raeuchle, W. Pohl, B. Blaich and J. Goubeau, Ber. Bunsenges. Phys. Chem., 75, 66 (1971).

353. S. I. Radchenko, V. N. Chistokletov and A. A. Petrov, Zh. Obshch. Khim., 35, 1735 (1965).

354. S. I. Radchenko, V. N. Chistokletov and A. A. Petrov, Zh. Org. Khim., 1, 51 (1965).

355. S. I. Radchenko and A. A. Petrov, Zh. Org. Khim., 1, 2115 (1965).

356. A. I. Razumov, B. G. Liorber, M. B. Gazizov and Z. M. Khammatova, Zh. Obshch. Khim., 34, 1851 (1964).

357. M. L. N. Reddy, M. R. Wiles and A. G. Massey, Nature, 217, 740 (1968).

357a. M. L. N. Reddy and D. S. Urch, Discuss. Faraday Soc., 47, 53 (1969).

358. L. Reichel and E. Kirschbaum, Ann. Chem., 523, 211 (1936).

359. L. Reichel and E. Kirschbaum, Chem. Ber., 76, 1105 (1943).

360. L. Reichel and K. Ilberg, Chem. Ber., 76, 1108 (1943).

361. H. Rheinboldt and E. Giesbrecht, J. Amer. Chem. Soc., 69, 2310 (1947).

362. H. Rheinboldt, "Houben-Weyl Methoden der Organischen Chemie", Vol. IX, E. Mueller, Ed., Georg Thieme Verlag, Stuttgart, 1955, p. 917.

363. H. Rheinboldt and G. Vicentini, Chem. Ber., 89, 624 (1956).

364. H. Rheinboldt and N. Petragnani, Chem. Ber., 89, 1270 (1956).

365. F. O. Rice and A. L. Glasebrook, J. Amer. Chem. Soc., 56, 2381 (1934).

366. F. O. Rice and A. L. Glasebrook, J. Amer. Chem. Soc., 56, 2472 (1934).

367. F. O. Rice and M. D. Dooley, J. Amer. Chem. Soc., 56, 2747 (1934).

368. H. Richter, Physik. Z., 44, 406 (1943).

368a. O. Riesser, Hoppe Seylor's Z. Physiol. Chem., 86, 415 (1913).

369. P. L. Robinson and K. R. Stainthorpe, Nature, 153, 24 (1944).

370. F. Rogoz, Dissertationes Pharm., 16, 157 (1964); Chem. Abstr., 62, 11722a.

371. E. Rohrbaech, Ann. Chem., 315, 9 (1901).

372. H. Roth, "Houben-Weyl Methoden der Organichen Chemie", Vol. II, E. Mueller, Ed., Georg Thieme Verlag, Stuttgart, 1953, p. 19.

373. H. Roth, "Houben-Weyl Methoden der Organischen Chemie", Vol. II, E. Mueller, Ed., Georg Thieme Verlag, Stuttgart, 1953, p. 176.

374. E. Rust, Chem. Ber., 30, 2828 (1897).

S

374a. I. D. Sadekov and V. I. Minkin, Khim. Geterotsikl. Soldin, 7, 138 (1971).

374b. I. D. Sadekov and V. I. Minkin, Doklad. Akad. Nauk SSSR, 197, 1094 (1971).

375. R. B. Sandin, F. T. McClure and F. Irwin, J. Amer. Chem. Soc., 61, 2944 (1939).

376. R. B. Sandin, R. G. Christiansen, R. K. Brown and S. Kirkwood, J. Amer. Chem. Soc., 69, 1550 (1947).

377. V. Sasisekharan, Proc. Indian Acad. Soc., Sect. A, 43, 224 (1956).

378. R. Savelli, Atti Accad. Naz. Lincei, Rend., Cl. Fis. Mat. Nat., 24, 151 (1936); Chem. Abstr., 31, 7466.

378a. E. D. Schermer and W. H. Baddley, J. Organometal. Chem., 27, 83 (1971).

378b. E. D. Schermer and W. H. Baddley, J. Organometal. Chem., 30, 67 (1971).

379. M. Schmidt and H. Schumann, Z. Naturforsch., 19b, 74 (1964).

379a. M. Schmidt and H. D. Block. Chem. Ber., 103, 3705 (1970).

380. O. Schmitz-DuMont and B. Ross, Angew. Chem., Int. Ed. Engl., 6, 1071 (1967).

381. W. Schneider and F. Wrede, Chem. Ber., 50, 793 (1917).

382. A. Schoenberg, E. Rupp and W. Gumlich, Chem. Ber., 66, 1932 (1933).

383. R. Schuhmann, J. Amer. Chem. Soc., 47, 356 (1925).

384. H. Schumann, K. F. Thom and M. Schmidt, Angew. Chem., 75, 138 (1963).

385. H. Schumann, K. F. Thom and M. Schmidt, J. Organometal. Chem., 2, 361 (1964).

386. H. Schumann and M. Schmidt, J. Organometal. Chem., 3, 485 (1965).

387. H. Schumann and K. F. Thom and H. Schmidt, J. Organometal. Chem., 4, 22 (1965).

388. H. Schumann, K. F. Thom and M. Schmidt, J. Organometal. Chem., 4, 28 (1965).

389. H. Schumann and M. Schmidt, Angew. Chem., 77, 1049, (1965).

390. H. Schumann and I. Schumann-Ruidisch, J. Organometal. Chem., 18, 355 (1969).

390a. H. Schumann and R. Weis, Angew. Chem., Int. Ed. Engl., 9, 246 (1970).

391. A. Scott, Proc. Chem. Soc., 20, 156 (1904).

392. J. Selbin, N. Ahmad, and J. M. Pribble, Chem. Commun., 1969, 759.

392a. J. Selbin, N. Ahmad and J. M. Pribble, J. Inorg. Nucl. Chem., 32, 3249 (1970).

392b. S. Sergi, F. Faraone and L. Silvestro, Inorg. Nucl. Chem. Letters, 7, 869 (1971).

392c. S. Sergi, F. Faraone, L. Silvestro and R. Pietropaolo, J. Organometal. Chem., 33, 403 (1971).

393. M. Shimose, Chem. News, 49, 157 (1884).

394. M. Shinagawa, H. Matsuo and H. Sunahara, Japan Analyst, 3, 204 (1954).

395. M. Shinagawa, Kagaku no Ryoiki, 10, 111 (1956); Chem. Abstr., 51, 17567e.

396. Yu. N. Shlyk, G. M. Bogolyubov and A. A. Petrov, Zh. Obshch. Khim., 38, 193 (1968).

397. Yu. N. Shlyk, G. M. Bogolyubov and A. A. Petrov, Zh. Obshch. Khim ., 38, 1199 (1968).

397a. K. P. Shresta and J. S. Thayer, J. Organometal. Chem., 27, 79 (1971).

398. G. Silvey, W. A. Hardy and C. H. Townes, Phys. Rev., 87, 236 (1952).

399. M. P. Simonnin, C. R. Acad. Sci., Paris, 257, 1075 (1963).

400. B. Singh and R. Krishen, J. Indian Chem. Soc., 12, 711 (1935).

401. C. W. Sink and A. B. Harvey, J. Chem. Soc.,D, 1969, 1023.

401a. C. W. Sink and A. B. Harvey, J. Mol. Structure, 4, 203 (1969).

402. M. K. Slattery, Phys. Rev., (2), 21, 378 (1923).

403. M. K. Slattery, Phys. Rev., (2), 25, 333 (1925).

404. C. P. Smyth, J. Amer. Chem. Soc., 60, 183 (1938).

405. R. Sochacka and A. Szushnik, Polska Akad. Nauk. Inst.Badan Jadrowych, No. 149/XIII, 1 (1960); Chem. Abstr., 54, 24481d.

406. G. R. Somalajulu, J. Chem. Phys., 33, 1514 (1960).

407. O. Steiner, Chem. Ber., 34, 570 (1901).

408. R. Steudel, Angew. Chem. Int. Ed. Engl., 6, 635 (1967).

409. E. T. Strom, B. S. Snowden, Jr., H. C. Custard, D. E. Woessner and J. R. Norton, J. Org. Chem., 33, 2555 (1968).

410. A. Stock and H. Blumenthal, Chem. Ber., 44, 1832 (1911).

411. A. Stock and P. Praetorius, Chem. Ber., 47, 131 (1914).

412. E. A. L. Suranyi and H. de Andrade, Jr., Bol Dept. Eng. Quim. EPUSP, No. 19, 31 (1964); Chem. Abstr., 64, 3001d.

413. J. F. Suttle and C. R. F. Smith, Inorg. Syn., 3, 140 (1950).

T

414. S. P. Tandon and K. Tandon, Indian J. Phys., 38, 460 (1964).

415. H. Taniyama, F. Miyoshi, E. Sakakibara and H. Uchida, Yakugaku Zasshi, 77, 191 (1957); Chem. Abstr., 51, 10407i.

416. H. Taniyama, F. Miyoshi and E. Sakakibara, Kyushu Yakugakkai Kaiho, No. 22, 57 (1968); Chem. Abstr. 71, 28990N.

417. E. T. Tao, Chem. Ind. (China), 10, 15 (1935); Chem. Abstr., 30, 2135.

418. J. S. Thayer, Inorg. Chem., 7, 2599 (1968).

419. E. W. Tillay, E. D. Schermer, and W. H. Baddley, Inorg. Chem., 7, 1925 (1968).

420. L. Tschugaeff and W. Chlopin, Chem. Ber., 47, 1269 (1914).

U

420a. G. P. Ustyugov, V. V. Taraskin, E. N. Vigdorovich, A. A. Titov, A. A. Kudryavtsev and L. A. Niselson, Nizkotemp. Termoelek. Mater, 1970, 5; Chem. Abstr., 76, 26736.

V

421. Y. P. Varshni, J. Chem. Phys., 28, 1081 (1958).

421a. G. Vasiliu, A. Gioaba and O. Maior, An. Univ. Bucuresti, Ser. Stiint Natur., Chim., 16, 57 (1967); Chem. Abstr., 71, 3339.

422. G. Vasiliu and A. Gioaba, Rev. Chim. (Bucharest), 19, 253 (1968).

422a. G. Vasiliu and A. Gioaba, Rev. Chim. (Bucharest), 20, 357 (1969).

423. R. H. Vernon, J. Chem. Soc., 1920, 86.

424. R. H. Vernon, J. Chem. Soc., 1920, 889.

425. R. H. Vernon, J. Chem. Soc., 1921, 687.

426. G. Vicentini, Chem. Ber., 91, 801 (1958).

427. G. Vicentini, E. Giesbrecht and L. R. M. Pitombo, Chem. Ber., 92, 40 (1959).

428. D. R. Vissers and M. T. Steindler, AEC Accession No. 43164, Rept. No. ANL-7142.

428a. M. Vobecky, V. D. Nefedov and E. N. Sinotova, Zh. Obshch. Khim., 33, 4023 (1963).

428b. M. Vobecky, V. D. Nefedov and E. N. Sinotova, AEC Accession No. 28445, Rept. No. UJV-1168/64.

429. M. Vobetsky, V. D. Nefedov and E. N. Sinomova, Zh. Obshch. Khim., 35, 1684 (1965).

429a. G. F. Voronin, Zh. Fiz. Khim., 45, 2100 (1971).

430. N. S. Vyazankin, M. N. Bochkarev and L. P. Sanina, Zh. Obshch. Khim., 36, 166 (1966).

431. N. S. Vyazankin, M. N. Bochkarev and L. P. Sanina, Zh. Obshch. Khim., 36, 1154 (1966).

432. N. S. Vyazankin, M. N. Bochkarev and L. P. Sanina, Zh. Obshch. Khim., 37, 1037 (1967).

433. N. S. Vyazankin, M. N. Bochkarev and L. P. Sanina, Zh. Obshch. Khim., 37, 1545 (1967).

434. N. S. Vyazankin, M. N. Bochkarev and L. P. Sanina, Zh. Obshch. Khim., 38, 414 (1968).

435. N. S. Vyazankin, L. P. Sanina, G. S. Kalina and M. N. Bochkarev, Zh. Obshch. Khim., 38, 1800 (1968).

W

436. E. L. Wagner, J. Chem. Phys., 43, 2728 (1965).

437. W. A. Waters, J. Chem. Soc., 1938, 1077.

438. T. Wentik, Jr., J. Chem. Phys., 29, 188 (1958).

439. T. Wentik, Jr., J. Chem. Phys., 30, 105 (1959).

439a. O. W. Wheeler and J. Trabal, Int. J. Appl. Radiat. Isotop., 21, 241 (1970).

439b. O. W. Wheeler, C. L. Gonzales and H. Lopez-Alonso, Int. J. Appl. Radiat. Isotop., 21, 244 (1970).

440. M. Wieber and H. U. Werther, Monatsh. Chem., 99, 1153 (1968).

441. F. D. Williams and F. X. Dunbar, J. Chem. Soc., D, 1968, 459.

442. G. Wittig and H. Fritz, Ann. Chem., 577, 39 (1952).

443. F. Woehler, Ann. Chem., 35, 111 (1840).

444. F. Woehler, Ann. Chem., 84, 69 (1852).

445. F. Woehler and J. Dean, Ann. Chem., 93, 233 (1855).

446. F. Woehler and J. Dean, Ann. Chem., 97, 1 (1856).

447. R. W. G. Wyckoff, "Crystal Structures," Interscience Publishers, New York, 1963, Second Edition, Vol. I, p. 36, 59.

447a. K. J. Wynne and P. S. Pearson, Inorg. Chem., 9, 106 (1970).

447b. K. J. Wynne and P. S. Philip, Inorg. Chem., 10, 1871 (1971).

447c. K. J. Wynne and P. S. Pearson, Inorg. Chem., 10, 2735 (1971).

447d. K. J. Wynne and P. S. Pearson, J. Chem. Soc., D, 1970, 556.

Y

448. C. H. Yao and C. E. Sun, J. Chin. Chem. Soc., 5, 22 (1937).

Z

449. F. Zeiser, Chem. Ber., 28, 1670 (1895).

450. R. A. Zingaro, J. Organometal. Chem., 1, 200 (1963).

451. R. A. Zingaro, B. H. Steeves and K. Irgolic, J. Organometal. Chem. 4, 320 (1965).

452. E. Zintl, J. Goubeau and W. Dullenkopf, Z. Phys. Chem. (Leipzig), A 154, 1 (1931).

453. W. G. Zoellner, Diss. Abstr., 9, 3139 (1959).

SUBJECT INDEX

The subject index has listed only those compounds, reactions and physical properties of compounds which are mentioned in the text. Compounds, which appear only in tables are not indexed.

In order to find information on a specific compound, the list of Tables should be consulted. The appropriate table will provide the mode of synthesis, the melting point and/or the boiling point, the yield and pertinent references for the compound in question, if it has been reported in the literature. Additional information, which might be contained in the text, can be retrieved with the help of the subject index.

Throughout this book general statements have been made in the text about the various classes of organic tellurium compounds. This information has not been incorporated into the tables. To find, for example, a general description of the behavior of diorganyl tellurium dihalides, the Table of Contents will give the pertinent page numbers. In addition entries such as "Diorganyl tellurium dihalides", "Diorganyl tellurium dichlorides", Dialkyl tellurium dihalides" and Diaryl tellurium dihalides" should be looked up in the subject index. Underlined page numbers refer to Tables, Figures or Reaction Schemes.

In arranging the entries alphabetically, the prefixes "Bis-, Cyclo-, Di-, Epi-, Iso-, Tetrakis-, Tri- and Tris-" determine the placement of each subject containing one of these prefixes. "Bis(2-naphthyl) telluride" is listed under "B". The prefix "Di" is used only when the organic radical is unsubstituted and does not have a positional number. The chapter on the nomenclature of organic tellurium compounds will be helpful in forming the names used in this book. The organic groups within a compound are arranged alphabetically.

The following abbreviations have been employed in the subject index:

compd(s).	compound(s)
conv.	conversion
dec.	decompose, decomposition
detn.	determination
ir	infrared
nmr	nuclear magnetic resonance
opt.	optical(ly)
refl.	reflux, refluxing
r.w.	reaction with
soln(s)	solution(s)
TLC	thin layer chromatography

A

Acetic anhydride
see also carboxylic acid anhydrides
-r.w. 4-alkoxyphenyl 2,4-dihydroxyphenyl tellurium dichlorides 186
-- $TeCl_4$ 33, 77
-- tellurophene 279
Acetone
see also ketones
-r.w. methylene bis(tellurium trichloride) 171
-- organyl tellurium trichlorides 43, 82
Acetonyl aryl tellurium dichlorides
-conv. to ArTeOOH 89
-conv. to Ar_2Te_2 99
-dec. by reducing agents 189
-from $ArTeCl_3$ and acetone 171
Acetylacetonyl tellurium trichlorides 67
-dec. by $K_2S_2O_5$ 97
-dec. by $Na_2S \cdot 9H_2O$ 98
-enol formation 69
Acetylene
-r.w. Al_2Te_3 279
Acrylonitrile
-r.w. H_2Te 42, 118
Acyl peroxides
-r.w. $[(C_2H_5)_3M]_2Te$ (M = group IV element) 252
Alcohols
-r.w. Al_2Te_3 42, 57, 106
Alkali metal halides
-r.w. R_3TeX 227
Alkali organyl tellurides 121
Alkanecarboditelluroic acid anhydride 42
Alkanetellurinic acids
-r.w. HI 80
Alkanetellurols <u>57</u>
-synthesis 42, 57
Alkanols
-r.w. Al_2Te_3 42, 57, 106
4-Alkoxyphenyl 2,4-diacetoxyphenyl tellurium dichloride 186
4-Alkoxyphenyl 2,4-dihydroxyphenyl tellurium dichloride
-dec. by reducing agents 189
-r.w. acetic anhydride 186
Alkyl aryl tellurides 102, 121, 131, 132, 133, <u>134</u>
Alkyl aryl tellurium dichlorides
-conv. to Ar_2Te_2 in refl. pyridine 99
Alkyl 2-formylphenyl telluride 131, <u>134</u>
Alkyl halides
see also individual compds.
-r.w. Na_2Te 106, 107
-- tellurium 148
Alkyl iodides
-r.w. tellurium 148
Alkyl phenyl tellurides 101, 136
Alkyl tellurium trichlorides 36, 67, <u>70</u>
-red. to R_2Te_2 97
Alkyl tellurium trifluorides 80
Alkyl tellurium trihalides 67, <u>70</u>
Alkyl tellurium triiodides <u>70</u>, 80
Alkyl vinylacetylenyl tellurides 129, 130
Alkylating agents
-r.w. Na_2Te 41, 106, 107
-- Na_2Te_2 41, 97, 106
Alkynyl sodium tellurides 245

Allenyl bromide
-r.w. $C_6H_5TeMgBr$ 131
Allenyl phenyl tellurides 131
9-Allyl-9-carboxyfluorene
-r.w. 2-chlorocyclohexyl tellurium trichloride 88
Aluminum telluride
-r.w. acetylene 279
-- alkanols 42, 106
-- dialkyl ethers 106
-- 1,4-dibromobutane 277
-- organic dihalides 42
-- pentamethylene dihalides 296
-- sodium succinate 279
Aluminum-tellurium compds. 246
2-Aminophenoxtellurine
- attempted resolution into opt. active compds. 320
Aminophenoxtellurines 311, 316
Amines
-r.w. $TeCl_4$ 35
Ammonia
-r.w. dimethyl tellurium diiodide 190
Antimony compds.
see Organyl antimony compds.
Antimony pentachloride
- adducts with organyl tellurium trichlorides 83
Arenediazonium chlorides
-r.w. Na_2Te 119
-- tellurium 28, 119, 185
-- $TeCl_4$ 185
Arenediazonium hexachloro-tellurates(IV) 185
Arenetellurinic acids
- from acetonyl or phenacyl aryl tellurium dichlorides 89
- nitration 88
Arenetellurols 58
Aromatic compds.
-r.w. $ArTeCl_3$ 43, 82, 171
-- $TeBr_4$ 35
-- $TeCl_4$ 34, 77, 106
Aromatic tellurenyl compds. 59
Arsenic-tellurium compds. 257
Aryl benzyl tellurides 121
Aryl butyl tellurides 129
Aryl chloromagnesium telluride 136
Aryl 2,4-dihydroxyphenyl tellurium dichlorides
- conv. to Ar_2Te_2 99
- dec. by reducing agents 189
Aryl 4-dimethylaminophenyl tellurium dichloride
- synthesis 171
Aryl lithium
see also organic lithium compds.
-r.w. tellurium 129, 245
Aryl lithium tellurides 129, 245
Aryl mercury chlorides
-r.w. $TeCl_4$ 39, 78, 184, 309
-- $RTeCl_3$ 82
-- 2-$C_{10}H_7TeI$ 64, 136
Aryl methyl tellurides 130, 131, 133, 135
Aryl organyl tellurides 129, 171
Aryl phenacyl tellurium dichlorides
- conv. to Ar_2Te_2 99
- dec. by reducing agents 189
- from $ArTeCl_3$ and acetophenone 171
Aryl sodium telluride 245
-r.w. carboxylic acid chlorides 129
-- β-chloropropionic acid 129
-- dimethyl sulfate 129

Aryl tellurium chlorides <u>60</u>, <u>62</u>, 136
- adducts with thiourea 59
-- structure 61
Aryl tellurium tribromides
-r.w. ketones 171
Aryl tellurium trichlorides 34, 67, <u>70</u>
- from ArHgCl and $TeCl_4$ 39
-r.w. 4-dimethylaminobenzene 171
-- ketones 171
-- olefins 82
-- resorcinol 171
- reduction by methanethiosulfonate 59
-- $Na_2S{\cdot}9H_2O$ 97
-- sulfites 97
-- thiourea 59
Aryl tellurium triiodides 171
Aryltelluro magnesium chloride 136
Azulene
-r.w. KTeCN 50

B

Base solutions (NaOH, KOH, carbonates)
-r.w. $2\text{-}C_{10}H_7TeI$ 64, 88, 100
-- dinitrophenoxtellurines 315
-- phenoxtellurine 320
-- R_2TeX_2 190, 191, 192
-- $R_2Te(OH)X$ 193
-- $R_2Te(OH)_2$ 146, 193
-- R_3TeX 223
-- 1-tellura-3,5-cyclohexanedione 300
-- tetrachlorotellurophene 286, 288
Benzaldehyde
-r.w. $(C_6H_5)_4Te$ 211, 240
Benzenetellurinic acid nitrate 88, 102
Benzenetellurol 58, 102
-r.w. $HgCl_2$ 102
Benzenetelluronic acid 368
Benzonitrile N-oxide
-r.w. vinylacetylenyl organyl tellurides 130
Benzotellurophene 289, <u>290</u>
- adducts with trinitrofluorenone, picric acid 290
- mass spectrum 363
- nitration 292, 294
-r.w. butyl lithium 294
-- halogens 289
-- methyl iodide 289
-- sulfur 294
Benzotellurophene 1,1-dibromide
- thermal dec. 294
Benzotellurophene 1,1-dichloride
- hydrolysis 294
Benzotellurophene 1,1-dihalides 289, <u>290</u>
Benzotellurophene 1,1-dioxide 294
-r.w. hydrohalic acids 294
Benzoyl butyl telluride 129
Benzyl chloride
-r.w. R_2Te 174, 208
Benzoyl peroxide
-r.w. R_2Te 140, 177
Benzyl dimethyl phenyl ammonium chloride
-r.w. Na_2Te 107
Benzyl iodide
-r.w. tellurium 148
Benzyl 4-methoxyphenyl telluride 141

- conv. to Ar_2Te_2 99
- conv. to $ArTeX_3$ and RX by halogens 173
Benzyl 2-naphthyl telluride 141
- conv. to Ar_2Te_2 99
- conv. to $ArTeX_3$ and RX by halogens 173
Benzyne
-r.w. diaryl ditellurides 138
Betaine
-r.w. Na_2TeO_3 106
Biphenyl
-r.w. $TeCl_4$ 35, 290
2,2-(2',2"-Biphenylylene)-4-pentenoic acid
-r.w. $RTeCl_3$ 173
2-Biphenylyl 2,2'-biphenylylene telluronium chloride 239, 240, 294
-r.w. butyl lithium 236
2-Biphenylyl 2,2'-biphenylylene telluronium hydroxide 239, 240, 294
4-Biphenylyl methyl tellurides 133, 138
-r.w. bromoacetyl chloride 131, 132
2-Biphenylyl tellurium trichloride
- intramolecular condensation 82
2,2'-Biphenylylene 2-(2'-bromobiphenylyl) telluronium bromide 239, 240, 294
2,2'-Biphenylylene 2-(2'-iodobiphenylyl) telluronium iodide 239, 240, 294
2,2'-Biphenylylene dibutyl tellurium 237
2,2'-Biphenylylene dimethyl tellurium 236
2,2'-Biphenylylene methyl telluronium iodide 291
-dec. 291
-r.w. 2,2'-dilithiobiphenyl 235
2,2'-Biphenylylene mercury
-r.w. tellurium 291
2,2'-Biphenylylene organyl telluronium salts 294, 295
- anion exchange reactions 294
2,2'-Biphenylylene sulfone
-r.w. tellurium 28, 290
2,2'-Biphenylylene telluride 237, 238
see also dibenzotellurophene
- adduct with picric acid 140
- polymeric 239, 323
-r.w. chloramine-T 193, 235
2,2'-Biphenylylene tellurium dichloride
-r.w. 2,2'-dilithiobiphenyl 235
2,2'-Biphenylylene tellur-4-methylphenylsulfonylimine 193
- hydrolysis 294
-r.w. 2,2'-dilithiobiphenyl 235
2,2'-Biphenylylene telluroxide 193
Bis(acetylacetonyl) tellurium dichloride
- enol form 333
- hydrogen bonding in 333
Bis(acetylacetonyl) tellurium dichlorides
- dec. by reducing agents 114
2,5-Bis(acetoxymercuric)tellurophene 286
Bis(2-benzothiazolyl) disulfides
-r.w. R_2Te_2 79
Bis(benzoylmethyl) tellurium dichloride
- dec. by reducing agents 114
Bis(4-biphenylyl) ditelluride
- thermal dec. 117

- mass spectrum 363, 365
Bis(4-biphenylyl) telluride 117
Bis(2,2'-biphenylylene) arsonium iodide
-r.w. $(C_4H_9)_4Te$ 238
Bis(2,2'-biphenylylene) butyl arsenic 238
Bis(2,2'-biphenylylene) butyl phosphorus 238
Bis(2,2'-biphenylylene) phosphonium iodide
-r.w. $(C_4H_9)_4Te$ 238
Bis(2,2'-biphenylylene) tellurium 234, 235, 237
- mass spectrum 362
-r.w. bromine, iodine 239, 294
-- butyl lithium 237, 238
-- HCl in ethanol 239, 294
-- methyl iodide 240
-- water 239, 294
- thermal dec. 239, 323
Bis(2-bromo-4-methoxyphenyl) ditelluride
-r.w. $NaBH_4$ 245
Bis(4-bromophenyl) ditelluride
- mass spectrum 363, 365
-r.w. $NaBH_4$ 245
{Bis[(carbethoxy)(triphenylphosphonio)methyl] tellurium dibromide} dibromide 37, 172
Bis(carbo-*l*-menthoxymethyl) telluride 117
-r.w. H_2O_2 197
Bis(carbomenthoxymethyl) telluroxide
- adduct with H_2O_2 197
Bis(2-carboxyethyl) telluride 118
Bis(carboxymethyl) tellurium dichloride 34, 170
Bis(2-carboxyphenyl) ditelluride 97
-r.w. $HCCl_2(OC_4H_9)$ 100
Bis(2-carboxyphenyl) telluride 119, 141
Bis(4-chlorobut-2-enyl) tellurium dichloride 172
- dec. by base 190
- thermal dec. 191
Bis(2-chlorocyclohexyl) tellurium dichloride
- dec. by reducing agents 114
- thermal dec. 191
Bis(3-chloro-2,4-dioxopentyl) tellurium dichloride
- enol form 333
- hydrogen bonding 333
Bis(2-chloroethyl) ether
-r.w. Na_2Te 305
Bis(2-chloroethyl) sulfide
-r.w. Na_2Te 320
Bis(2-chloroethyl) tellurium dichloride
- irritant 378
- thermal dec. 191
Bis(2-chloroformylphenyl) ditelluride 100
Bis(4-chlorophenyl) ditelluride
- structure 91, 357
Bis(4-chlorophenyl) ether
-r.w. $TeCl_4$ 35
Bis(2-chloro-2-phenylvinyl) tellurium dichloride 184
Bis(2-chloropropyl) tellurium dichloride
- dec. by base 190
- dec. by heat 191
- dec. by reducing agents 189

Bis(2-cyanoethyl) telluride 42, 118
Bis(N,N-diisopropylthiocarbazoyl) ditelluride 52
Bis(4-dimethylaminophenyl) tellurium dichloride
- dec. by HNO_2 190
Bis(4,4'-dimethyl-2,2'-biphenylylene) tellurium 235
- mass spectrum 362
- nmr 241
-r.w. bromine, iodine 294
-- HCl 294
-- water 294
- thermal dec. 239
Bis(2,4-dimethylphenyl) telluride
-r.w. methyl iodide 208
Bis(2,5-dimethylphenyl) telluride
- TLC 367
Bis(2,4-dimethylphenyl) tellurium dichloride
- dec. by KOH, K_2CO_3 solns. 191
Bis(3,3-diphenyl-2-oxo-1-oxacyclopentylmethyl) tellurium dichloride 184
Bis(2,2-diphenylvinyl) telluride
- ir spectrum 333
- mass spectrum 362
Bis(2,2-di-*p*-tolylvinyl) telluride
- ir spectrum 333
Bis(2-naphthyl) ditelluride 64, 88, 99, 100
Bis(1-naphthyl) tellurium dibromide
- reduction by C_2H_5MgI 115
Bis(4-ethoxyphenyl) ditelluride 99, 101
- dec. by methyl iodide 102, 186, 223
-r.w. Ar_2TeBr_2 119
-- benzyne 138
-- diazomethane 103, 138
Bis(4-ethoxyphenyl) telluride 119
Bis(4-ethoxyphenyl) tellurium dibromide
-r.w. 2<u>N</u> NaOH 192
Bis(4-ethoxyphenyl) tellurium dichloride
- dec. by conc. HNO_3 144, 190
- dec. by reducing agents 114
- from $ArTeCl_3$ and ArH 171
-r.w. 2<u>N</u> NaOH 192
Bis(4-ethoxyphenyl) telluroxide 192
Bis(3-fluorophenyl) ditelluride
- mass spectrum 363, <u>365</u>
Bis(4-fluorophenyl) ditelluride
- mass spectrum 363, <u>365</u>
Bis(haloalkyl) tellurium dichlorides
- dec. by base 190
- from $TeCl_4$ and olefins 172
- thermal dec. 191
2,5-Bis(2'-hydroxyisopropyl)-tellurophene 289
Bis(4-hydroxy-2-methylphenyl) tellurium dichloride
- conv. to Ar_3TeCl 146, 223
- from cresol and $TeOCl_2$ 223
2,5-Bis(hydroxymethyl)tellurophene 289
Bis(2-iodoethyl) ether
-r.w. tellurium 308
Bis(iodomethyl) telluride 323
Bis(iodomethyl) tellurium diiodide
- from CH_2I_2 and Te 148
Bis(4-methoxyphenyl) ditelluride 99
- dec. by HNO_3 102
-r.w. Ar_2TeBr_2 119
-- bis(thiocarbamoyl) disulfides 79

-- diazomethane 103, 138
-- $Fe(CO)_2(NO)_2$ 269
-- mercuric halides 274
-- methyl iodide 102, 186, 223
- thermal stability 101
Bis(4-methoxyphenyl) ether
-r.w. $TeCl_4$ 35
Bis(4-methoxyphenyl) telluride 119
- TLC 367
Bis(4-methoxyphenyl) tellurium dibromide
-r.w. C_6H_5MgBr 137
Bis(4-methoxyphenyl) tellurium dichloride
- TLC 367
Bis(4-methoxyphenyl) tellurium difluoride 143, 174
Bis(4-methoxyphenyl) tellurium diiodide 186
Bis(methylene diphenyl telluronium) dibromide 175, 208
- thermal dec. 232
Bis(2-methylphenyl) ditelluride
-r.w. $NaBH_4$ 245
Bis(4-methylphenyl) ditelluride
-r.w. $NaBH_4$ 245
Bis(2-methylphenyl) methyl telluronium iodide
- loss of methyl iodide 231
Bis(2-methylphenyl) telluride
-r.w. $Hg(OH)NO_3$ 274
- TLC 367
Bis(3-methylphenyl) telluride
- TLC 367
Bis(4-methylphenyl) telluride 114, 119
- contg. ^{125m}Te 119
-r.w. $FeCl_3$, $CuCl_2$, $HgCl_2$ 174
- TLC 367
Bis(4-methylphenyl) tellurium dibenzoate 175
Bis(4-methylphenyl) tellurium dibromide
- dipole moment 144, 356
- TLC 367
Bis(4-methylphenyl) tellurium chloride hydroxide
- dipole moment 357
Bis(2-methylphenyl) tellurium dichloride
- TLC 367
Bis(3-methylphenyl) tellurium dichloride
- TLC 367
Bis(4-methylphenyl) tellurium dichloride 175
- contg. ^{125m}Te 148
- dipole moment 144, 356
- from Ar_2Te and $FeCl_3$, $CuCl_2$, $HgCl_2$ 174
-r.w. Ar_2Po 114
- TLC 367
Bis(4-methylphenyl) tellurium difluoride 174
Bis(4-methylphenyl) tellurium diiodide
- TLC 367
Bis(4-methylphenyl) telluroxide
- binary systems with Ar_2SO, Ar_2Se, Ar_2SO_2, Ar_2SeO_2 192
- dipole moment 192
Bis(1-naphthyl) telluride 115
- TLC 367
Bis(1-naphthyl) tellurium difluoride 174
Bis(1-naphthyl) tellurium diiodide
- TLC 367
Bis(perfluorophenyl) 239

Bis(perfluorophenyl) ditelluride 98
Bis(perfluorophenyl) telluride
27, 116, 117, 148, 239
-r.w. selenium 141
Bis(perfluorophenyl) thallium bromide
-r.w. tellurium 27, 117
Bis(2-phenoxyphenyl) ditelluride
307
Bis(4-phenoxyphenyl) ditelluride
-r.w. HNO_3 102
Bis(4-phenoxyphenyl) tellurium
dichloride
- dec. by KOH, K_2CO_3 solns. 191
- from $ArTeCl_3$ and ArH 171
Bis(phenyltelluro) mercury
102, 273
1,2-Bis(phenyltelluro)benzene 103
-r.w. iodine 59, 101
Bis(phenyltelluro) hexacarbonyldiiron
- mass spectrum 362
Bis(2-thienyl) tellurium dichloride
- dec. by KOH, K_2CO_3 solns. 191
Bis(thiocarbamoyl) disulfides
-r.w. R_2Te_2 79, 100
Bis(triethylgermyl) telluride
252
- deuterium acceptor 337
- ir spectrum 333
-r.w. dicyclohexyl percarbonate
252
-- halogens 252
-- triethylstannane 252
Bis(triethylsilyl) telluride 251
- deuterium acceptor 337
- ir spectrum 333
-r.w. acyl peroxides 252
-- halogens 252
-- sulfur 253
-- triethylgermane 252
Bis(triethylstannyl) telluride 252
- deuterium acceptor 337
- ir spectrum 333
-r.w. acyl peroxides 252
-- halogens 252
-- selenium 253
-- sulfur 253
Bis(trifluoromethyl) ditelluride 97
Bis(2,4,6-trimethylphenyl) methyl
telluronium iodide 208
Bis(2,4,6-trimethylphenyl) telluride
-r.w. mercuric halides 274
-- methyl iodide 208
Bis(trimethylsilyl) telluride
- ir spectrum 337
- complex with Cr and W carbonyls
257
-- ir spectrum 337
-r.w. silver iodide 252
Bis(triphenylgermyl) telluride 247, 248
- ir spectrum 337
Bis(triphenylplumbyl) telluride
247, 248
- ir spectrum 337
Bis(triphenylstannyl) telluride
247, 248
- ir spectrum 337
Boron-tellurium compds. 83, 212, 246
Boron tribromide
-r.w. R_3TeX 212
Boron trichloride
-r.w. $RTeCl_3$ 83
Boron trihalides
-r.w. R_2Te 246
Bromine
-r.w. CSTe 54
-- $(CH_3)_2Te_2$ 101
-- tellurium 17

Bromoacetyl chloride
-r.w. aryl methyl tellurides 131, 132
4-Bromobutyl cyclotetramethylene telluronium bromide 278
4-Bromobutyl cyclotetramethylene telluronium iodide 279
α-Bromocarboxylic acids
-r.w. diorganyl tellurides 208
2-(β-Bromoethyl)benzyl bromide
-r.w. Na_2Te 305
2-Bromo-4-methoxyphenyl sodium telluride 245
4-Bromophenyl sodium telluride 245
α-Bromopropionic acid
-r.w. diphenyl telluride 175
Bromotrifluoromethane
-r.w. tellurium 19
Butadiene
-r.w. $TeCl_4$ 172
Butanetellurinic acid 84
- adducts with diorganyl telluroxides 196
-r.w. HI 80
1-Butene
-r.w. $TeCl_4$ 36, 172
Butyl bromide
-r.w. Na_2Te_2 97
-- diphenyl telluride 175, 208
Butyl carbethoxymethyl telluride 231
Butyl carbethoxymethyl telluroxide
- adduct with butanetellurinic acid 196
Butyl carbomenthoxymethyl telluride 84, 231
-r.w. benzoyl peroxide 177
Butyl carbomenthoxymethyl tellurium dibenzoate 177
Butyl carbomenthoxymethyl telluroxide
- adduct with butanetellurinic acid 196
tert-Butyl chloride
-r.w. Na_2Te 107
Butyl dichloroformate
-r.w. bis(2-carboxyphenyl) ditelluride 100
Butyl 2-formylphenyl telluride
- oxidation by Ag_2O 141
Butyl lithium
-r.w. benzotellurophene 294
-- bis(2,2'-biphenylylene) tellurium 237, 238
-- $(C_6H_5)_3TeBr$ 211, 236
-- tellurium 129, 245
-- $TeCl_4$ 116, 236
-- tellurophene 279
-- tetraphenyl tellurium 238
-- tellurium halides 230, 236, 238
Butyl lithium telluride 129, 245
Butyl tellurium triiodide 80
- from R_2TeI_2 and HI 190

C

Cadmium-tellurium compds. 273
Cadmium iodide
- adduct with dimethyl telluride 273
Carbethoxymethyl dibutyl telluronium bromide
- thermal dec. 231
Carbethoxymethyl dipentyl telluronium bromide
- thermal dec. 231
Carbethoxymethyl diphenyl telluronium bromide
-r.w. silver oxide/water 223

Carbethoxymethyl 2-formylphenyl telluronium bromide
 - thermal dec. 135
Carbethoxymethyl pentyl telluride 84, 231
 -r.w. benzoyl peroxide 177
Carbethoxymethyl pentyl tellurium dibenzoate 177
Carbethoxymethylenetriphenyl-phosphorane
 -r.w. $TeBr_4$ 36, 172
Carbo-ℓ-menthoxymethyl dibutyl telluronium bromide 118, 229
 - thermal dec. 231
Carbomenthoxymethyl pentyl telluride 84
Carbomethoxymethyl diphenyl telluronium bromide
 - cleavage of alkyl bromide 231
 -r.w. silver oxide/water 223
Carbon disulfide
 -r.w. tellurium 53
Carbon ditelluride 53
Carbon monoxide
 -r.w. tellurium 53
Carbon monoselenide
 -r.w. tellurium 53
Carbon monosulfide
 -r.w. tellurium 53
Carbon monotelluride 54
Carbon oxide telluride 52
Carbon selenide telluride 53
 - ir spectrum 54, 55
Carbon sulfide telluride 53
 - ir spectrum 54, 55
 -r.w. bromine 54
 -- mercury 54
2-Carboxybenzenetellurinic acid
 - reduction to ditelluride 97
2-Carboxybenzotellurophene 142, 289
 - decarboxylation 289
Carboxylic acids
 -r.w. P_2Te_5 42
Carboxylic acid anhydrides
 -r.w. $TeCl_4$ 33, 77, 170
Carboxylic acid chlorides
 -r.w. ArTeNa 245
Carboxymethyl 2-carboxyphenyl telluride 141, 290
 - cyclization 290
Carboxymethyl diphenyl telluronium hydroxide 223
Carboxymethyl 2-formylphenyl methyl telluronium bromide
 - cleavage of methyl bromide 231
Carboxymethyl 2-formylphenyl telluride
 - cyclization 142, 289
Carboxymethyl tellurium trichloride 34
2-Carboxyphenoxtellurine
 - attempted resolution into opt. active compds. 320
2-Carboxyphenyl methyl telluride
 - toxicity 379
4-(4'-Carboxy-2'-quinolyl)phenyl tellurium trichloride
 - dec. by $K_2S_2O_5$ 97
 - dec. by $Na_2S \cdot 9H_2O$ 98
Chloral
 -r.w. H_2Te 119
Chloramine-T
 -r.w. dibenzotellurophene 193, 235
Chlorine
see also Halogen
 -r.w. tellurium 17
Chloroacetic acid
 -r.w. Na_2Te_2 97
2-Chloroalkyl tellurium trichlorides

- thermal dec. 83
2-Chlorocyclohexyl tellurium trichloride
- dec. by $K_2S_2O_5$ 97
- dec. by $Na_2S \cdot 9H_2O$ 98
- halogen exchange 79
-r.w. 9-alkyl-9-carboxyfluorene 88
-- 2,2-diphenylpentenoic acid 82
- reactivity towards cyclohexene 36
2-Chlorocyclohexyl tellurium triiodide 79
2-Chloro-1,2-diphenylvinyl tellurium trichloride
- intramolecular condensation 82, 289
2-Chloroethyl tellurium trichloride
- structure 67, 361
Chloroform
-r.w. $(C_6H_5)_4Te$ 211, 240
Chloroform-*d* hydrogen bonded to $[(C_2H_5)_3M]_2Te$ (M = Si, Ge, Sn) 337
4-Chloro-4-methyldiphenyl ether
-r.w. $TeCl_4$ 35
4-Chlorophenacyl halide
-r.w. telluroisochroman 227, 305
2-Chloro-8-methylphenoxtellurine 316
- adduct with phenoxtellurine 316
3-Chloro-2-phenylbenzotellurophene 290
3-Chloro-2-phenylbenzotellurophene 1,1-dichloride 289
-r.w. $Na_2S \cdot 9H_2O$ 290
2-Chloro-2-phenylvinyl tellurium trichloride 184
- disproportionation into $TeCl_4$, R_2TeCl_2 184
β-Chloropropionic acid
-r.w. ArTeNa 129
2-Chloropropyl tellurium trichloride 82
Chlorosilanes
-r.w. C_2H_5TeLi 251
2-Chlorostyryl tellurium trichloride
- dec. by reducing agents 97
4-Chloro-1-tellura-3,5-cyclohexanedione 1,1-dichloride
- ir spectrum 326
Choline hydrochloride
-r.w. Na_2TeO_3 106
Chromatography of organic Te compds. 367
Chromium hexacarbonyl
-r.w. bis(trimethylsilyl) telluride 257
Conductivity of organic Te compds. 140, 233, 297, 357
Cresol
-r.w. $TeOCl_2$ 78, 171, 212, 223
Cyclohexene
-r.w. $TeCl_4$ 36, 78, 172
Cyclopentamethylene 5-halopentyl telluronium halides 296
- thermal dec. 296
Cyclopentamethylene hydrogen telluronium ion 199

D

Diacetonyl tellurium dichlorides
- dec. by reducing agents 114
Diacetylenes
-r.w. Na_2Te 41, 279
Dialkyl diaryl tellurium
- thermal stability 236, 238
Dialkyl ditellurides 91, 92
see also Diorganyl ditellurides

- reactions 100, 104
-r.w. halogens 100
-- transition metal compds. 103
- synthesis 97, 104
Dialkyl ethers
-r.w. Al_2Te_3 106
Dialkyl tellurides 105, 108
see also Diorganyl tellurides
- from Al_2Te_3 and ROH 42, 106
- from Na_2Te and alkyl halides 106, 107
- from R_2TeCl_2 and reducing agents 106, 107
- from tetraalkyl tellurium 38
- reactions 120, 139
-r.w. H_2O_2 141
-- $KMnO_4$ 193
-- organic halides 139, 208
-- oxygen 141, 193, 197
- synthesis 120
Dialkyl tellurium dichlorides 36
Dialkyl tellurium dihalides
see also Diorganyl tellurium dihalides
- from R_2TeO and HX 196
- reduction to tellurides 106, 107
- stability 145
Dialkyl tellurium dihydroxides
- thermal rearrangement 146, 212
Dialkyl tellurium diiodides 24
Dialkyl tellurium dinitrates 196
Dialkyl telluroketones 242
Dialkyl telluroxides 193
- oxidation 193
-r.w. hydrohalic acids 196
-- nitric acid 196
- stability 193
Diaminophenoxtellurines 311, 316
- diazonium salts 316
Diaryl ditellurides 91, 92
see also Diorganyl ditellurides
- from acetonyl, phenacyl and 2,4-dihydroxyphenyl aryl tellurium dichlorides and reducing agents 99
- from $RArTeCl_2$ in refluxing pyridine 99
- from tellurium and Grignard reagents 98
- from tellurium and organic lithium compds. 98
- from tellurium and organic radicals 97
- from "TeX_2" and Grignard reagents 98
- mass spectra 363, 365
- reactions 100, 104
-r.w. aryl lithium 245
-- bis(2-benzothiazolyl) disulfide 79
-- bis(thiocarbamoyl) disulfide 79, 100
-- diazomethane 103
-- Grignard reagents 101, 136
-- halogens 100
-- phenyl lithium 101
-- sodium borohydride 102, 121, 129 245
-- transition metal compds. 103, 258, 266, 269, 270, 274
- synthesis 104
- thermal dec. 101, 117
Diarylethylenes
- adducts with $TeCl_2$ 36
-r.w. $TeCl_4$ 26
Diaryl mercury
see also organic mercury compds.
-r.w. tellurium 28, 106, 115, 119, 2

Diaryl tellurides
see also Diorganyl tellurides
- from Ar_2Te_2 by thermal dec. 101
- from Ar_2TeCl_2 and CH_3MgI 114
- from Ar_2TeX_2 and ArMgBr 115
- from tellurium and diaryl mercury 115
- reactions 120, 139
-- affecting the organic moiety 129
-r.w. organic halides 139, 208
- unsymmetric 115
Diaryl tellurium dichlorides 28, 34
see also Diorganyl tellurium dichlorides
-r.w. Grignard reagents 211
- reduction by ArMgBr 115, 189
- reduction by CH_3MgI 137
- unsymmetric 171
Diaryl tellurium dihalides 144
see also Diorganyl tellurium dihalides
- antibacterial action 378
- stability 145, 146
Diazomethane
-r.w. diaryl ditellurides 103, 138
Diazotetraphenylcyclopentadiene
-r.w. diphenyl telluride 139, 232
Dibenzotellurophene 28, 290, 291, 292
see also 2,2'-biphenylylene telluride
- adducts with $HgCl_2$, picric acid, 1,3,5-trinitrobenzene 291
- from trimethyl telluronium iodide and 2,2'-dilithiobiphenyl 230
-r.w. chloramine-T 193
- halogens 291
-- methyl iodide 291
- telluronium salts of 294, 295
Dibenzotellurophene 5,5-dibromide 35
Dibenzotellurophene 5,5-dichloride 35
Dibenzotellurophene 5,5-dihalides 290, 291, 292
- reduction 291
Dibenzotellurophene p-tolylsulfonimide 193
Dibenzotellurophene 5-oxide 193
Dibenzothiophene 294
Dibenzoylmethane
-r.w. $TeCl_4$ 31
Dibenzoylmethyl tellurium trichloride 31
Dibenzyl telluride 107, 141
- platinum complexes 270
- stability 141
Dibenzyl tellurium diiodide
- from tellurium and benzyl iodide 148
vic-Dibromides
-r.w. R_2Te_2 44, 103, 139
-- R_2Te 139, 174, 176
1,4-Dibromobutane
-r.w. Al_2Te_3 277
-- Na_2Te 277
-- 1-telluracyclopentane 278
1,2-Dibromoethane
-r.w. diphenyl telluride 175
Dibutyl ditelluride 97
Dibutyl telluride 238, 294
- adduct with BX_3 246
-- trimethyl aluminum 246
-- trimethyl gallium 246
- from ^{132}Te and diaryl mercury 119

- oxidation to butyric acid 196
-r.w. $Mn(CO)_5X$ 267
-- $[Ru(CO)_2X_2]_n$ 270
- ^{132}Te disintegration 142
Dibutyl tellurium diiodide
-r.w. HI 190
Dibutyl telluroxide
- adduct with butanetellurinic acid 196
Dichlorodifluoromethane
-r.w. tellurium 19
Dichloromethane
-r.w. $(C_6H_5)_4Te$ 211, 240
2,8-Dichlorophenoxtellurine 10,10-dihalides
- reduction by $K_2S_2O_5$ 315
2,5-Dideuteriotellurophene 286
2-(Diethoxymethyl)phenyl dimethyl telluronium iodide 208
- saponification 223
Diethyl allylphosphonite
-r.w. tellurium 255
Diethyl (allyl)tellurophosphonate 255
Diethyl ditelluride
- mass spectrum 363, 364
-r.w. boron tribromide 246
-- lithium 102, 245, 251
-- triethylgermyl lithium 102
Diethyl telluride 105, 106
- from triethyl telluronium chloride and diethyl zinc 229, 234
- hydrogen bonded to phenol, indole 337
- mass spectrum 363, 364
-r.w. air 197
-- group IV hydrides 142, 250
-- $Re_2(CO)_8Cl_2$ 269
-- $[Rh(CO)_2Cl]_2$ 270, 271
- platinum complexes 270
- palladium complexes 270
Diethyl tellurium dihalides
- conductivities 357
Diethyl tellurium diiodide
- dec. in solution 145, 146
- from tellurium and ethyl iodide 148
-r.w. diphenyl mercury 273
Diethyl telluroketone 243
Diethyl tellurone 197
Diethyl Zinc
-r.w. $TeCl_4$ 199, 206, 234
- triethyl telluronium chloride 43, 229, 234
Diethylsilyl ethyl telluride 251
Di(ethyltelluro)silane
- ir spectrum 337
2,8-Difluorophenoxtellurine
-r.w. sulfur 320
2,8-Difluorophenoxtellurine 10,10-dihalides
-r.w. $K_2S_2O_5$ 315
-- sulfur 320
Digermyl telluride 352
- ir spectrum 337
Dihexadecyl telluride 107
1,4-Diiodobutane
-r.w. Li_2Te 282
- tellurium 277, 279
Diiodomethane
-r.w. tellurium 148, 323
1,3-Diketones
-r.w. $TeCl_4$ 29, 67, 148, 297
2,2-Diiodooctafluorobiphenyl
-r.w. tellurium 291
2,2'-Dilithiobiphenyl
-r.w. R_2TeCl_2 235

-- $Te(OCH_3)_4$ 43, 234
-- $Te(OCH_3)_6$ 43, 234
-- tellurimines 235
-- telluronium halides 230, 235
-- $TeCl_4$ 234, 235
-- triphenyl telluronium chloride 230
1,4-Dilithiobutadiene
-r.w. $TeCl_4$ 282
3,7-Dimethyldibenzotellurophene 291
Dimethyl ditelluride 91
- from tellurium and $CH_3\cdot$ 97
- ir, Raman spectrum 326, 327
-r.w. bromine 101
- vibrational analysis 91, 327
Dimethyl ditelluride-d_6
- ir, Raman spectrum 326, 327
Dimethyl 2-formylphenyl telluronium iodide 223
- loss of methyl iodide 231
Dimethyl hydrogen telluronium chloride 199
Dimethyl 4-methoxyphenyl telluronium iodide 223
- loss of methyl iodide 231
-r.w. $Na_2S\cdot 9H_2O$ 135, 232
Dimethyl phenyl telluronium iodide 208
- loss of methyl iodide 231
Dimethyl selenide telluride 257
Dimethyl sulfate
-r.w. ArTeNa 129
Dimethyl sulfite
-r.w. $TeCl_4$ 77
Dimethyl telluride 105, 106, 116, 118, 138
- adduct with $HgCl_2$ 379
- adduct with HCl 140
- adduct with trinitrobenzene 140
- conductivity in liquid HCl 140, 357
- flash photolysis 142, 277
- from K_2TeO_3 and molds 379
- ir, Raman spectrum 326, 327
- Palladium complexes 270
- Platinum complexes 270
- protonated 199
-r.w. cadmium iodide 273
-- hydrogen peroxide 197
-- liquid HCl 199
-- silver iodide 273
- toxicity 378
Dimethyl tellurium dibromide
- ir, Raman spectrum 326, 327
Dimethyl tellurium dichloride
- action on nerve cells 379
- conductivity in liquid HCl 357
- ir, Raman spectrum 326, 327
- structure 361
- titration with base 192
Dimethyl tellurium difluoride 143
Dimethyl tellurium dihalides
- conductivity in liquid HCl 357
- magnetic susceptibilities 356
Dimethyl tellurium dihydroxide 89
-r.w. hydrogen peroxide 197
Dimethyl tellurium diiodide
- adducts with NH_3 190
- contg. ^{127}Te, ^{129}Te 148
- crystallographic data 357
- dec. ni solution 145, 146
- dipole moment 356
- from tellurium and methyl iodide 148
- ir, Raman spectrum 326, 327
- structure 356
Dimethyl tellurium dinitrate
- magnetic susceptibility 356

Dimethyl tellurium hydroxide nitrate
 - titration with base 192
Dimethyl telluroketone 243
Dimethyl tellurone 197
Dimethyl telluroxide 192
 - basicity 192
4-Dimethylaminobenzene
 -r.w. $ArTeCl_3$ 43, 82, 171
1,3-Dimethylbenzimidazoline 28, 242
 -r.w. tellurium 28, 242
4,4'-Dimethyl-2,2'-diphenylylene mercury
 -r.w. tellurium 291
4,4'-Dimethyl-2,2'-biphenylylene tellurium dichloride
 -r.w. 4,4'-dimethyl-2,2'-dilithio-biphenyl 235
Dimethyl(ethyltelluro)silane
 - ir spectrum 337
2,8-Dimethylphenoxtellurine
 -r.w. sulfur 320
Dimethylsilyl ethyl telluride 251
2,4-Dimethyl-1-tellura-3,5-cyclohexanedione
 - toxicity 378
4,4-Dimethyl-1-tellura-3,5-cyclohexanedione
 -r.w. hydroxylamine sulfate 300
2,6-Dimethyl-1-tellura-3,5-cyclohexanedione
 - bactericide 369
 - toxicity 378
1,3-Dimethyl-2-telluroxobenzimidazoline 28, 242
2,8-Dinitrophenoxtellurine
 - adducts with 2-nitrophenoxtellurine 316
Dinitrophenoxtellurine 10,10-dinitrates 311, 315
Dinitrophenoxtellurines 310, 311
 - dec. by KOH 315
 -r.w. tin/HCl 316
Diorganyl ditellurides 91, 92
see also Dialkyl, Diaryl ditellurides
 - as ligands 247, 258
 - from Na_2Te_2 and organic halides 41
 - modification of the organic moiety 100
 - oxidation to tellurinic acids 84
 - properties 91, 92
 - reactions 104
 -r.w. air 102
 -- bis(thiocarbamoyl) disulfides 79, 100
 -- diphosphines 103, 255
 -- Grignard reagents 43, 58, 101, 245
 -- halogens 78, 100
 -- hydrogen peroxide 102
 -- iron carbonyls 269
 -- mercuric halides 103
 -- methyl iodide 44
 -- nitric acid 88
 -- phenyl lithium 101
 -- $RTeCl_3$ 43, 82, 103
 -- R_2TeBr_2 103, 190
 -- R_2SeBr_2 103
 -- tellurium tetrachloride 103, 247
 -- transition metal compds. 103, 247, 258, 266, 269, 270, 274
 -- *vic*-dibromides 44, 103
 -- uranium pentachloride 103, 274
 - synthesis 91, 104
 - thermal dec. 44, 101, 117
Diorganyl mercury
 -r.w. diphenyl ditelluride 101, 136

Diorganyl selenium dibromides
-r.w. R_2Te_2 103
Diorganyl methyl telluronium iodides 139
- from R_2TeI_2 and methyl iodide 44
Diorganyl tellurides
see also Dialkyl, Diaryl tellurides
- as ligands 257, 258
- from Na_2Te and organic halides 44, 107
- from R_2TeCl_2 and sulfites 107, 114
- from R_2TeCl_2 and Na_2S, $LiAlH_4$, Zn, $SnCl_2$, $Na_2S_2O_3$ 114
- from R_2TeCl_2 and CH_3MgI 114, 189
- from R_2TeCl_2 and R_2Po 189
- from R_2Te_2 by thermal dec. 44, 117
- from R_2Te_2 and Grignard reagents 43
- from R_2Te_2 and R_2TeBr_2 190
- from $TeCl_4$ and Grignard reagents 37, 38
- from $TeCl_4$ and organic lithium compds. 40
- modification of the organic moiety 118, 131, 132
- reactions 120, 139
-r.w. Ar_3BiF_2 139, 174
-- benzoyl chloride 174
-- benzoyl peroxide 140, 177
- α-halocarboxylic acids 174
-- halogens 139, 143, 173
-- hydrogen peroxide 140
-- mercuric halides 274, 275
-- nitric acid 140, 177
-- organic halides 44, 139, 199, 208, 224
-- oxygen 140
-- oxygen(air) and HCl or HBr 177
-- R_2TeX_2 140
-- SO_2X_2, $SOCl_2$ 139, 174
-- *vic*-dibromides 139, 174, 176
- purification 114
- symmetric 105
-- synthesis 107, 120
- unsymmetric 121, 122, 189
-- synthesis 121
Diorganyl tellurium compounds, R_2TeXY 179, 187
Diorganyl tellurium dibenzoates 140, 177
Diorganyl tellurium dibromides
- from $ArTeBr_3$ and ArH 171
- from Ar_2Te and *vic*-dibromides 174, 176
- from Ar_2Te_2 and *vic*-dibromides 44, 103, 139
- hydrolysis 177, 187
-r.w. R_2Te_2 103, 190
Diorganyl tellurium dichlorides
see also Dialkyl, Diaryl tellurium dichlorides and dihalides
- from $ArTeCl_3$ and ketones 171
- from $RTeCl_3$ and ArH 82, 171
- from $RTeCl_3$ and unsaturated compds. 82
- from $RTeCl_3$ and ArHgCl 82
- from $RTeCl_3$ and R_2Te_2 82, 103
- from R_2Te and $SOCl_2$ 139
- from R_2Te and transition metal chlorides 139
- from R_2Te_2 and $TeCl_4$ 103
- from $TeCl_4$ and Grignard reagents 37, 38
- from $TeCl_4$ and ketones 69
- hydrolysis 177, 187

-r.w. Grignard reagents 199
Diorganyl tellurium difluorides 139
- from R_2Te and tris(4-methylphenyl) bismuth difluoride 139, 174
Diorganyl tellurium dihalides 143
see also Diorganyl tellurium dichlorides, Dialkyl and Diaryl tellurium chlorides and halides
- from $ArTeCl_3$ and ArH 171
- from $RTeCl_3$ and olefins 173
- from $RTeCl_3$ and organic compds 43
- from $RTeCl_3$ and R_2Te_2 43
- from R_2Te and air/HX 177
- from R_2Te and halogens 139, 173
- from R_2Te and *vic*-dibromides 174, 176
- from Te and alkyl iodides 148
- from $TeCl_4$ and acetic anhydride 170
- from $TeCl_4$ and ArH 170
- from $TeCl_4$ and ketones 148
- from $TeCl_4$ and 1,3-diketones 148
- from $TeOCl_2$ and cresols 171
- halogen exchange reactions 177
- reactions 149, 188
-r.w. base 192
-- Grignard reagents 43, 114, 189, 211
-- R_2Po 189
-- R_2Te 140
-- silver oxide/water 192
- reduction to R_2Te by sulfites 107, 114, 137, 143
-- by $Na_2S{\cdot}9H_2O$ 114, 137, 185
-- by $LiAlH_4$, $SnCl_2$, CH_3MgI, $Na_2S_2O_3$ 114
-- by Zn 114, 137
-- by Grignard reagents 43, 114, 189, 211
- symmetric 147, 150
- structure 143, 144, 357, 358, 361
- synthesis 147, 149
- unsymmetric 147, 161, 171, 173, 184
-- resolution into opt. active compds. 227, 228
Diorganyl tellurium dihydroxides 150, 177, 184, 187, 192
- dehydration 192
-r.w. NaOH 193
Diorganyl tellurium diiodide
-r.w. iodine 188
-- water 177, 187
Diorganyl tellurium dinitrates 140, 177
Diorganyl tellurium halide hydroxides 177, 187
- anhydrides of 187
- from R_2TeO and HX 188
- halogen exchange 188
-r.w. NaOH 193
-- $R_2Te(OH)_2$ 188
-- silver oxide/water 193
Diorganyl tellurium iodide triiodide 188
Diorganyl tellurones 196
-r.w. HI 197
- structure 197
Diorganyl telluroxides 140, 175, 177, 184, 192, 194
- adducts with RTeOOH 196
- basicity 192
- enol form 196
- oxidation to tellurinic acids 196
-r.w. HX 188
- synthesis 192, 193

Dipentyl telluride 84, 105
Diphenyl acetylene
-r.w. $TeCl_4$ 36, 78
Diphenyl ditelluride 58, 138
- complex with UCl_5 274
- dissociation into radicals 101
- from Te and $C_6H_5\cdot$ 97
- from $Te(CN)_2$ and $(C_6H_5)_3Bi$ 51
- mass spectrum 363, 365
- molecular mass 101
-r.w. bis(thiocarbamoyl) disulfides 79
-- dialkyl mercury 101, 136
-- Grignard reagents 246
-- mercuric halides 274
-- mercury 102, 273
-- nitric acid 88, 102
-- sodium 58, 102
-- sodium borohydride 245
-- $[\pi\text{-cpFe(CO)}_2]_2$ 269
-- $[\pi\text{-cpMo(CO)}_3]_2$ 257, 266
-- $[Ru(CO)_4]_3$ 270
- thermal dec. 117
Diphenyl ether
-r.w. 4-phenoxyphenyl tellurium trichloride 171, 307
-- $TeCl_4$ 35, 307
Diphenyl methyl telluronium iodide 118, 175
- conductivity 357
- loss of methyl iodide 230
Diphenyl selenide 141, 192
- binary system with telluride 107
-r.w. potassium amide 142
Diphenyl sulfide
- binary system with telluride 107
-r.w. $TeCl_4$ 35
Diphenyl telluride 116, 117, 118, 137, 138, 232, 322
- adduct with HCl 140
- binary systems with diphenyl sulfide, selenide 107
- contg. ^{132}Te, from Te and Ar_2Hg 119
-- disintegration 142
- contg. ^{125m}Te, from $Ar_3{}^{125}Sb$ 119
- determination of Pd 368
- effect on syphylis in rabbits 379
- from Ar_2Te_2 by thermal dec. 117
- from Ar_3TeCl and ArMgBr 229
- from R_2TeBr_2 and $LiAlH_4$ 191
- neutron irradiation 142
- platinum complexes 270
-r.w. α-bromopropionic acid 175
-- butyl bromide 175, 208
-- diazotetraphenylcyclopentadiene 139, 232
-- 1,2-dibromoethane 175
-- ethyl iodide 208
-- ethyl iodoacetate 175
-- HCl 140
-- methyl iodide 175
-- potassium amide 142
-- sulfur 141
-- $AuCl_3$ 273
-- $Co_2(CO)_8$ 257
-- $Cr(CO)_6$ 257
-- $Fe(CO)_4X_2$ 269
-- $Fe_2(CO)_2(NO)_2$ 269
-- $Fe_3(CO)_{12}$ 269
-- $Hg(OH)NO_3$ 274
-- $Mn(CO)_5X$ 266
-- $Mn_2(CO)_{10}$ 266
-- $Ni(CO)_4$ 257
-- $[Ru(CO)_2X_2]_n$ 270
- TLC 367

Diphenyl tellurium bis(d-bromocamphorsulfonate) 229
Diphenyl tellurium d-bromocamphorsulfonate hydroxide 229
Diphenyl tellurium dibromide 175, 231
- crystallographic data 357
-r.w. lithium aluminum hydride 191
-- 2-methylphenyl magnesium bromide 137
- optically active 229
- structure 361
Diphenyl tellurium dichloride
- conductivity 357
- contg. ^{125m}Te 148
- dipole moment 143
- from $Ar_3{}^{125}SbCl_2$, $Ar_2{}^{125}SbCl_3$ 148
- from Te and diphenyliodonium chloride 185
-r.w. silver salts of dicarboxylic acids 184, 324
- TLC 367
Diphenyl tellurium difluoride 143
- dipole moment 143
Diphenyl tellurium dihalides
-r.w. halide ions 189
-- diphenyl telluride 189
Diphenyl tellurium diiodide 175, 193
- conductivity 357
- dipole moment 143
-r.w. iodine 189
- TLC 367
Diphenyl tellurium hydroxide iodide
- conv. to Ar_2TeO and Ar_2TeI_2 193
Diphenyl telluronium tetraphenylcyclopentadienylide 139
- mass spectrum 362
Diphenyl telluroxide 193
Diphenylacetylene
-r.w. $TeCl_4$ 36, 78
1,4-Diphenylbutadiene
-r.w. $TeCl_4$ 36
2,2-Diphenyl-5-chloromercuri-4-pentanolactone
-r.w. $TeCl_4$ 40, 184
Diphenyliodonium chloride
-r.w. tellurium 28, 119, 185
Diphenylnitrilimine
-r.w. acetylenyl organyl tellurides 130
N-α-Diphenylnitrone
-r.w. acetylenyl organyl tellurides 130
5-(3,3-Diphenyl-2-oxo-1-oxacyclopentyl)-methyl 4-ethoxyphenyl tellurium diiodide 99
- conv. to Ar_2Te_2 by $Na_2S \cdot 9H_2O$ 99
5-(3,3-Diphenyl-2-oxo-1-oxacyclopentyl)-methyl 2-naphthyl telluride 64, 65
-r.w. iodine 64, 65
5-(3,3-Diphenyl-2-oxo-1-oxacyclopentyl)-methyl 2-naphthyl tellurium diiodide 64, 65
3,3-Diphenyl-2-oxo-5-tetrahydrofurylmeth 2-naphthyl telluride 138
Diphosphines
-r.w. R_2Te_2 103, 255
Diphosphorus pentatelluride
-r.w. carboxylic acids 42
2,2-Diphenyl-4-pentenoic acid
-r.w. 2-chlorocyclohexyl tellurium trichloride 82
-- $2\text{-}C_{10}H_7TeI$ 64
-- $RTeCl_3$ 173
-- $TeCl_4$ 36, 172

Dipole moments of organic Te compds. 83, 143, 144, 192, 233, 256, 356, 357
Dipropyl telluride
 - complexes with Pd, Pt 270
Dipropyl telluroketone 243
Disilyl telluride 250
 - ir spectrum 337
 -r.w. germyl bromide 252
 - HI 253
 -- light 253
 -- oxygen 253
Disulfur dichloride
 -r.w. tellurium 17
1,1'-Ditellura-3,3'-bicyclopentyl 279
Ditellurium decafluoride 17
Ditellurium dicyanide 52
Ditelluromethane 323

E

5,10-Epitelluroarsanthrene 257
 - mass spectrum 362
 - structure 257
Epitellurobenzene 277, 363
5,10-Epoxyarsanthrene
 -r.w. H_2Te 256
4-Ethoxy-2-oxo-3-pentenyl tellurium trichloride 300
4-Ethoxyphenyl 4-ethoxyphenyltelluromethyl telluride 138
4-Ethoxyphenyl 2-(4-ethoxyphenyltelluro)phenyl telluride 138
4-Ethoxyphenyl 4-methoxyphenyl tellurium dichloride
 - from 4-ethoxyphenyl tellurium trichloride and ArH 171
4-Ethoxyphenyl tellurium trichloride
 -r.w. alkoxybenzenes 171
 -- dimethylaminobenzene 171
 -- ketones 171
 -- resorcinol 171
Ethyl chloride
 -r.w. 1-tellura-3,5-cyclohexanedione 1,1-dichlorides 300
Ethyl ethylsilyl telluride 251
Ethyl iodide
 -r.w. diorganyl tellurides 208
 -- tellurium 148
Ethyl iodoacetate
 -r.w. diphenyl telluride 175
Ethyl lithium telluride 102, 245
 -r.w. chlorosilanes 251
 -- triethylgermyl bromide 251
_l_-Ethyl methyl 4-methoxyphenyl telluronium _d_-bromocamphorsulfonate 227
 -r.w. potassium iodide 227
Ethyl methyl 4-methoxyphenyl telluronium iodide 227
Ethyl phenyl telluride 121, 137
Ethyl silyl telluride 251
 - ir spectrum 337
Ethyl tellurium tribromide
 - adduct with thiourea 83
 - ir spectrum 340
Ethyl tellurium trichloride
 - adduct with $SbCl_5$ 83
 - adduct with thiourea 83
 - ir spectra of adducts 340
Ethyl tellurium triiodide
 - adduct with thiourea 83
 -- ir spectrum 340
Ethyl triethylgermyl telluride 251
Ethyl triethylsilyl telluride 251
Ethyl vinylacetylenyl telluride
 -r.w. 1,3-diphenylnitrilimine 130
 -- N-α-diphenylnitrone 130

1,2-Ethylene bis(diphenyl telluronium bromide) 175
- dec. to Ar_2TeBr_2 174
4-Ethyl-1-tellura-3,5-cyclohexanedione
- therapeutic value 378
- toxicity 378
5-(Ethyltelluroethynyl)-2,3-diphenylisoxazolidine 130
5-(Ethyltelluroethynyl)-1,3-diphenyl-2-pyrazoline 130
- ir spectrum 333

F

Flash photolysis of dimethyl telluride 142, 277
Fluorine
-r.w. tellurium 17
4-Fluorodiphenyl ether
-r.w. $TeCl_4$ 35
Fluorosulfonic acid
-r.w. 1-telluracyclohexane 199
2-Formylphenyl methyl telluride
-r.w. silver nitrate 100
-- silver oxide 141

G

Gallium-tellurium compds. 246
Germanium-tellurium compds. 247
Gold-tellurium compds. 258, 273
Germyl bromide
-r.w. disilyl telluride 252
Grignard reagents
-r.w. $ArTeCl_3$ 186
-- Ar_2TeCl_2 114, 115, 137, 189, 199, 211
-- Ar_3TeCl 229, 230
-- $2\text{-}C_{10}H_7TeI$ 43, 64, 136
-- R_2TeCl_2 43, 114, 115, 137, 189, 211
-- R_2Te_2 43, 101, 136, 245, 246
-- tellurium 26, 58, 98, 106, 117, 137, 245
-- TeX_2 38, 98, 138
-- $TeCl_4$ 37, 38, 116, 186, 199, 206

H

Halogen exchange
- in $RTeX_3$ 79, 80
- in RTeX 61
- in R_3TeX 208, 209, 210, 294
Halogens
-r.w. benzotellurophene 289
-- benzyl aryl tellurides 173
-- bis(triorganyl -group IV element) tellurides 252
-- dibenzotellurophene 291
-- 1-oxa-4-telluracyclohexane 305
-- phenoxtellurine 315
-- R_2Te_2 78, 100
-- R_2Te 139, 143, 173
-- 1-telluracyclohexane 296
-- 1-tellura-3,5-cyclohexanedione 300
-- 1-telluracyclopentane 278, 279
-- telluroformaldehyde 242
-- tellurophene 286
-- tetraorganyl tellurium 239, 294
-- 1-thia-4-telluracyclohexane 320
Hexachloro-1,3-butadiene 26, 282
Hexamethoxy tellurium
-r.w. 2,2'-dilithiobiphenyl 43, 234
Hydrides of Si, Ge, Sn
-r.w. diethyl telluride 142, 250
Hydrochloric acid
see hydrogen chloride
Hydrogen chloride
-r.w. dimethyl telluride 140, 199
-- diphenyl telluride 140

-- 1-tellura-2,5-cyclohexanedione 300
-- 1-tellura-3,5-cyclohexanedione 1,1-dichlorides 300
-- tetraorganyl tellurium 239, 294
Hydrogen halides
-r.w. R_2TeO 177, 196
Hydrogen iodide
-r.w. $C_4H_9TeI_3$ 190
-- $(C_4H_9)_2TeI_2$ 190
-- $(H_3Si)_2Te$ 253
-- RTeOOH 80, 89
-- R_2TeO_2 197
-- trialkyltelluronium alkyltetrahydroxytellurate(IV) 212
Hydrogen peroxide
- adduct with diorganyl telluroxides 197
-r.w. diorganyl ditellurides 102
-- diorganyl tellurides 140, 141, 197
-- phenoxtellurine 197, 198
-- $R_2Te(OH)_2$ 197
-- 1-telluracyclohexane 197
-- telluroisochroman 197
Hydrogen telluride 58, 136
-r.w. acrylonitrile 42, 118
-- benzil 243
-- benzophenone 243
-- chloral 119
-- 5,10-epoxyarsanthrene 256
-- N-isothiocyanatodiisopropylamine 52
-- ketones 42, 242
Hydrogen triethylsilyl telluride 251
Hydrolysis
- 2,2'-biphenylylene tellur-4-methylphenylsulfonylimine 193
- R_2TeX_2 177, 187, 228
- R_2TeI_2 177, 187
- $RTeX_3$ 80, 82, 83, 84, 88
- RTe(O)X 89
- R_4Te 40, 208, 211, 238, 239, 294
Hydroxy 4-methylphenyl phenyl telluronium
- d-bromocamphorsulfonate, opt. active 228
- hydroxide 228
4-Hydroxybenzenetellurinic acid
- nitration 88
4-Hydroxy-3,5-dinitrobenzenetellurinic acid nitrate 88
4-(1'-Hydroxy-2'-hydroxymethyl-2'-dichloroacetamidoethyl)biphenylyl methyl telluride
- toxicity 379
4-(1'-Hydroxy-2'-hydroxymethyl-2'-dichloroacetamidoethyl)phenyl methyl telluride
- toxicity 379
Hydroxylamine sulfate
-r.w. 1-telluracyclohexanediones 300, 304
4-Hydroxy-3-nitrobenzenetellurinic acid 88
4-Hydroxyphenyl tellurium trichloride 79
- dec. by $K_2S_2O_5$ 97
- dec. by $Na_2S\cdot 9H_2O$ 98
- halogen exchange 79
4-Hydroxyphenyl tellurium triiodide 79

I

Indole
- hydrogen bond to diethyl telluride 337

Iodine
-r.w. diorganyl tellurides 59, 64, 65, 101
-- phenoxtellurine 315
-- R_2TeI_2 188, 189
-- tellurium 19
Iodine bromide
-r.w. tellurium 19
Iron-tellurium compds. 259, 269
Iron carbonyls
-r.w. tetraphenyltellurophene 1,1-dibromide 288
Iron compds.
- complexes with tellurides, ditellurides 259, 269
Isonitriles
-r.w. tellurium 50
Isopropenylethynyl methyl telluride
- dec. by mercuric nitrate 141
Isopropenylethynyl sodium telluride 245
N-Isothiocyanatodiisopropylamine
-r.w. H_2Te 52

K

Ketones
-r.w. $ArTeBr_3$ 171
-- $ArTeCl_3$ 43, 82, 171
-- H_2Te 42, 242, 243
-- $TeCl_4$ 32, 69, 77, 148, 170

L

Lead-tellurium compds. 244, 247
2-Lithiotellurophene 286, 287
-r.w. alkylating agents 286
Lithioalkanes
see also organic lithium compds.
-r.w. $TeCl_4$ 40, 199, 236
-r.w. R_3TeI 236
Lithium
-r.w. diethyl ditelluride 102, 245, 251
Lithium aluminum hydride
-r.w. R_2TeX_2 114, 191
Lithium pentaphenyltellurate(IV) 234
Lithium-tellurium compds. 244
Lithium telluride 250
-r.w. silyl halides 250
-- 1,4-diiodobutadiene 282
Lithium triphenylgermyl telluride 245, 248
-r.w. $(C_6H_5)_3MCl$ (M = Ge, Sn, Pb) 247
Lithium triphenylplumbyl telluride 245, 248
-r.w. $(C_6H_5)_3MCl$ (M = Ge, Sn, Pb) 247
Lithium triphenylstannyl telluride 245, 248
-r.w. $(C_6H_5)_3MCl$ (M = Ge, Sn, Pb) 247

M

Magnesium-tellurium compds. 244, 245
Magnetic susceptibility of organic Te compds. 356
Maleic anhydride
-r.w. tetraphenyltellurophene 286
-- tetraphenyltellurophene 1,1-dibromide 286
Manganese-tellurium compds. 266
Mercuric acetate
-r.w. tellurophene 286
Mercuric chloride
-r.w. benzenetellurol 102
-- dibenzotellurophene 291
Mercuric halides
- complexes with diorganyl ditellurides 274

- complexes with diorganyl tellurides 64, 103, 273, 274, 275, 379
-r.w. R_3TeX 212
Mercuric hydroxide nitrate
-r.w. diorganyl tellurides 274
Mercuric nitrate
-r.w. methyl alkynyl tellurides 141
Mercury
-r.w. CSTe 54
-- diphenyl ditelluride 102, 273
Mercury tellurium compds. 244, 273, 274, 275
Metal halides
-r.w. diorganyl tellurides 139, 174
-- R_3TeX 212
Metal ions
- detn. with organic Te compds. 368
Methanetellurinic acid 89
Methantellurol
- ir, Raman spectrum 325, 327
4-Methoxybenzenetellurinic acid iodide 84
4-Methoxyphenyl dimethyl telluronium iodide 118
4-Methoxyphenyl 4-methoxyphenyl-telluromethyl telluride 138
4-Methoxyphenyl methyl telluride
- adduct with picric acid 140
4-Methoxyphenyl tellurium trichloride 79
- adduct with $SbCl_5$ 83
-- ir spectrum 340
- adduct with thiourea 83
-- ir spectrum 340
- halogen exchange 79
- hydrolysis 80
4-Methoxyphenyl tellurium trihydroxide 80
4-Methoxyphenyl tellurium triiodide
- hydrolysis 84
4-Methoxyphenyl tellurium tris(dithiocarbamate) 79, 80
Methyl iodide
-r.w. benzotellurophene 289
-- dibenzotellurophene 291
-- diorganyl ditellurides 44, 102, 186, 223
-- diorganyl tellurides 175, 208
-- sodium selenide, telluride 257
-- tellurium 148
-- tetraorganyl tellurium 240
-- tetraphenyltellurophene 288
-- 1-thia-4-telluracyclohexane 323
Methyl lithium
-r.w. $TeCl_4$ 206, 236
-- $(CH_3)_3TeI$ 230, 236
Methyl methylethynyl telluride
- dec. by mercuric nitrate 141
ℓ-Methyl 4-methylphenyl phenyl telluronium
- d-α-bromocamphorsulfonate 224
- iodide 224
d-Methyl 4-methylphenyl phenyl telluronium
- d-10-camphorsulfonate 224
- iodide 224
Methyl 2-methylphenyl phenyl telluronium iodide 224
Methyl 4-methylphenyl phenyl telluronium iodide 224
-r.w. silver d-α-bromocamphorsulfonate 224
-- silver d-10-camphorsulfonate 224

Methyl phenyl telluride 138
- from "TeX_2" and CH_3MgBr/ C_6H_5MgBr 138
-r.w. bromoacetyl chloride 131, 132
- uv spectrum 141
Methyl tellurium iodide 145
Methyl tellurium tribromide 67, 80, 101
- adduct with thiourea 83
- adduct with tetramethylthiourea 83
Methyl tellurium trichloride 67
- adduct with $SbCl_5$ 83
-- ir spectrum 340
- association in solution 67
-r.w. boron trichloride 83
Methyl tellurium trifluoride 80
Methyl tellurium trihalides
- adducts with $(CH_3)_3TeX$ 146, 147
- adducts with tetramethylthiourea, ir spectrum 340
Methyl tellurium triiodide 80
- adduct with $(CH_3)_3TeI$ 146
-- crystallographic data 357
-- structure 146
- adduct with thiourea 83
- halogen exchange 80
-r.w. potassium iodide 82
Methyl vinylacetylenyl telluride
-r.w. benzonitrile N-oxide 130
-- N-α-diphenylnitrone 130
-- 1,3-diphenylnitrilimine 130
Methylene bis(acetonyl tellurium dichloride) 171
Methylene bis(tellurium trichloride) 34
-r.w. acetone 171
-- $K_2S_2O_5$ 323
Methylene ditelluride 323
Methylene radicals
-r.w. tellurium 322
Methylethynyl sodium telluride 245
3-Methyl-4-hydroxyphenyl tellurium trichloride 78
2-Methylphenol
-r.w. $TeOCl_2$ 78
2-Methylphenoxtellurine 10,10-dichloride 309
8-Methylphenoxtellurine 10,10-dichloride 316
-r.w. $TeCl_4$ 316
4-Methylphenyl phenyl ether
-r.w. $TeCl_4$ 316
2-Methylphenyl phenyl telluride 137
4-Methylphenyl phenyl telluride 137
4-Methylphenyl phenyl tellurium bis(d-α-bromocamphorsulfonate)
- hydrolysis 228
- opt. active 228
4-Methylphenyl phenyl tellurium dibromide
- opt. active 228
-r.w. silver d-α-bromocamphorsulfonate 228
4-Methylphenyl phenyl tellurium diiodide 228
- opt. active 228
4-Methylphenyl phenyl telluroxide
- attempted resolution into opt. active compds. 227
2-Methylphenyl sodium telluride 245
4-Methylphenyl sodium telluride 245
4-Methylphenyl tellurium trichloride
- TLC 367
Methyltelluracyclopropane 277
Methyltelluro radical 142

5-(Methyltelluroethynyl)-1,3-diphenyl-2-pyrazoline 130
- ir spectrum 333
5-(methyltelluroethynyl)-3-phenyl-2-isoxazoline 130
1-Methyltelluro-3-methyl-3-buten-1-yne
- ir spectrum 333
- nmr spectrum 356
Methyltetraiodotellurate(IV) anion
- structure 361
Mineral acids
-r.w. tellurophene 286
Molybdenum-tellurium compds. 258, 259, 266

N

Naphthalenetellurinic acid anhydride 64, 88
2-Naphthyl phenyl telluride
- adduct with HgClI 64
--r.w. $Na_2S \cdot 9H_2O$ 64
2-Naphthyl tellurium iodide
- adduct with triphenylphosphine 64
-- dec. to Ar_2Te_2 100
- reactions 66
-r.w. aqueous NaOH 64, 88, 100
-- 2,2-diphenyl-2-allylacetic acid 64, 138
-- Grignard reagents 43, 64, 136
-- phenyl mercury chloride 64, 136
-- triethyl phosphite 64, 100
-- triethylamine 64, 100
- synthesis 59
1-Naphthyl tellurium trichloride
-r.w. olefins 173
2-Naphthyl tellurium trichloride
-r.w. olefins 173
Nitric acid
-r.w. arenetellurinic acids 88
-- benzotellurophene 292, 294
-- diorganyl ditellurides 88, 102
-- diorganyl tellurides 140, 177
-- phenoxtellurine 309, 310, 317
-- R_2TeX_2 144, 190
-- R_2TeO 196
Nitrodiphenyl ethers
-r.w. $TeCl_4$ 309
Nitromethane
-r.w. $TeCl_4$ 34, 77
Nitromethyl tellurium trichloride 34, 77
Nitrophenoxtellurines 310, 311
-r.w. tin/HCl 316
Nitrophenoxtellurine 10,10-dinitrates 311, 315
2-Nitrophenoxtellurine
- adduct with 2,8-dinitrophenoxtellurine 316
Nitrous acid
-r.w. bis(4-dimethylaminophenyl) tellurium dichloride 190
Nomenclature of organic tellurium compds. 4

O

Octafluorodibenzotellurophene 291
Olefins
-r.w. $ArTeCl_3$ 82, 173
-- $TeBr_4$ 26, 172
-- $TeCl_4$ 36, 78, 172
Organic Compds.
-r.w. tellurium 25
-- $TeCl_4$ 29, 30
Organic dihalides
-r.w. Al_2Te_3 42

-- tellurium 26
Organic halides
-r.w. Na_2Te 44, 107
-- Na_2Te_2 41
-- R_2Te 44, 139, 199, 208, 224
-- tellurium 24, 148
Organic lithium compds.
-r.w. R_2Te_2 101, 245
-- R_3TeX 230, 234, 236
-- tellurium 26, 98, 121, 122, 245
-- $TeCl_4$ 40, 116, 199, 206, 208, 234, 236
Organic radicals
-r.w. tellurium 27, 97, 116, 138, 322
Organic mercury compds.
-r.w. tellurium 28, 106, 115, 119, 291
-- $TeCl_4$ 39, 78, 184, 309
-- $RTeCl_3$ 43, 82, 135, 185
-- $2\text{-}C_{10}H_7TeI$ 64, 136
Organic tellurium compds.
- as ligands in transition metal complexes 257
-- ir spectra 338
-- nmr spectra 356
- biological activity 369
- chromatography 367
- contg. radioactive tellurium isotopes 142, 148
- contg. a tellurium-metal bond 244
- dipole moments 143, 144, 192, 356, 357
- heterocyclic 277
- ir, Raman 325, <u>327</u>, <u>330</u>, <u>334</u>
- magnetic sysceptibilities 356
- mass spectra 362
- nmr 341
- nomenclature 4
- parachor 356
- patents 380
- physicochemical investigations 325
- polymeric 323
- reviews 3
- structures 356
- synthesis from tellurium 24, <u>25</u>
-- from organic tellurium compds. 43, <u>45</u>
- uv, visible spectra 341, <u>342</u>
Organyl antimony compds. (^{125}Sb)
- disintegration to Te compds. 119, 148, 206, 236
Organyl mercury chlorides
see also Organic mercury compds.
-r.w. $RTeCl_3$ 43, 82, 135, 185
Organyl tellurium tribromides <u>70</u>
- from R_2Te_2 and bromine 79
- hydrolysis 83, 84
Organyl tellurium trichlorides 67, <u>70</u>
see also individual compds.
- adducts with thioureas 83
- halogen exchange 79
- hydrolysis 83, 84
- intramolecular condensation 82
-r.w. aromatic compds. 43, 82
-- dialkyl ditellurides 82, 103
-- dialkyl tellurides 43
-- Grignard reagents 186
-- ketones 43, 82
-- olefins 173
-- organyl mercury chlorides 43, 8
135, 185
-- sulfites, sodium sulfide 79, 80, 97
- reduction to RTeCl 59, 82

- synthesis 67, 68, 69, 77, 78
Organyl tellurium trifluorides 80
Organyl tellurium trihalides 70
- halogen exchange 79
- hydrolysis 84
- reactions 80, 81
- synthesis 67, 68
Organyl tellurium triiodides
- from R_2Te_2 and iodine 70, 79
- hydrolysis 84
Organyl tellurium tris(dithiocarbamate) 70, 79
- thermochromic properties 79
Organyl tellurocyanates 49, 50
Organyltelluro magnesium bromide 58, 137
Organyltellurophosphines 103, 255
1-Oxa-4-telluracyclohexane 305, 308
- dec. in CCl_4 307
- reactions 308
-r.w. halogens 305
1-Oxa-4-telluracyclohexane 4,4-dichloride 305
-r.w. $K_2S_2O_5$ 307
1-Oxa-4-telluracyclohexane 4,4-diiodide 307
Oxygen (air)
-r.w. disilyl telluride 253
-- R_2Te 140, 141, 177, 193, 197
-- R_2Te_2 102

P

Palladium
- determination with diphenyl telluride 368
Palladium complexes
- with dialkyl tellurides 259, 270, 271
Pentamethylene bis(cyclopentamethylene telluronium halide) 296
- thermal dec. 296
α,ω-Pentamethylene dihalides
-r.w. Al_2Te_3 296
-- Na_2Te 296
-- 1-telluracyclohexane 296
-- tellurium 296
Pentanetellurinic acid 84
Perfluorodibenzotellurophene 291
Perfluorphenyl iodide
-r.w. tellurium 148
Perfluorophenyl lithium
-r.w. $TeCl_4$ 41, 116, 236
Perfluorophenyl tellurium trichloride
-r.w. zinc 98
Phenacyl bromide
-r.w. diaryl tellurides 208
Phenol
-hydrogen bonded to diethyl telluride 337
Phenoxtellurine 307, 311
- adduct with 2-chloro-8-methylphenoxtellurine 316
- adducts with phenoxtellurine derivatives 317, 319
- adducts with trinitrobenzene, picric acid, picryl chloride 317, 318
-- ir spectra 317
- 2,8-disubstituted 309, 311
- electrochemical oxidation 319
- from 2-phenoxyphenyl tellurium trichloride 43
- nitration 309, 310
- product of oxidation with H_2O_2 198
- reactions 311
-r.w. base 320

-- bromine, iodine 315
-- hydrogen peroxide 197
-- $Mn(CO)_5X$ 266
-- nitric acid 309, 310, 317
-- pyrocatechol dichloromethylene ether 315
-- sulfuric acid 317
- 2-substituted 309, 311
- triplet-triplet energy transfer 320
Phenoxtellurine 10,10-dichloride 35, 307, 311, 315
-r.w. $K_2S_2O_5$ 315
Phenoxtellurine 10,10-dihalides 315
- anion exchange 316
Phenoxtellurine 10,10-dinitrate 315
- complex with reduced species 317, 319
-r.w. $K_2S_2O_5$ 315
Phenoxtellurine 10-hydroxide 10-nitrate 315
-r.w. $K_2S_2O_5$ 315
Phenoxtellurine sulfates 317
2-Phenoxyphenyl tellurium trichloride
- intramolecular condensation 43, 82, 307
-r.w. $K_2S_2O_5$ 307
4-Phenoxyphenyl tellurium trichloride
- intramolecular rearrangement 43, 82, 307
-r.w. diphenyl ether 171, 307
Phenyl bromomagnesium telluride 58, 245, 246
-r.w. allenyl bromide 131
-- propargyl bromide 131
-- trimethylsilyl chloride 250
Phenyl lithium
-r.w. Ar_2TeCl_2 234
-- Ar_3TeCl 234
-- $TeCl_4$ 234
Phenyl propargyl telluride 129
- isomerization 131
Phenyl sodium telluride 245
Phenyl tellurium halides
- thiourea adducts, structure 61, 63
Phenyl tellurium trichloride
- contg. ^{127}Te 83
- dipole moment 83
-r.w. olefins 173
- TLC 83, 367
Phenyltellurium tris(dithiocarbamates) 79
Phenyl trimethylsilyl telluride 250
Phenylethynyl sodium telluride 245
Phenyltelluro magnesium bromide
- see phenyl bromomagnesium telluride
Phenyltelluro mercury chloride 58, 102
Phosgene
-r.w. tellurophene 279
Phosphorus-tellurium compds. 255
Picric acid
-r.w. benzotellurophene 290
-- dibenzotellurophene 140, 291
-- diorganyl tellurides 140
-- phenoxtellurine 317, 318
-- R_3TeX 209
Picryl chloride
-r.w. phenoxtellurine 317, 318
Platinum complexes
- with dialkyl tellurides 259, 270, 271
Polonium compds. 174, 189, 367
Potassium amide
-r.w. diphenyl selenide and telluride 142

Potassium cyanide
-r.w. tellurium 46, 47
Potassium diethyl tellurophosphate 255
Potassium diethyl phosphite
-r.w. tellurium 255
Potassium halides
-r.w. R_3TeX 209
Potassium hexacyanoferrate(II)
-r.w. tellurium 46
Potassium disulfite
- dec. of acetonyl tellurium compds. 97, 114, 189
- dec. of 4-hydroxyphenyl tellurium trichloride 97
-r.w. 1-oxa-4-telluracyclohexane 4,4-dichloride 307
-- phenoxtellurine 10,10-dihalides 315
-- RTeOOH 89, 97
-- $RTeCl_3$ 80, 79, 97, 307, 323
-- R_2TeX_2 99, 106, 107, 114, 137, 143, 291, 315
-- 1-telluracyclohexane 1,1-dihalides 297
-- 1-tellura-3,5-cyclohexanedione 1,1-dichlorides 300
Potassium iodide
-r.w. $RTeI_3$ 82
-- tetramethylene bis(cyclotetramethylene telluronium iodide) 279
Potassium methyltetraiodotellurate(IV) 82
Potassium permanganate
-r.w. diorganyl tellurides 193
Potassium telluride 47, 105
Potassium tellurite
- conv. to dimethyl telluride by molds 379
Potassium tellurocyanate
-r.w. azulene 50
- synthesis 46, 47
Potassium tetratelluride 47
Propargyl bromide
-r.w. $C_6H_5TeMgBr$ 131
Pyrocatechol dichloromethylene ether
-r.w. phenoxtellurine 315
Pyrrole
-r.w. $TeBr_4$ 35

Q

Quinones
-r.w. $TeBr_4$ 35

R

Resorcinol
-r.w. $ArTeCl_3$ 171
-- $TeCl_4$ 170
Rhenium complexes
- with diethyl telluride 259, 269
Rhodium complexes
- with diethyl telluride 259, 270, 271
Rongalite 21
Ruthenium complexes
- with diphenyl ditelluride, telluride, dibutyl telluride 239, 270

S

Selenanthrene 142
Selenium-tellurium compounds 257
Selenium
-r.w. diorganyl tellurides 141
-- bis(triethylstannyl) telluride 252
Silyl halides
-r.w. lithium telluride 250
Silver d-α-bromocamphorsulfonate
-r.w. telluronium salts 224, 227, 228, 305, 306

Silver d-10-camphorsulfonate
-r.w. telluronium salts 224
Silver complexes
- with dimethyl telluride 273
Silver cyanide
-r.w. $TeBr_4$ 50
Silver dicarboxylates
-r.w. R_2TeX_2 184, 324
Silver halides
-r.w. R_3TeX 209
Silver iodide
-r.w. bis(trimethylsilyl) telluride 252
Silver nitrate
-r.w. 2-formylphenyl methyl telluride 100
Silver oxide
-r.w. alkyl 2-formylphenyl telluride 141
-- carbalkoxymethyl diphenyl telluronium bromide 223
-- $R_2Te(OH)X$ 193
-- R_2TeX_2 192
-- R_3TeX 209, 223
Silver thiocyanate
-r.w. $TeBr_4$ 51
Sodium
-r.w. diphenyl ditelluride 58, 102
-- tellurium 19, 21, 282
Sodium acetylides
-r.w. tellurium 26, 121, 245
Sodium borohydride
-r.w. diorganyl ditellurides 102, 121, 129, 245
Sodium ditelluride 19, 21
-r.w. alkylating agents 41, 91, 97
-- butyl bromide 97
-- chloroacetic acid 97
-- diacetylenes 41
Sodium hydrogen sulfite
-r.w. tellurophene 1,1-dihalides 286
Sodium hydroxide
see base solutions
Sodium picrate
-r.w. R_3TeX 209
Sodium selenide telluride
-r.w. methyl iodide 257
Sodium succinate
-r.w. Al_2Te_3 279
Sodium sulfide nonahydrate
- reduction of benzotellurophene dichlorides 290
-- $RTeCl_3$ 79, 97, 80
-- RTeOOH 89, 97
-- R_2TeX_2 99, 106, 107, 114, 137, 185 291
-- $R_2Te \cdot HgX_2$ 64
-- R_3TeI 135, 232
-- thiophenoxtellurine 10,10-dichlorid 321
- dec. of acetonyl tellurium compds. 98, 114, 189
- dec. of 4-hydroxyphenyl tellurium trichloride 98
Sodium telluride 19, 21
- from tellurium and sodium in NH_3 21, 282
- from tellurium, NaOH and Rangalite in H_2O 21
-r.w. alkylating agents 41, 106, 107
-- arenediazonium chlorides 119
-- benzyl dimethyl phenyl ammonium chloride 107
-- bis(2-chloroethyl) ether 305
-- bis(2-chloroethyl) sulfide 320

-- 2-(β-bromoethyl)benzyl bromide 305
-- *tert*-butyl chloride 107
-- diacetylenes 41, 279
-- 1,4-dibromobutane 277
-- pentamethylene dihalides 296
-- 1,2,3,4-tetrakis(iodomethyl)-butane 279
Sodium tellurate
-r.w. betaine 106
-- choline hydrochloride 106
Sodium-tellurium compds. 244, 245
Sodium thiosulfate
- reduction of R_2TeX_2 114
Sodium vinylethynyl telluride 245
Sulfur
-r.w. benzotellurophene 294
-- bis(triethysilyl) telluride 253
-- 2,8-difluorophenoxtellurine 320
-- 2,8-dimethylphenoxtellurine 320
-- diphenyl telluride 141
Sulfur dioxide
-r.w. 1-tellura-3,5-cyclohexane-dione 1,1-dichlorides 300
-- 1-telluracyclopentane 1,1-diiodide 279
Sulfuric acid
-r.w. phenoxtellurine 317
Sulfuric acid-$\underline{d}_2$
-r.w. tellurophene 279
Sulfuryl chloride
-r.w. R_2Te 139, 174
-- tellurium 17

T

1-Telluracycloalkanes
- from Al_2Te_3 and organic dihalides 42
1-Telluracycloalkanes 1,1-dihalides 26
1-Telluracyclobutane 277
1-Telluracyclohexane 277, 296, 298
- protonated 199
- reactions 298
-r.w. fluorosulfonic acid 199
-- hydrogen peroxide 197
-- halogens 296
-- pentamethylene dihalides 296
1-Telluracyclohexane-4,4-$\underline{d}_2$
- nmr 356
1-Telluracyclohexane-3,3,5,5-$\underline{d}_4$
- nmr 356
- protonated 356
1-Telluracyclohexane 1,1-dibromide
- conformation 356
- nmr 341
1-Telluracyclohexane 1,1-dihalides 296, 298
- conductivities 297
-r.w. $K_2S_2O_5$ 297
- uv spectra 297
1-Telluracyclohexane 1,1-dioxide 197
1-Tellura-3,5-cyclohexanedione 296, 297, 301
- bactericide 369
- dec. by HCl, KOH 300
- ir spectrum 326
- oximes 300, 304
-r.w. bromine, iodine 300
-- hydroxylamine sulfate 300
- structure 362
1-Tellura-3,5-cyclohexanedione 1,1-dibromides 304
1-Tellura-3,5-cyclohexanedione 1,1-dichlorides 31, 277, 297, 301
- dec. by SO_2 300

-r.w. ethyl chloride, HCl in $CHCl_3$ 300
-- $K_2S_2O_5$ 300
1-Tellura-3,5-cyclohexanedione 1,1-diiodides 304
2,4-R_2-1-Tellura-3,5-cyclohexanedione
-r.w. hydroxylamine sulfate 300
1-Telluracyclopentane 277, 279, 280
- conductivity of derivatives 357
- mass spectrum 363
-r.w. 1,4-dibromobutane 278
-- halogens 278, 279
- vapor phase chromatography 368
1-Telluracyclopentane 1,1-dibromide 278
- conductivity 357
1-Telluracyclopentane 1,1-diiodide 279
- conductivity 357
-r.w. sulfur dioxide 279
1-Telluracyclopropane 277
1-Tellura-4-ethyl-3,5-cyclohexanedione 1,1-dichloride
- dec. by HCl 300
Telluranthrene 28, 118, 322
1-Tellura-2-oxa-cyclohexane-5-one 297
Tellurenyl compounds 59, 62, 99
- exchange of X in RTeX 61
- preparation 59, 60
Tellurides 105
see individual compds.
- R_3M-Te-$M'R_3$ (M,M' = group IV element) 248, 254
- R_3M-Te-R (M = group IV element) 247, 253
-- ir spectra 253
Tellurinic acids 84, 85, 90, 140, 141
- adducts with R_2TeO 84, 196
- aromatic, nitration 88
- from $RTeX_3$ and NaOH 83
- from R_2Te_2 on oxidation 84
- from R_2TeO on oxidation 196
-r.w. HI 80, 89
-- reducing agents 89, 97
Tellurinic acid anhydrides 85
- from $RTeX_3$ and NaOH 82, 88
- reduction 89
Tellurinic acid chlorides 85
- hydrolysis 89
Tellurinic acid halides 84, 85
- from acetonyl and phenacyl aryl tellurium dichlorides on hydrolysis 89
- from $RTeX_3$ and water 83
- hydrolysis 88
Tellurinic acid nitrates 85, 88
- hydrolysis 89
Tellurium
- abundance 13
- amorphous 14
- applications 16
- crystalline 15
- determination in organic compds. 366
- manufacture 13
- properties 15
-r.w. alkyl iodides 148
-- arenediazonium chlorides 28, 119, 185
-- aryl lithium 129, 245
-- benzyl iodide 148
-- 2,2'-biphenylylene mercury 291
-- biphenylylene sulfone 28, 290
-- bis(2-iodoethyl) ether 308
-- bis(perfluorophenyl) thallium bromide 27, 117

-- bromine 17
-- bromotrifluoromethane 19
-- butyl lithium 129, 245
-- carbon disulfide 53
-- carbon monoxide 53
-- carbon monoselenide 53
-- carbon monosulfide 53
-- chlorine 17
-- diaryl mercury 28, 106, 115, 119, 291
-- dichlorodifluoromethane 19
-- diethyl allylphosphonite 255
-- 1,4-diiodobutane 277, 279
-- diiodomethane 148, 323
-- 2,2'-diiodooctafluorobiphenyl 291
-- 1,3-dimethylbenzimidazoline 28, 242
-- 4,4'-dimethyl-2,2'-biphenylylene mercury 291
-- diphenyliodonium chloride 28, 119, 185
-- disulfur dichloride 17
-- ethyl iodide 148
-- fluorine 17
-- Grignard reagents 26, 58, 98, 106, 117, 137, 245
-- hexachloro-1,3-butadiene 26, 282
-- inorganic compds. 18
-- iodine 19
-- iodine bromide 19
-- isonitriles 50
-- methyl iodide 148
-- methylene radical 322
-- organic compds. 25
-- organic dihalides 26
-- organic halides 24, 148
-- organic lithium compds. 26, 98, 121, 122, 245
-- organic radicals 27, 97, 116, 138
-- pentamethylene dihalides 296
-- perfluorophenyl iodide 148
-- potassium cyanide 46, 47
-- potassium diethyl phosphite 255
-- potassium hexacyanofenate(II) 46
-- sodium 19
-- sodium acetylides 26, 121, 245
-- sulfuryl chloride 17
-- tellurium tetrabromide 19
-- tellurium tetrachloride 19
-- tetraethylammonium cyanide 49
-- tetramethylammonium cyanide 49
-- tetraphenyl tin 28, 118, 322
-- tetraphenylarsonium cyanide 49
-- thianthrene 5,5,10,10-tetroxide 28, 290
-- thionyl chloride 17
-- triethylmetal hydrides 250
-- triorganylphosphines 255
-- triphenylgermyl lithium 245
-- triphenylplumbyl lithium 245
-- triphenylstannyl lithium 245
-- separation from other elements 13
- structure 15
Tellurium bis(diselenodiethylphosphinate) 257
Tellurium bis(thioselenodiethylphosphinate) 257
- structure 257
Tellurium-carbon bond
- bond dipole moment 357
- bond length 357
- formation 24
Tellurium dibromide 19, 20
-r.w. Grignard reagents 138
Tellurium dichloride 19, 20
- adducts with diarylethylenes 36
Tellurium dicyanide 47, 50

- ir spectrum 51
-r.w. triphenylbismuthine 51
Tellurium dihalides 19, 20
-r.w. Grignard reagents 38, 98, 115
- stability 39
Tellurium dioxide 23, 191
Tellurium di(thiocyanate) 51
Tellurium-halogen bond
- bond length 357
Tellurium hexafluoride 17
Tellurium-iodine, fused
-r.w. mixture of CH_3MgI and C_6H_5MgI 138
Tellurium-metal bonds 244
Tellurium oxychloride 78
-r.w. 2-methylphenol 78
-- cresols 171, 212
Tellurium-palladium compds. 262, 270
- Te-Pd vibration 338
Tellurium-phosphorus compds. 255
- Te-P stretching vibration 341
Tellurium-platinum compds. 263, 276
- Te-Pt stretching vibration 338
Tellurium tetrabromide
- preparation 17
- properties 20
-r.w. anthraquinone 35
-- benzoquinone 35
-- biphenyl 35, 290
-- carbethoxymethylenetriphenylphosphorane 36, 172
-- ethylene 36
-- naphthoquinone 35
-- olefins 172
-- phenetole 35
-- pyrrole 35
-- silver cyanide 50
-- silver thiocyanate 51
-- tellurium 19
Tellurium tetrachloride
- addition to carbon-carbon multiple bonds 36, 78, 172, 323
- preparation 17
- properties 20
-r.w. acetic anhydride 33, 77
-- amines 35
-- arenediazonium chlorides 185
-- aromatic compounds 34, 77, 106
-- aryl mercury chlorides 39, 78, 184
-- biphenyl 35, 290
-- bis(4-chlorophenyl) ether 35
-- bis(4-methoxyphenyl) ether 35
-- butadiene 172
-- 1-butene 36, 172
-- butyl lithium 116, 236
-- carboxylic acid anhydrides 33, 77, 170
-- 4-chloro-4'-methyldiphenyl ether 35
-- cyclohexane 36, 78, 172
-- diarylethylenes 30
-- dibenzoylmethane 31
-- diethyl zinc 199, 206, 234
-- diisobutylene 36
-- 1,3-diketones 29, 67, 148, 297
-- 2,2'-dilithiobiphenyl 234, 235
-- 1,4-dilithiobutadiene 282
-- dimethyl sulfite 77
-- diphenyl ether 35, 307
-- diphenyl sulfide 35
-- diphenylacetylene 36, 78
-- 1,4-diphenylbutadiene 36
-- 2,2-diphenyl-5-chloromercuric-4-pentanolactone 40, 184

-- 2,2-diphenyl-4-pentenoic acid 36, 172
-- ethylene 36, 172
-- 4-fluorodiphenyl ether 35
-- Grignard reagents 37, 116, 186, 199, 206
-- ketones 32, 69, 170
-- lithioalkanes 40, 199, 236
-- methyl lithium 206, 236
-- 4-methylphenyl phenyl ether 316
-- nitrodiphenyl ethers 309
-- nitomethane 34, 77
-- organic lithium compds. 40, 116, 199, 206, 208, 234 236
-- organic compds. 29, 30
-- perfluorophenyl lithium 41, 116, 236
-- phenyl lithium 234
-- phenylacetylene 36, 78
-- propene 36, 78, 172
-- resorcinol 170
-- 2-(4'-R-phenoxyphenyl) mercury chloride 309
-- 4,4'-R_2-diphenyl ether 309
-- stilbene 36
-- styrene 36
-- tellurium 19
-- 1,1,3,3-tetramethylacetone 33, 77
-- toluene 34, 78
-- xylenes 34, 78
Tellurium tetrafluoride 17
Tellurium tetrahalides
see also individual compds.
- preparation 17, 18
- properties 20
-r.w. organometallic compds. 37
-- diorganyl ditellurides 103
Tellurium-tin compds. 247
- Te-Sn stretching vibration 337
Tellurium ylides 232, 233
Telluroaldehydes 242
Tellurocyanate ion
- physical properties 47
- ir spectrum 48, 49
Tellurocyanates 46
Tellurocyanic acid 46
Telluroformaldehyde 28, 138, 242, 322, 323
- mass spectrum 242
-r.w. bromine, iodine 242
Telluroisochroman 198, 296, 305, 306
- reactions 306
-r.w. hydrogen peroxide 197
- telluronium salt with 4-chlorophenacyl halide 227, 305
-- resolution into ℓ-R_3Te^+ d-bromocamphorsulfonate$^-$ and d-d salt 227, 305, 306
-- opt. active picrates 227, 306
Telluroketones 242
- from H_2Te and ketones 42, 242
Tellurols 56, 58
- preparation 57
- properties 56, 57
Telluropentathionate 46
Tellurophene 1,1-dihalides
-r.w. $NaHSO_3$ 286
Tellurophene(s) 279, 283, 287, 289
- aromaticity 286
- ir spectra 333
- nucleophilic substitution 286
-r.w. acetic anhydride 279
-- butyl lithium 279
-- D_2SO_4/CH_3OD 286
-- halogens 286

-- mercuric acetate 286
-- mineral acids 286
-- phosgene 279
- 2,5-substituted 41, 279, 287
-- from diacetylenes and Na_2Te 41, 279
Tetraalkyl tellurium 238
see also Tetraorganyl tellurium
- hydrolysis 238
- thermal stability 236, 238
Tetrabutyl tellurium 230, 236, 237
-r.w. bis(2,2'-biphenylylene)-arsonium, phosphonium iodide 238
- thermal dec. 238
Tetrachlorotellurophene 210, 282
- dec. by base 286, 288
Tetraethylammonium cyanide
-r.w. tellurium 49
Tetraethylammonium tellurocyanate 49
1,2,3,4-Tetrakis(iodomethyl)butane
-r.w. Na_2Te 279
Tetrakis(perfluorophenyl)tellurium 236
- thermal dec. 116, 238, 239
Tetramethoxy tellurium
-r.w. 2,2'-dilithiobiphenyl 43, 234
Tetramethyl tellurium 230, 236
Tetramethylammonium cyanide
-r.w. tellurium 49
Tetramethylammonium tellurocyanate 49
Tetramethylene bis(cyclotetramethylene telluronium bromide) 278
-r.w. KI 279
- thermal dec. 279
Tetramethylene bis(cyclotetramethylene telluronium iodide) 279
Tetramethylthiourea
-r.w. $RTeBr_3$ 83, 340
Tetraorganyl tellurium
see also individual compds.
- from R_2TeCl_2 and Grignard reagents 43
- from $TeCl_4$ and RLi 40, 116
- hydrolysis 40, 208, 211, 238
- structure 241
- thermal stability 38, 115, 116, 238
Tetraphenyl tellurium
- contg. ^{125m}Te 236
- from Ar_2TeCl_2 and C_6H_5Li 234
- from Ar_3TeCl and C_6H_5Li 234
- from $^{125}Sb(C_6H_5)_5$, $^{125}Sb(C_6H_5)_4Cl$ 236
- from $TeCl_4$ and C_6H_5Li 234
- hydrolysis 239
-r.w. benzaldehyde 211, 240
-- butyl lithium 238
-- $CHCl_3$, CH_2Cl_2 211, 240
-- triphenyl boron 211, 240
- thermal dec. 116, 117, 238
Tetraphenyl tin
-r.w. tellurium 28, 118, 322
Tetraphenylarsonium cyanide
-r.w. tellurium 49
Tetraphenylarsonium tellurocyanate 49
Tetraphenyltellurophene 282
-r.w. maleic anhydride 286
-- methyl iodide 288
Tetraphenyltellurophene 1,1-dibromide 286
-r.w. iron carbonyls 288
-- maleic anhydride 286
Thianthrene 5,5,10,10-tetroxide
-r.w. tellurium 28, 290

1-Thia-4-telluracyclohexane 320
-r.w. halogens 320
-- methyl iodide 321
- telluronium salt with methyl iodide 321
1-Thia-4-telluracyclohexane 4,4-dihalides 320
1-Thia-4-telluracyclohexane 4,4-diiodide
- structure 321, 361
Thionyl chloride
-r.w. R_2Te 139, 174
-- tellurium 17
Thiophenoxtellurine 321
- from 2-thiophenoxyphenyl tellurium trichloride 43, 321
Thiophenoxtellurine 10,10-dichloride 321
-r.w. sodium sulfide 321
2-Thiophenoxyphenyl tellurium trichloride
- intramolecular condensation 43, 82
Thiourea
- adducts with ArTeCl 59, 61, 63
-r.w. $RTeX_3$ 59, 83
Tin/hydrochloric acid 247
- reduction of nitrophenoxtellurines 316
Tin dichloride
-r.w. R_2TeX_2 114
Tin-tellurium compds. 247
Toluene
-r.w. $TeCl_4$ 34
Transition metal compds.
-r.w. ditellurides 103, 247, 258, 269
- tellurides 257, 258, 267, 269, 270, 271, 273
Trialkyl telluronium alkyltetrahydroxytellurate(IV) 212
-r.w. HX 212
Trialkyl telluronium hydroxides 238
Trialkyl telluronium iodides
-r.w. lithioalkanes 236
Trialkyl telluronium salts
- from R_2TeO upon heating 193, 212
Triarylbismuth difluoride
-r.w. R_2Te 139, 174
Tribromotelluromethyl methyl sulfite 77
Tributyl telluronium bromide 211
Tributyl telluronium iodide 230
- from $(C_6H_5)_3TeX$ and butyl lithium 230
-r.w. butyl lithium 230, 236
Trichlorotelluromethyl methyl sulfite 77
Triethylmetal(group IV) hydrides
-r.w. tellurium 250
Triethyl phosphite
-r.w. 2-$C_{10}H_7TeI$ 64, 100
Triethyl telluronium chloride
- from $TeCl_4$ and diethyl zinc 206, 234
- polarography 368
-r.w. diethyl zinc 43, 229, 234
Triethylamine
-r.w. 2-$C_{10}H_7TeI$ 64, 100
Triethylgermane
-r.w. bis(triethylsilyl) telluride 252
Triethylgermyl bromide
-r.w. C_2H_5TeLi 251
Triethylgermyl lithium
-r.w. diethyl ditelluride 102
Triethylgermyl triethylsilyl telluride
- hydrogen bonded to $CDCl_3$ 337

Triethylgermyl triethylstannyl telluride
- hydrogen bonded to $CDCl_3$ 337

Triethylsilyl triethylstannyl telluride
- hydrogen bonded to $CDCl_3$ 337

Triethylstannane
-r.w. bis(triethylgermyl) telluride 252

Tri(ethyltelluro)silane
- ir spectrum 337

Trimethyl aluminum
-r.w. dibutyl telluride 246

Trimethyl gallium
-r.w. dibutyl telluride 246

Trimethyl onium compds.
- hydrogen exchange in D_2O 232
- hyperconjugation 232

Trimethyl telluronium bromide
- ir spectrum of BBr_3 adduct 340

Trimethyl telluronium halides
- compds. with CH_3TeX_3 146, 147
- from R_3Te^+ $RTe(OH)_4^-$ and HX 212

Trimethyl telluronium hydroxide
- conductivity 233

Trimethyl telluronium iodide
- adduct with CH_3TeI_3 146
-- crystallographic data 257
-- structure 146
- from $TeCl_4$ and methyl lithium 206
-r.w. 2,2'-dilithiobiphenyl 236
-- methyl lithium 230, 236

Trimethyl telluronium methyltetrahalotellurates(IV) 149, 197
- medullary stimulant 378

Trimethyl telluronium salts
- catalytic activity 233
- hydrogen exchange in D_2O 232
- hyperconjugation 232
- thermal dec. 230

Trimethylsilyl chloride
-r.w. $C_6H_5TeMgBr$ 250

1,3,5-Trinitrobenzene
- adduct with dibenzotellurophene 291
- adduct with dimethyl telluride 140

Trinitrofluorenone
- adduct with benzotellurophene 290
- adduct with phenoxtellurine 317, 318

Trioctylphosphine telluride
- dipole moment 256

Triorganyl telluronium bromides
- conv. to chlorides with AgCl 209
- conv. to iodides 209

Triorganyl telluronium chlorides
- conv. to bromides and iodides 209
- from R_2TeCl_2 and Grignard reagents 189
- from $TeCl_4$ and Grignard reagents 37, 38
- from $TeCl_4$ and RLi 40

Triorganyl telluronium halides 199
- from R_2Te and RX 44, 139, 208
-r.w. organic lithium compds. 230
-- picric acid, sodium picrate 209
-- potassium halides 209
-- silver oxide/H_2O 209
- solubility 209
- stability 232
- thermal dec. 131, 230

Triorganyl telluronium hydroxides 209
- basicity 209

-r.w. acids 209
Triorganyl telluronium iodides 206, 209
- conv. to chlorides, bromides with AgX 209
Triorganyl telluronium picrates 209
Triorganyl telluronium salts 199, 207
- anion exchange 208, 210
- from $TeCl_4$ and Grignard reagents 206
- modification of organic moiety 223
- of the type R_3Te^+ 200
-- $R_2R'Te^+$ 211
-- $RR'R''Te^+$ 224
-r.w. boron tribromide 212
-- mercuric halides 212
-- other metal chlorides 212
-- zinc chloride 212
Triorganylphosphine
-r.w. tellurium 255
Triorganylphosphine tellurides 255, 256
- ir spectrum 341
Triphenyl boron
-r.w. $(C_6H_5)_4Te$ 211, 240
Triphenyl telluronium bromide
-r.w. butyl lithium 211, 236
Triphenyl telluronium chloride 118
- analytical applications 233, 368
- conductivity 357
- contg. ^{125m}Te from ^{125}Sb compds. 206
- dipole moment 233
- from Ar_4Te and $CHCl_3$, CH_2Cl_2 or C_6H_5CHO 211, 240
- polarography 368
-r.w. 2,2'-dilithiobiphenyl 230
-- Grignard reagents 230
-- phenyl lithium 234
- TLC 367
Triphenyl telluronium fluoride
- dipole moment 233
Triphenyl telluronium halides
-r.w. butyl lithium 230
Triphenyl telluronium iodide
- conductivity 357
- detn. of Bi, Co 368
- dipole moment 233
-r.w. butyl lithium 238
-- $Na_2S \cdot 9H_2O$ 232
Triphenyl telluronium salts
- analytical applications 233
Triphenyl telluronium tetraphenylcyclopentadienylide 233
Triphenyl telluronium tetraphenyl borate 211, 240
Triphenylbismuthine
-r.w. $Te(CN)_2$ 51
Triphenylgermyl chloride
-r.w. R_3MLi (M = Ge, Sn, Pb) 247
Triphenylgermyl lithium
-r.w. tellurium 245
Triphenylphosphine
-r.w. $2-C_{10}H_7TeI$ 64
Triphenylplumbyl chloride
-r.w. R_3MLi (M = Ge, Sn, Pb) 247
Triphenylplumbyl lithium
-r.w. tellurium 245
Triphenylstannyl chloride
-r.w. R_3MLi (M = Ge, Sn, Pb) 247
Triphenylstannyl lithium
-r.w. tellurium 245
1,3,5-Tritelluracyclohexane 242, 322, 323

Tritelluroformaldehyde 242, 322, 323
Tris(dimethylamino)phosphine telluride
 - ir spectrum 341
Tris(4-hydroxy-2-methylphenyl) telluronium chloride 146, 223
 -r.w. NaOH, Na_2CO_3 223
Tris(2-hydroxy-5-methylphenyl) telluronium chloride 212, 213
Tris(4-hydroxy-2-methylphenyl) telluronium hydroxide
 - sodium salt 223
Tris(2-hydroxy-5-methylphenyl) telluronium oxotrichlorotellurate(IV) 212
 - conv. to telluronium chloride 223
Tris(4-methylphenyl) telluronium chloride
 - contg. ^{125m}Te from ^{125m}Sb compds. 206
 - TLC 367
Tris(perfluorophenyl) telluronium chloride 40
Tungsten hexacarbonyl
 -r.w. bis(trimethylsilyl) telluride 257

U

Uranium-tellurium compds. 244, 274
Uranium pentachloride
 - complex with diphenyl ditelluride 274

V

Vernon's β-base 146, 212
Vinylethynyl sodium telluride 245

W

Water
 -r.w. bis(2,2'-biphenylylene) tellurium 239, 294
 -- $RTeCl_3$ 83

X

Xylenes
 -r.w. $TeCl_4$ 34, 78

Y

Ylides 232

Z

Zinc
 -r.w. R_2TeX_2 114
 -- $RTeX_3$ 98
Zinc chloride
 -r.w. R_3TeX 212